GCSE

Mathematics

Foundation Level

Complete Revision and Practice

Contents

Throughout this book you'll see grade stamps like these: Ⓓ Ⓕ Ⓔ Ⓒ
You can use these to focus your revision on easier or harder work.
But remember — to get a top grade you have to know **everything**, not just the hardest topics.

Contents

Published by CGP

From original material by Richard Parsons

Updated by: Ceara Hayden, Paul Jordin, Sharon Keeley-Holden, Simon Little, Alison Palin, Andy Park, Sam Pilgrim, Caley Simpson, Ruth Wilbourne

Contributors: Rosie Hanson

With thanks to Glenn Rogers and Mark Moody for the proofreading

ISBN: 978 1 84146 480 2

Groovy website: www.cgpbooks.co.uk
Printed by Elanders Ltd, Newcastle upon Tyne.
Jolly bits of clipart from CorelDRAW®

Calculating Tips

Welcome to GCSE Maths — not always fun, but stuff you have to learn. GCSE Maths is tested by two or three exams. Thankfully there are some nifty exam tricks you only have to learn once, which could get you marks in all your exams. Read on...

BODMAS — <u>B</u>rackets, <u>O</u>ther, <u>D</u>ivision, <u>M</u>ultiplication, <u>A</u>ddition, <u>S</u>ubtraction

<u>BODMAS</u> tells you the <u>ORDER</u> in which these operations should be done:
Work out <u>Brackets</u> first, then <u>Other</u> things like squaring, then <u>Divide</u> / <u>Multiply</u> groups of numbers before <u>Adding</u> or <u>Subtracting</u> them.

EXAMPLES:

1. Work out $7 + 9 \div 3$
1) Follow BODMAS — do the <u>division</u> first... $7 + 9 \div 3$
2) ...then the <u>addition</u>: $= 7 + 3$
   $= 10$

If you don't follow BODMAS, you get:
$7 + 9 \div 3$
$= 16 \div 3$
$= 5.333...$ ✗

2. Calculate $15 - 7^2$
1) The square is an 'other' so that's first: $15 - 7^2$
2) Then do the <u>subtraction</u>: $= 15 - 49$
   $= -34$

3. Find $(5 + 3) \times (12 - 3)$
1) Start by working out the <u>brackets</u>: $(5 + 3) \times (12 - 3)$
2) And now the <u>multiplication</u>: $= 8 \times 9$
   $= 72$

Don't Be Scared of *Wordy Questions*

A lot of the marks on your exam are for answering <u>wordy</u>, <u>real-life</u> questions. For these you don't just have to do <u>the maths</u>, you've got to work out what the question's <u>asking you to do</u>. <u>Relax</u> and work through them <u>step by step</u>.

1) <u>READ</u> the question <u>carefully</u>. Work out <u>what bit of maths</u> you need to answer it.
2) <u>Underline the INFORMATION YOU NEED</u> to answer the question — you might not have to use <u>all</u> the numbers they give you.
3) Write out the question <u>IN MATHS</u> and answer it, showing all your <u>working</u> clearly.

EXAMPLE:

A return car journey from Carlisle to Manchester uses $\frac{4}{7}$ of a tank of petrol.
It costs £56 for a full tank of petrol. How much does the journey cost?

1) The "$\frac{4}{7}$" tells you this is a <u>fractions</u> question. (Fractions questions are covered on page 22.)

2) You need <u>£56</u> (the cost of a full tank) and $\frac{4}{7}$ (the fraction of the tank used). It doesn't matter where they're driving from and to.

3) You want to know $\frac{4}{7}$ of £56, so in maths: $\frac{4}{7} \times £56 = £32$

Don't forget the units in your final answer — this is a question about cost in pounds, so the units will be £.

Brackets, Other, Division, Multiplication, Addition, Subtraction

It's really important to check your working on BODMAS questions. You might be certain you did the calculation right, but it's surprisingly easy to make a slip.

Calculating Tips

This page covers some really important stuff about using calculators.

Know Your **Buttons** ©

Look for these buttons on your calculator — they might be a bit different on yours.

Ans This uses your <u>last answer</u> in your current calculation. Super useful.

∛☐ The <u>cube root</u> button. You might have to press <u>shift</u> first.

S⇔D Flips your answer from a <u>fraction or root</u> to a <u>decimal</u> and vice versa.

x⁻¹ The <u>reciprocal</u> button.

The <u>RECIPROCAL</u> of a number is <u>1 DIVIDED BY IT</u>.
- So the reciprocal of 2 is ½, and the reciprocal of ¼ is 4 (1 ÷ ¼).
- <u>0</u> doesn't have a reciprocal (because you <u>can't</u> divide by 0).
- A <u>number</u> multiplied by its <u>reciprocal</u> is <u>1</u> (e.g. 2 × ½ = 1, 4 × ¼ = 1).
- <u>Dividing</u> by a <u>number</u> is the same as <u>multiplying</u> by its <u>reciprocal</u>, (i.e. ÷ 2 is the same as × ½).

BODMAS on Your Calculator Ⓓ

A BODMAS question on the <u>calculator paper</u> will be packed with <u>tricky decimals</u> and possibly a <u>square root</u>. You <u>could</u> do it on your calculator in one go, but that runs the risk of losing a precious mark.

EXAMPLE:

Work out $\dfrac{2.48 - 0.79}{\sqrt{9.2 + 6.35}}$.

Write down all the figures on your calculator display.

Do it in stages and write down each step:

1 Work out the number inside the <u>square root sign</u>: $\dfrac{2.48 - 0.79}{\sqrt{9.2 + 6.35}}$

9.2 **+** **6.35** **=**

$= \dfrac{2.48 - 0.79}{\sqrt{15.55}}$

2 Use the answer to work out the <u>bottom</u> of the fraction: **√☐** **Ans** **=**

Write the answer down and store it in the <u>memory</u> by pressing: **STO** **M+**

3 Now work out the <u>top</u> of the fraction:

2.48 **−** **0.79** **=**

$= \dfrac{1.69}{3.943348831}$

$= 0.4285697443$

4 Do the division:

1.69 **÷** **RCL** **M+** **=**

This gets the value of the bottom of the fraction out of the memory.

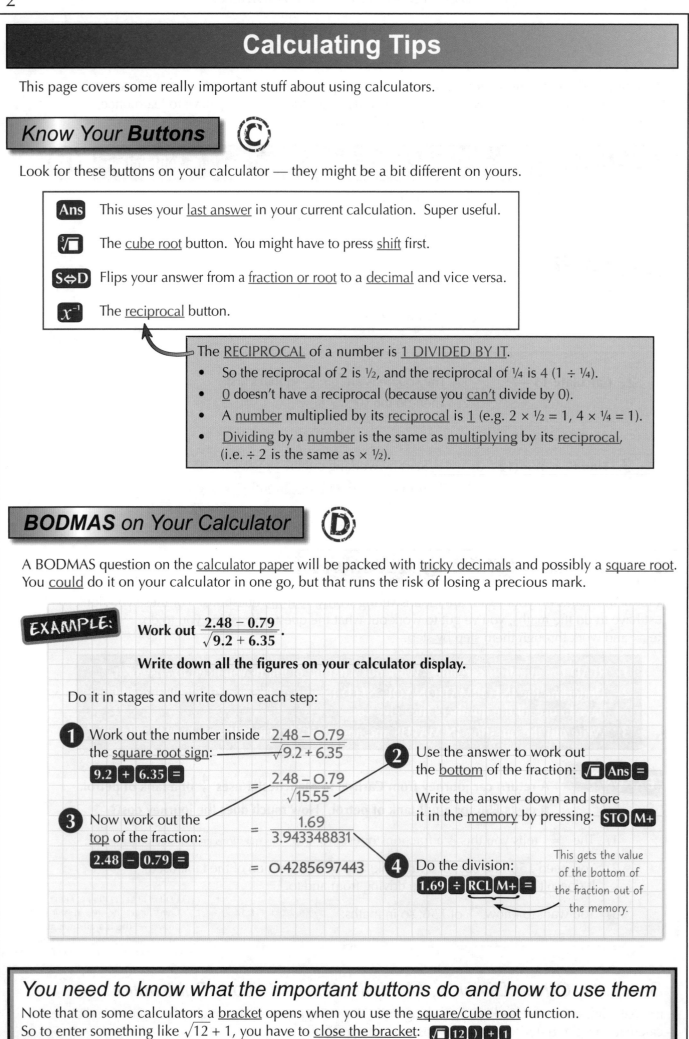

You need to know what the important buttons do and how to use them

Note that on some calculators a <u>bracket</u> opens when you use the <u>square/cube root</u> function.
So to enter something like $\sqrt{12} + 1$, you have to <u>close the bracket</u>: **√☐** **12** **)** **+** **1**

Ordering Numbers and Place Value

Here's a nice easy page to get you going.
You need to be able to: 1) <u>Read big numbers</u>, 2) <u>Write them down</u>, 3) <u>Put numbers in order</u>.

Always Look at *Big Numbers* in Groups of *Three* Ⓖ

EXAMPLE: **Write the number 2 351 243 in words.**

1) The number has spaces which break it up into groups of 3:

So many <u>MILLION</u> ⟶ 2 351 243 ⟵ And <u>THE REST</u>

So many <u>THOUSAND</u>

2) So this is: Two million, three hundred and fifty-one thousand, two hundred and forty-three

Putting Numbers in *Order of Size* Ⓕ

EXAMPLE: **Write these numbers in ascending order:**   ⟵ <u>Ascending</u> order just means smallest to largest.
12   84   623   32   486   4563   75   2143

1) First put them into groups, the ones with fewest digits first:

2-digit	3-digit	4-digit
12 84 32 75	623 486	4563 2143

2) Then just put each separate group in order of size:

12 32 75 84	486 623	2143 4563

For <u>decimals</u>: 1) Do the <u>whole-number bit first</u>, then the bit <u>after the decimal point</u>.
2) With numbers between <u>0 and 1</u>, first <u>group them</u> by the number of 0s at the start. The group with the <u>most 0s</u> at the start comes <u>first</u>.

EXAMPLE: **Write these numbers in order, from smallest to largest:**
0.531 0.098 0.14 0.0026 0.7 0.007 0.03

In decimals, like in whole numbers, the value of the digits decreases from left to right.

0.256
tenths / thousandths
hundredths

1) These are all between 0 and 1, so group them by the number of 0s at the start:

2 initial 0s	1 initial 0	no initial 0s
0.0026 0.007	0.098 0.03	0.531 0.14 0.7

2) Once they're in groups, just order them by comparing the first non-zero digits. (If the first digits are the same, look at the next digit along instead.)

0.0026 0.007   0.03 0.098   0.14 0.531 0.7

Ordering numbers is easy if you know your place values

There's nothing too tricky about putting numbers in order of size, as long as you remember the tips above. With decimals, look at the whole-number bit first, then at each decimal place in turn.

Addition and Subtraction

With a non-calculator paper coming up, I'd imagine you'd like to learn some methods for doing sums with just a pen and paper. Here they are...

Adding

1) Line up the <u>units</u> columns of each number.
2) Add up the columns from <u>right to left</u>.
3) <u>Carry over</u> any spare tens to the next column.

EXAMPLE: **Add together 292, 484 and 29.**

1) 292
   484
 + 29
     5
     1
Units lined up
2 + 4 + 9 = 15
— write 5 and carry the 1

2) 292
   484
 + 29
    O5
    2 1
9 + 8 + 2 + carried 1 = 20
— write O and carry the 2

3) 292
   484
 + 29
   8O5
   2 1
2 + 4 + carried 2 = 8

Subtracting

1) Line up the <u>units</u> columns of each number.
2) Working <u>right to left</u>, subtract the <u>bottom</u> number from the <u>top</u> number.
3) If the top number is <u>smaller</u> than the bottom number, <u>borrow</u> 10 from the left.

EXAMPLE: **Work out 693 − 665.**

1) 693
 − 665
Units lined up
You can't do 3 − 5, so borrow 10 from the left.

2) 6̶9̶3̶ (8 13)
 − 665
   O28
13 − 5 = 8
8 − 6 = 2
6 − 6 = O

And with **Decimals**... Ⓕ

The <u>method's just the same</u>, but start instead by lining up the <u>decimal points</u>.

EXAMPLES:

1. Work out 3.74 + 24.2 + 0.6.

1) 3.74
   24.20
 + 0.60
     54
      1
Decimal points lined up
It often helps to write in extra zeros to make all the decimals the same length
7 + 2 + 6 = 15 — write 5 and carry the 1

2) 3.74
   24.20
 + 0.60
   28.54
      1
3 + 4 + O + carried 1 = 8

2. Bob has £8, but spends 26p on chewing gum. How much is left?

1) £8.00
 − £0.26
Decimal points lined up
O is smaller than 6, so you can't do O − 6.

2) £8.00 (7 10)
 − £0.26
Borrow 10...

3) £8.00 (7 10 10, 9)
 − £0.26
   £7.74
...then borrow 10 again
10 − 6 = 4
9 − 2 = 7
7 − O = 7

It's vital you can add and subtract without using a calculator

Practise these methods until you're sure you've got them sussed — you're bound to need them in the exam. And watch out for questions where you need to give the units in your answer too.

Multiplying by 10, 100, etc.

You really should know the stuff on this page because:
a) it's <u>nice and simple</u>, and b) they're likely to <u>test you on it</u> in the exam.

1) To **Multiply** Any Number by **10** Ⓖ

Move the decimal point <u>ONE</u> place <u>BIGGER</u> and if it's needed, <u>ADD A ZERO</u> on the end.

E.g.   23.6 × 10 = 2 3 6

485 × 10 = 4 8 5 0

45.678 × 10 = 4 5 6 . 7 8

2) To **Multiply** Any Number by **100** Ⓖ

Move the decimal point <u>TWO</u> places <u>BIGGER</u> and <u>ADD ZEROS</u> if necessary.

E.g.   296.5 × 100 = 2 9 6 5 0

34 × 100 = 3 4 0 0

2.543 × 100 = 2 5 4 . 3

3) To **Multiply** by **1000** or **10 000**, the same rule applies: Ⓕ

Move the decimal point so many places <u>BIGGER</u> and <u>ADD ZEROS</u> if necessary.

E.g.   341 × 1000 = 3 4 1 0 0 0

2.3542 × 10 000 = 2 3 5 4 2

You always <u>MOVE</u> the <u>DECIMAL POINT</u> this much:
<u>1 place for 10,</u>   <u>2 places for 100,</u>
<u>3 places for 1000,</u>   <u>4 for 10 000</u>   etc.

4) To **Multiply** by Numbers like **20, 300, 8000** etc. Ⓔ

<u>MULTIPLY</u> by <u>2</u> or <u>3</u> or <u>8</u> etc. <u>FIRST</u>, then move the decimal point so many places <u>BIGGER</u> (↷) according to how many noughts there are.

EXAMPLE: **Calculate 234 × 200.**

1) First multiply by 2...   234 × 2 = 468
2) ...then move the decimal point 2 places   468 × 100 = 46800

Multiplying — move the decimal point to the right

This stuff might seem easy, but you still need to practise it. So turn the page and write down some examples using each of the four rules in the dark blue boxes above.

Dividing by 10, 100, etc.

This is <u>pretty easy</u> stuff too. Just <u>make sure you know it</u> — that's all.

1) To **Divide** Any Number by **10** Ⓖ

Move the decimal point <u>ONE</u> place <u>SMALLER</u> and if it's needed, <u>REMOVE ZEROS</u> after the decimal point.

E.g. $23.6 \div 10 = 2.36$

$340 \div 10 = 34$

$45.678 \div 10 = 4.5678$

2) To **Divide** Any Number by **100** Ⓖ

Move the decimal point <u>TWO</u> places <u>SMALLER</u> and <u>REMOVE ZEROS</u> after the decimal point.

E.g. $296.5 \div 100 = 2.965$

$340 \div 100 = 3.4$

$2543 \div 100 = 25.43$

3) To **Divide** by **1000** or **10 000**, the same rule applies: Ⓕ

Move the decimal point so many places <u>SMALLER</u> and <u>REMOVE ZEROS</u> after the decimal point.

E.g. $341 \div 1000 = 0.341$

$23\ 500 \div 10\ 000 = 2.35$

You always <u>MOVE</u> the <u>DECIMAL POINT</u> this much:
<u>1 place for 10</u>,   <u>2 places for 100</u>,
<u>3 places for 1000</u>,   <u>4 for 10 000</u>   etc.

4) To **Divide** by Numbers like **40**, **300**, **7000** etc. Ⓔ

<u>DIVIDE</u> by <u>4</u> or <u>3</u> or <u>7</u> etc. <u>FIRST</u>, then move the decimal point so many places <u>SMALLER</u> (i.e. to the left).

EXAMPLE: Calculate 960 ÷ 300.

1) First divide by 3... $960 \div 3 = 320$
2) ...then move the decimal point 2 places smaller. $320 \div 100 = 3.2$

Dividing — move the decimal point to the left

Four more rules for you to learn, but this time it's dividing. Make sure you can recognise when to use rule 4 — it should save you time in the exam.

Multiplying Without a Calculator

You need to be really happy doing multiplications <u>without</u> a calculator — you'll definitely need to do it in your non-calculator exam.

Multiplying **Whole Numbers**

There are lots of methods you can use for this. Three popular ones are shown below.
Just make sure <u>you can do it</u> using whichever method <u>you prefer</u>...

EXAMPLE: **Work out 46 × 27**

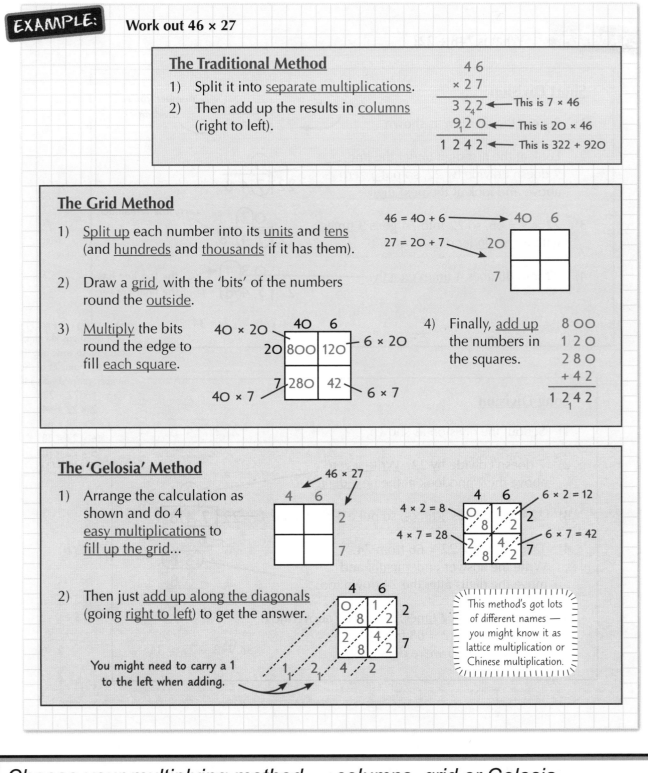

The Traditional Method

1) Split it into <u>separate multiplications</u>.
2) Then add up the results in <u>columns</u> (right to left).

$$\begin{array}{r} 46 \\ \times\ 27 \\ \hline 3\,2\,2 \\ 9\,2\,0 \\ \hline 1\,2\,4\,2 \end{array}$$

This is 7 × 46
This is 20 × 46
This is 322 + 920

The Grid Method

1) <u>Split up</u> each number into its <u>units</u> and <u>tens</u> (and <u>hundreds</u> and <u>thousands</u> if it has them).

2) Draw a <u>grid</u>, with the 'bits' of the numbers round the <u>outside</u>.

3) <u>Multiply</u> the bits round the edge to fill <u>each square</u>.

$46 = 40 + 6$
$27 = 20 + 7$

40 × 20
20 | 800 | 120 | 6 × 20
7 | 280 | 42 | 6 × 7
40 × 7

4) Finally, <u>add up</u> the numbers in the squares.

$$\begin{array}{r} 800 \\ 120 \\ 280 \\ +\ 42 \\ \hline 1\,2\,4\,2 \end{array}$$

The 'Gelosia' Method

1) Arrange the calculation as shown and do 4 <u>easy multiplications</u> to <u>fill up the grid</u>...

46 × 27

4 × 2 = 8
4 × 7 = 28
6 × 2 = 12
6 × 7 = 42

2) Then just <u>add up along the diagonals</u> (going <u>right to left</u>) to get the answer.

You might need to carry a 1 to the left when adding.

This method's got lots of different names — you might know it as lattice multiplication or Chinese multiplication.

Choose your multiplying method — columns, grid or Gelosia

Once you've chosen your favourite method for multiplying without a calculator, stick to it, and make sure you get lots of practice at using it before your non-calculator exam.

Dividing Without a Calculator

OK, time for <u>dividing</u> now. Just remember, if you don't learn one of these <u>basic methods</u>, you'll find yourself in real trouble in the exam...

Dividing **Whole Numbers** (E)

There are two common ways to do <u>division</u> — <u>long division</u> and <u>short division</u>.
Here's an example done using <u>both methods</u> so you can compare them. <u>Use</u> the method you find easier.

EXAMPLE: **What is 748 ÷ 22?**

Short Division

number you're
dividing by

number you're
dividing

22 | 7 4 8

1) Set out the division as shown.

2) Look at the first digit under the line.
7 doesn't divide by 22, so <u>put a zero</u>
above and look at the <u>next digit</u>.

22 | 7 4 8

3) 22 × 3 = 66, so 22 into 74 goes <u>3 times</u>,
with a <u>remainder</u> of 74 – 66 = 8.

carry the remainder

22 | 7 4 8 8

4) 22 into 88 goes <u>4 times exactly</u>.

the top line has
the final answer

22 | 7 4 8 8

So 748 ÷ 22 = 34

> For questions like this,
> it's useful to write out the
> first few multiples of the
> number you're dividing by,
> e.g. 1 × 22 = 22
> 2 × 22 = 44
> 3 × 22 = 66
> 4 × 22 = 88
> 5 × 22 = 110...

Long Division

1) Set out the division as shown.

22 | 7 4 8

2) 7 doesn't divide by 22. <u>Write a zero</u>
above the 7 and look at the <u>next digit</u>.

3) 22 into 74 goes <u>3 times</u>, so put a <u>3</u> above the 4.

22 | 7 4 8
– 6 6
8 8
– 8 8
0

4) <u>Take away</u> 3 × 22 = 66 from 74.
Write the answer <u>underneath</u>, and
move the digits after the 74 down too.

5) 22 into 88 goes <u>4 times</u>, so put a <u>4</u> above the 8.
<u>Take away</u> 4 × 22 = 88 from 88.
That leaves 0, so we're done.

So 748 ÷ 22 = 34

Long or short division — both give the same result

It doesn't matter which method you use to divide whole numbers. Have a go at doing a question or two with each method and decide which one you prefer. Once you've decided — practise, practise, practise.

Multiplying and Dividing with Decimals

You might get a nasty non-calculator question on multiplying or dividing using decimals. Luckily, these aren't really any harder than the whole-number versions. You just need to know what to do in each case.

Multiplying **Decimals**

1) Start by <u>ignoring</u> the decimal points. Do the multiplication using <u>whole numbers</u>.
2) Count the <u>total</u> number of digits after the <u>decimal points</u> in the original numbers.
3) Make the answer have the <u>same number</u> of decimal places.

EXAMPLE:

Work out 4.6 × 2.7

We know this 'cos we worked it out on page 7.

1) Do the whole-number multiplication: $46 × 27 = 1242$
2) Count the digits after the decimal points: 4.6 × 2.7 has <u>2 digits</u> after the decimal points, so the answer will have 2 digits after the decimal point.
3) Give the answer the same number of decimal places: $4.6 × 2.7 = 12.42$

Dividing a **Decimal** by a **Whole Number**

For these, you just set the question out like a whole-number division <u>but</u> put the <u>decimal point</u> in the answer <u>right above</u> the one in the question.

EXAMPLE: **What is 52.8 ÷ 3?**

Put the decimal point in the answer above the one in the question

3 into 5 goes once, carry the remainder of 2

3 into 22 goes 7 times, carry the remainder of 1

3 into 18 goes 6 times exactly

So 52.8 ÷ 3 = 17.6

Dividing a **Number** by a **Decimal**

Two-for-one here — this works if you're dividing a whole number by a decimal, or a decimal by a decimal.

EXAMPLE: **What is 36.6 ÷ 0.12?**

1) The trick here is to write it as a fraction: $36.6 ÷ 0.12 = \dfrac{36.6}{0.12}$
2) Get rid of the decimals by multiplying top and bottom by 100 (see p5): $= \dfrac{3660}{12}$
3) It's now a decimal-free division that you know how to solve:

12 into 3 won't go so carry the 3

12 into 36 goes 3 times exactly

12 into 6 won't go so carry the 6

12 into 60 goes 5 times exactly

So 36.6 ÷ 0.12 = 305

To divide decimals by decimals, first turn them into whole numbers

Multiply your decimals by 10, 100, etc. to get rid of the decimal points. This will give you an equivalent fraction. But be careful to multiply both the top and bottom by the same amount.

Warm-up and Worked Exam Questions

Doing sums without a calculator becomes easier the more you practise. These warm-up questions will help to get your brain in gear. Work through them without using your calculator.

Warm-up Questions

1) Write these numbers in words: a) 1 234 531   b) 23 456   c) 3402
2) Write this down as a number: Fifty-six thousand, four hundred and twenty-one.
3) Put these numbers in order of size: 23 493 87 1029 3004 345 9
4) Write these numbers in ascending order: 0.37 0.008 0.307 0.1 0.09 0.2
5) When Ric was 10 he was 142 cm tall. Since then he has grown 29 cm. How tall is he now in centimetres?
6) I have 3 litres of water and drink 1.28 litres. How many litres are left?
7) Work out: a) 12.3×100 b) 2.4×20
8) Work out: a) $2.45 \div 10$ b) $4000 \div 800$
9) Work out the following:
   a) 28×12 b) 104×8 c) 3.2×56 d) 0.6×10.2
10) Work out the following:
   a) $96 \div 8$ b) $242 \div 2$ c) $33.6 \div 0.6$ d) $45 \div 1.5$

Worked Exam Question

This first question already has the answers filled in. Have a careful read through the working and handy hints before you have a go at the exam questions on the next two pages.

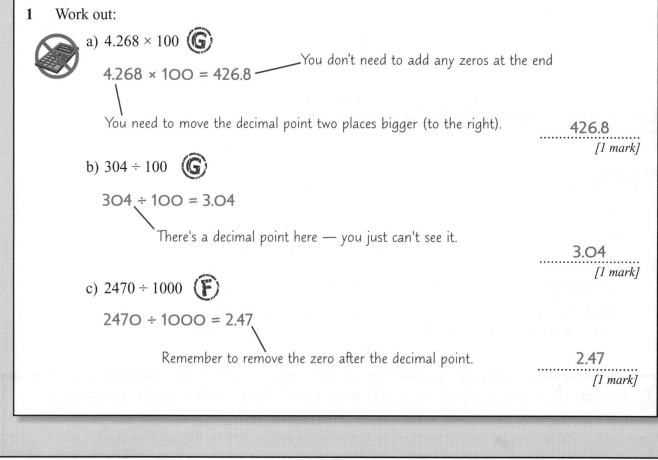

1 Work out:

a) 4.268×100 Ⓖ

You don't need to add any zeros at the end

$4.268 \times 100 = 426.8$

You need to move the decimal point two places bigger (to the right).

426.8
[1 mark]

b) $304 \div 100$ Ⓖ

$304 \div 100 = 3.04$

There's a decimal point here — you just can't see it.

3.04
[1 mark]

c) $2470 \div 1000$ Ⓕ

$2470 \div 1000 = 2.47$

Remember to remove the zero after the decimal point.

2.47
[1 mark]

Exam Questions

2 Answer each of the following. (G)

a) Write 5079 in words.

five thousand and seventy nine

[1 mark]

b) Write six thousand, one hundred and five as a number.

6,105

[1 mark]

c) What is the value of the figure 9 in 13 692?

90

[1 mark]

3 Jamie has 522 stickers. He gives 197 to his brother and 24 to his sister. (G)
How many stickers does he have left?

$$197 + 24 = 221$$

$$\begin{array}{r} 522 \\ -221 \\ \hline 301 \end{array}$$

301

[2 marks]

4 Sue and Alan meet Mark in a juice bar. (F)
Mark offers to buy a round of drinks.

Mark wants a Passion Fruit Punch and
Sue and Alan both want a Tutti Frutti.

Mark pays with a £10 note.
How much change will he get?

Juice Bar Price List	
St Clements:	£2.80
Cranberry Crush:	£2.90
Tutti Frutti:	£2.40
Passion Fruit Punch:	£2.15

$$\begin{array}{r} 2.15 \\ +\ 2.40 \\ 2.40 \\ \hline £6.95 \end{array}$$

$$\begin{array}{r} 10.00 \\ -06.95 \\ \hline 3.05 \end{array}$$

£ *3.05*

[2 marks]

5 Put these numbers in order of size, from high to low. (F)

53.30   35.60   35.54   52.91   35.06

53.30 , *52.91* , *35.60* , *35.54* , *35.06*

[1 mark]

6 The mileages of four cars are given below. (F)
Put the distances in order, starting with the lowest.

98 653   100 003   98 649   100 010

98,649 , *98,653* , *100003* , *100,010*

[1 mark]

Exam Questions

7 Work out: 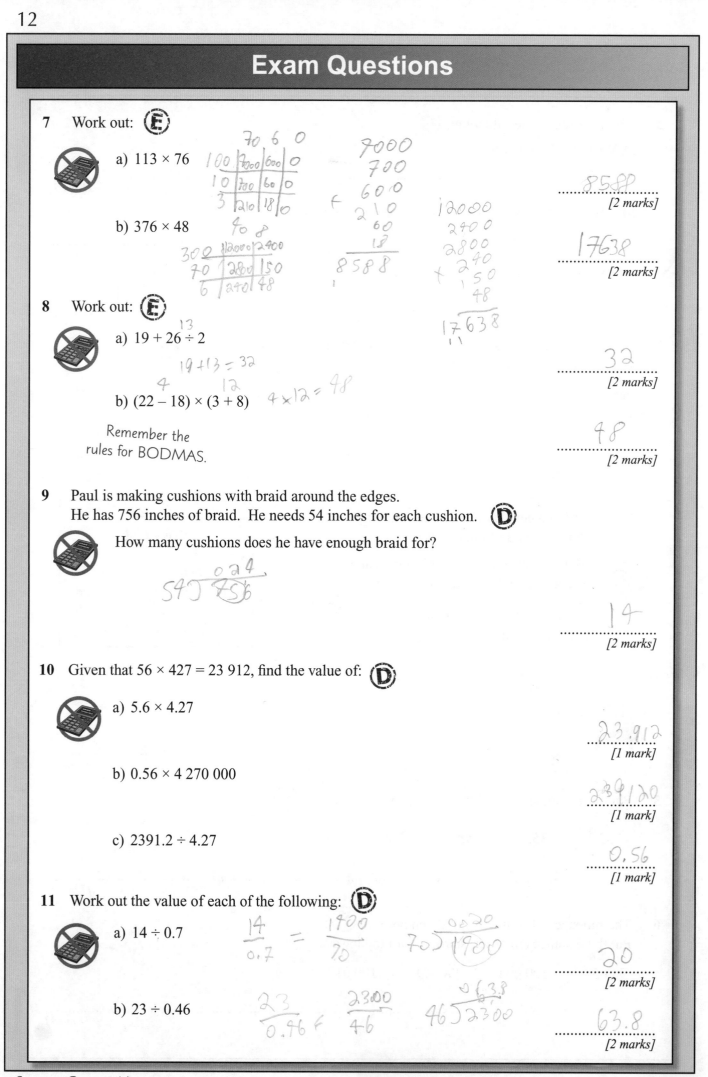 (E)

a) 113 × 76

	70	6	0
100	7000	600	0
10	700	60	0
3	210	18	0

7000
700
600
210
18

8588

.... 8588

[2 marks]

b) 376 × 48

	40	8
300	12000	2400
70	2800	150
6	240	48

12000
2400
2800
240
150
48

17638

.... 17638

[2 marks]

8 Work out: (E)

a) 19 + 26 ÷ 2

13

19 + 13 = 32

.... 32

[2 marks]

b) (22 − 18) × (3 + 8)

4   12

4 × 12 = 48

Remember the rules for BODMAS.

.... 48

[2 marks]

9 Paul is making cushions with braid around the edges.
He has 756 inches of braid. He needs 54 inches for each cushion. (D)

How many cushions does he have enough braid for?

54) 756   024

.... 14

[2 marks]

10 Given that 56 × 427 = 23 912, find the value of: (D)

a) 5.6 × 4.27

.... 23.912

[1 mark]

b) 0.56 × 4 270 000

.... 2391200

[1 mark]

c) 2391.2 ÷ 4.27

.... 0.56

[1 mark]

11 Work out the value of each of the following: (D)

a) 14 ÷ 0.7

$$\frac{14}{0.7} = \frac{1400}{70}$$

70) 1900   0020

.... 20

[2 marks]

b) 23 ÷ 0.46

$$\frac{23}{0.46} \leftarrow \frac{2300}{46}$$

46) 2300   063.8

.... 63.8

[2 marks]

Negative Numbers

Numbers less than zero are <u>negative</u>. You should be able to <u>add</u>, <u>subtract</u>, <u>multiply</u> and <u>divide</u> with them.

Adding and *Subtracting* with Negative Numbers $\textcircled{F}$

Use the <u>number line</u> for <u>addition</u> and <u>subtraction</u> involving negative numbers:

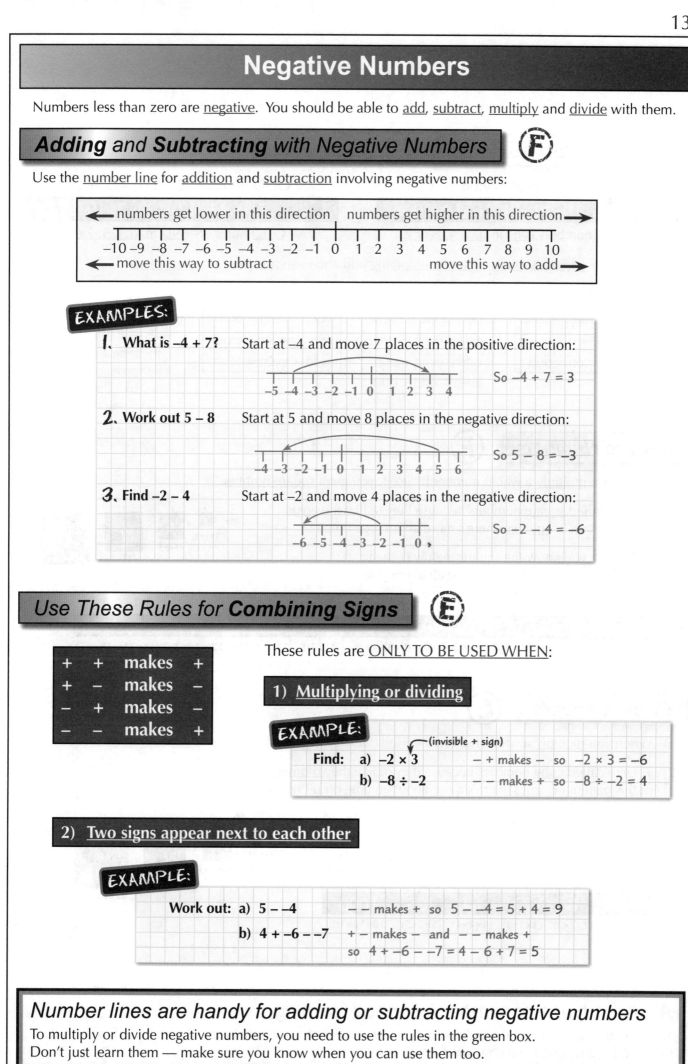

← numbers get lower in this direction   numbers get higher in this direction →

–10 –9 –8 –7 –6 –5 –4 –3 –2 –1 0 1 2 3 4 5 6 7 8 9 10

← move this way to subtract   move this way to add →

EXAMPLES:

1. **What is –4 + 7?**   Start at –4 and move 7 places in the positive direction:

–5 –4 –3 –2 –1 0 1 2 3 4

So –4 + 7 = 3

2. **Work out 5 – 8**   Start at 5 and move 8 places in the negative direction:

–4 –3 –2 –1 0 1 2 3 4 5 6

So 5 – 8 = –3

3. **Find –2 – 4**   Start at –2 and move 4 places in the negative direction:

–6 –5 –4 –3 –2 –1 0

So –2 – 4 = –6

Use These Rules for *Combining Signs* $\textcircled{E}$

+	+	makes	+
+	–	makes	–
–	+	makes	–
–	–	makes	+

These rules are <u>ONLY TO BE USED WHEN</u>:

1) Multiplying or dividing

EXAMPLE:

(invisible + sign)

Find: a) **–2 × 3**   – + makes – so –2 × 3 = –6
     b) **–8 ÷ –2**   – – makes + so –8 ÷ –2 = 4

2) Two signs appear next to each other

EXAMPLE:

Work out: a) **5 – –4**   – – makes + so 5 – –4 = 5 + 4 = 9
     b) **4 + –6 – –7**   + – makes – and – – makes +
                so 4 + –6 – –7 = 4 – 6 + 7 = 5

Number lines are handy for adding or subtracting negative numbers

To multiply or divide negative numbers, you need to use the rules in the green box.
Don't just learn them — make sure you know when you can use them too.

Special Types of Number

You need to know all the types of number on this page. They're each <u>special</u> in their very own way.

Even and *Odd* Numbers (F)

EVEN numbers all divide by 2

2 4 6 8 10 12 14 16 18 20 ...

All <u>EVEN</u> numbers <u>END</u> in <u>0, 2, 4, 6 or 8</u>

ODD numbers don't divide by 2

1 3 5 7 9 11 13 15 17 19 21 ...

All <u>ODD</u> numbers <u>END</u> in <u>1, 3, 5, 7 or 9</u>

These <u>rules</u> for <u>adding, subtracting and multiplying</u> odd and even numbers are <u>always true</u>:

Adding	Subtracting	Multiplying
odd + odd = even	odd − odd = even	odd × odd = odd
even + even = even	even − even = even	even × even = even
odd + even = odd	odd − even = odd	odd × even = even
	even − odd = odd	

Don't stress too hard trying to remember these rules — if you're not sure, try doing a calculation with some odd or even numbers. The answer will tell you the rule.

Square Numbers (F)

1) When you <u>multiply</u> a whole number by <u>itself</u>, you get a <u>square number</u>.

2) They're called <u>square</u> numbers because they're like the <u>areas</u> of this pattern of <u>squares</u> (there's more about area on p92):

3) Make sure you know the squares below <u>by heart</u> — they could come up on a non-calculator paper.

1^2	2^2	3^2	4^2	5^2	6^2	7^2	8^2	9^2	10^2	11^2	12^2	13^2	14^2	15^2
1	4	9	16	25	36	49	64	81	100	121	144	169	196	225
(1×1)	(2×2)	(3×3)	(4×4)	(5×5)	(6×6)	(7×7)	(8×8)	(9×9)	(10×10)	(11×11)	(12×12)	(13×13)	(14×14)	(15×15)

Cube Numbers (E)

1) When you <u>multiply</u> a whole number by <u>itself</u>, then by itself <u>again</u>, you get a <u>cube number</u>.

2) They're called <u>cube</u> numbers because they're like the <u>volumes</u> of this pattern of <u>cubes</u> (there's more about volume on p99):

3) You need to know some cubes <u>by heart</u> too — these are the ones to learn:

1^3	2^3	3^3	4^3	5^3	10^3
1	8	27	64	125	1000
(1×1×1)	(2×2×2)	(3×3×3)	(4×4×4)	(5×5×5)	(10×10×10)

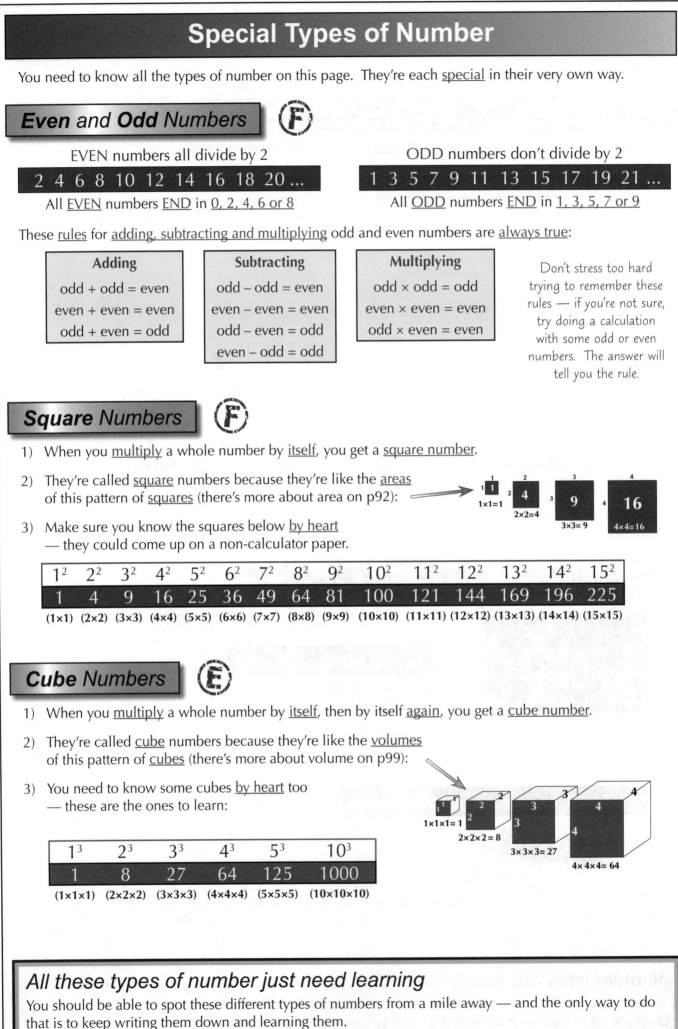

All these types of number just need learning

You should be able to spot these different types of numbers from a mile away — and the only way to do that is to keep writing them down and learning them.

Prime Numbers

There's one more special set of numbers you need to know about — the prime numbers...

PRIME Numbers Don't Divide by Anything

Prime numbers are all the numbers that DON'T come up in times tables:

$$2 \quad 3 \quad 5 \quad 7 \quad 11 \quad 13 \quad 17 \quad 19 \quad 23 \quad 29 \quad 31 \quad 37 \quad ...$$

The only way to get ANY PRIME NUMBER is: $1 \times$ ITSELF

E.g. The only numbers that multiply to give 7 are: 1×7
      The only numbers that multiply to give 31 are: 1×31

EXAMPLE: **Show that 24 is not a prime number.**

Just find another way to make 24 other than 1×24: $2 \times 12 = 24$

24 divides by other numbers apart from 1 and 24, so it isn't a prime number.

Five Important Facts

1) 1 is NOT a prime number.

2) 2 is the ONLY even prime number.

3) The first four prime numbers are 2, 3, 5 and 7.

4) Prime numbers end in 1, 3, 7 or 9 (2 and 5 are the only exceptions to this rule).

5) But NOT ALL numbers ending in 1, 3, 7 or 9 are primes, as shown here:
   (Only the circled ones are primes.)

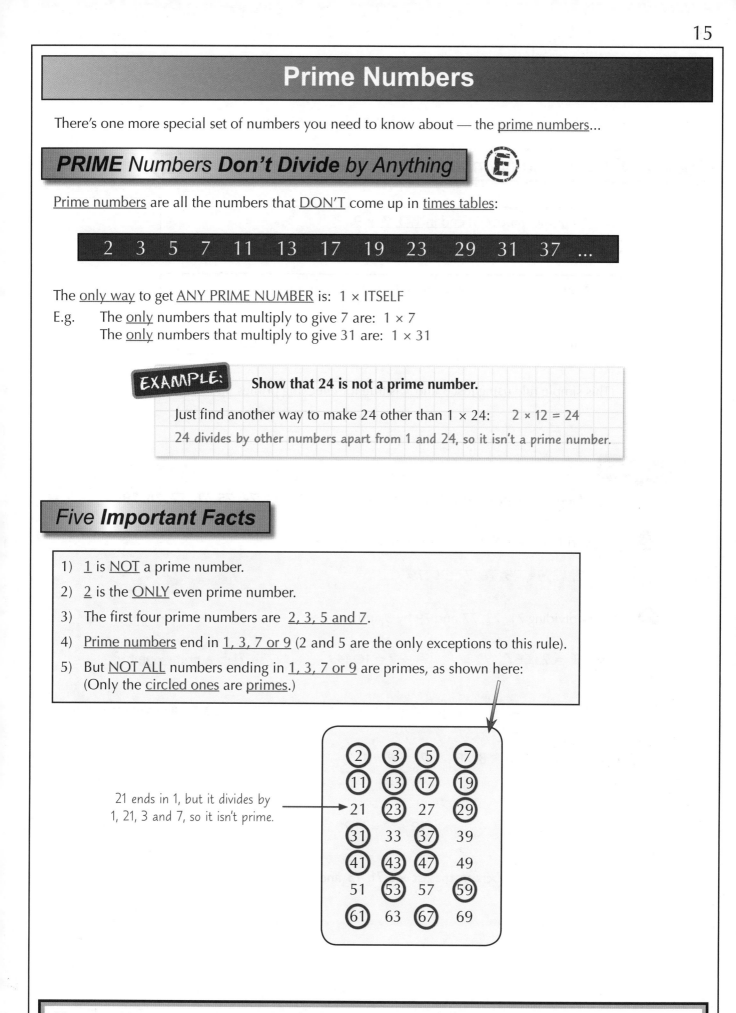

21 ends in 1, but it divides by 1, 21, 3 and 7, so it isn't prime.

Remember — prime numbers don't come up in times tables

You'll save time if you can answer prime number questions straight off, without having to test for primes. So memorise as many primes as you can from the diagram above.

Prime Numbers

Now you know what prime numbers are, it's time to learn how to find them...

How to **FIND** Prime Numbers — a very simple method

1) <u>All primes</u> (above 5) <u>end in 1, 3, 7 or 9</u>.
   So ignore any numbers that don't end in one of those.

2) Now, to find which of them <u>ACTUALLY ARE</u> primes
   you only need to <u>divide each one by 3 and 7</u>.

   If it <u>doesn't</u> divide exactly by either 3 or 7 then <u>it's a prime</u>.

(This simple rule <u>using just 3 and 7</u> is true for checking primes <u>up to 120</u>.)

EXAMPLE:

Find all the prime numbers in this list: 71, 72, 73, 74, 75, 76, 77, 78, 79

1 First, get rid of anything that doesn't end in 1, 3, 7 or 9:

   71, ~~72~~, 73, ~~74~~, ~~75~~, ~~76~~, 77, ~~78~~, 79

2 Now try dividing 71, 73, 77 and 79 by 3 and 7.

   $71 \div 3 = 23.667$ $71 \div 7 = 10.143$ so <u>71 is a prime number</u>

   $73 \div 3 = 24.333$ $73 \div 7 = 10.429$ so <u>73 is a prime number</u>

   $77 \div 3 = 25.667$ BUT: $77 \div 7 = 11$ — 11 is a whole number,
   so <u>77 is NOT a prime</u>, because it divides by 7.

   $79 \div 3 = 26.333$ $79 \div 7 = 11.286$ so <u>79 is a prime number</u>

   So the prime numbers in the list are **71, 73 and 79**.

To test for prime numbers, divide by 3 and 7
You have to be able to recognise prime numbers. The first few are easy enough to remember by heart, but when it comes to bigger numbers the only way is to use the prime number test above.

Multiples, Factors and Prime Factors

Hmm, the words above look <u>important</u>. Don't panic, explanations and examples are below.

Multiples and Factors E

The <u>MULTIPLES</u> of a number are just its <u>times table</u>.

EXAMPLE: **Find the first 8 multiples of 13.**
You just need to find the first 8 numbers in the 13 times table:
13   26   39   52   65   78   91   104

The <u>FACTORS</u> of a number are all the numbers that <u>divide into it</u>.

There's a method that guarantees you'll find them all:

1) Start off with 1 × the number itself, then try 2 ×, then 3 × and so on, listing the pairs in rows.

2) Try each one in turn. Cross out the row if it doesn't divide exactly.

3) Eventually, when you get a number <u>repeated</u>, <u>stop</u>.

4) The numbers in the rows you haven't crossed out make up the list of factors.

EXAMPLE: **Find all the factors of 24.**

Increasing by 1 each time

1 × 24
2 × 12
3 × 8
4 × 6
5 ×
6 × 4

So the <u>factors of 24</u> are: 1, 2, 3, 4, 6, 8, 12, 24

Finding Prime Factors — The Factor Tree C

<u>Any number</u> can be broken down into a string of prime numbers all multiplied together — this is called '<u>expressing it as a product of prime factors</u>'.

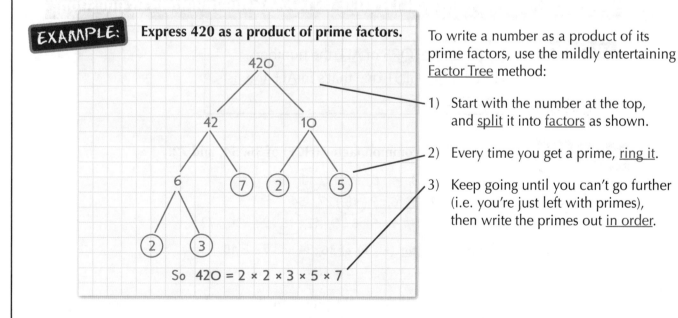

EXAMPLE: **Express 420 as a product of prime factors.**

420
42   10
6   (7) (2)   (5)
(2) (3)

So 420 = 2 × 2 × 3 × 5 × 7

To write a number as a product of its prime factors, use the mildly entertaining <u>Factor Tree</u> method:

1) Start with the number at the top, and <u>split</u> it into <u>factors</u> as shown.

2) Every time you get a prime, <u>ring it</u>.

3) Keep going until you can't go further (i.e. you're just left with primes), then write the primes out <u>in order</u>.

Follow the methods above to find factors and prime factors

It doesn't matter how you split the numbers when drawing a factor tree, as long as each pair of numbers multiplies to give the number above. For instance, we could have first split 420 into 21 and 20. The important thing is that you keep going until you only have prime numbers left.

LCM and HCF

Two big fancy names but don't be put off — they're both <u>real easy</u>.

LCM — 'Lowest Common Multiple' Ⓒ

'<u>Lowest Common Multiple</u>' — sure, it sounds kind of complicated, but all it means is this:

The <u>SMALLEST</u> number that will <u>DIVIDE BY ALL</u> the numbers in question.

METHOD:
1) <u>LIST</u> the <u>MULTIPLES</u> of <u>ALL</u> the numbers.
2) Find the <u>SMALLEST</u> one that's in <u>ALL the lists</u>.
3) That's the <u>LCM</u>.

The LCM is sometimes called the Least (instead of 'Lowest') Common Multiple.

EXAMPLE: **Find the lowest common multiple (LCM) of 12 and 15.**

Multiples of 12 are: 12, 24, 36, 48, (60,) 72, 84, 96, ...

Multiples of 15 are: 15, 30, 45, (60,) 75, 90, 105, ...

So the <u>lowest common multiple</u> (LCM) of 12 and 15 is **60**.

HCF — 'Highest Common Factor' Ⓒ

'<u>Highest Common Factor</u>' — all it means is <u>this</u>:

The <u>BIGGEST</u> number that will <u>DIVIDE INTO ALL</u> the numbers in question.

METHOD:
1) <u>LIST</u> the <u>FACTORS</u> of <u>ALL</u> the numbers.
2) Find the <u>BIGGEST</u> one that's in <u>ALL the lists</u>.
3) That's the <u>HCF</u>.

EXAMPLE: **Find the highest common factor (HCF) of 36, 54, and 72.**

Factors of 36 are: 1, 2, 3, 4, 6, 9, 12, (18,) 36

Factors of 54 are: 1, 2, 3, 6, 9, (18,) 27, 54

Factors of 72 are: 1, 2, 3, 4, 6, 8, 9, 12, (18,) 24, 36, 72

So the <u>highest common factor</u> (HCF) of 36, 54 and 72 is **18**.

Just <u>take care</u> listing the factors — make sure you use the <u>proper method</u> (as shown on the previous page) or you'll miss one and blow the whole thing out of the water.

LCM and HCF — learn what the names mean

LCM and HCF questions shouldn't be too bad as long as you know exactly what's meant by each of the terms. Then you just multiply or divide to find the multiples or factors.

Warm-up and Worked Exam Questions

These warm-up questions will test whether you've learned the facts from the last few pages. Keep practising any you get stuck on, before moving on.

Warm-up Questions

1) Work out: a) –4 × –3   b) –4 + –5 + 3   c) (3 + –2 – 4) × (2 + –5)   d) 120 ÷ –40
2) Choose from the numbers 1, 2, 3, 4, 5, 6, 7, 8, 9, 10.
   Which numbers are: a) even?   b) odd?   c) square?   d) cube?
3) Which of the following numbers are prime? 30, 31, 32, 33, 34, 35, 36, 37, 38, 39, 40.
4) Explain why 27 is not a prime number.
5) Find all the factors of 40.
6) Find the prime factors of 40.
7) Find the lowest common multiple of 4 and 5.
8) Find the highest common factor of 36 and 96.

Worked Exam Question

Time for another exam question with the answers filled in. Understanding this solution will help you with the exam questions that follow and in the exam itself.

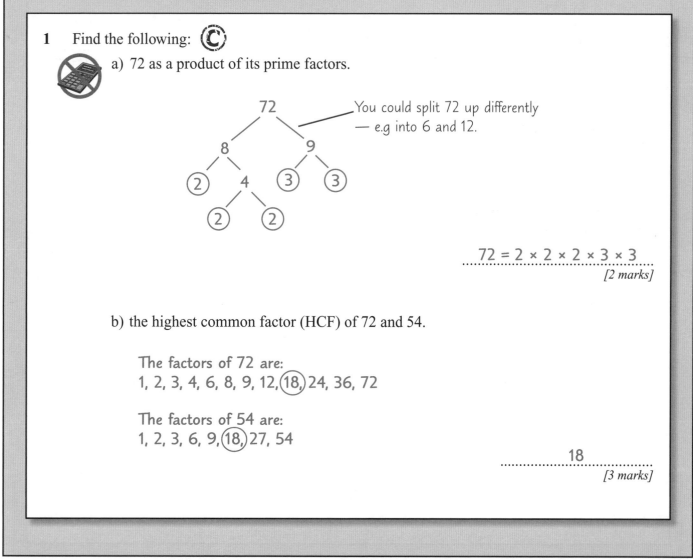

1   Find the following: (C)

a) 72 as a product of its prime factors.

72

You could split 72 up differently — e.g into 6 and 12.

8   9

(2)   4   (3) (3)

(2)   (2)

72 = 2 × 2 × 2 × 3 × 3
....
[2 marks]

b) the highest common factor (HCF) of 72 and 54.

The factors of 72 are:
1, 2, 3, 4, 6, 8, 9, 12, (18,) 24, 36, 72

The factors of 54 are:
1, 2, 3, 6, 9, (18,) 27, 54

18
....
[3 marks]

Exam Questions

2 Choose a number from the list which matches each description.

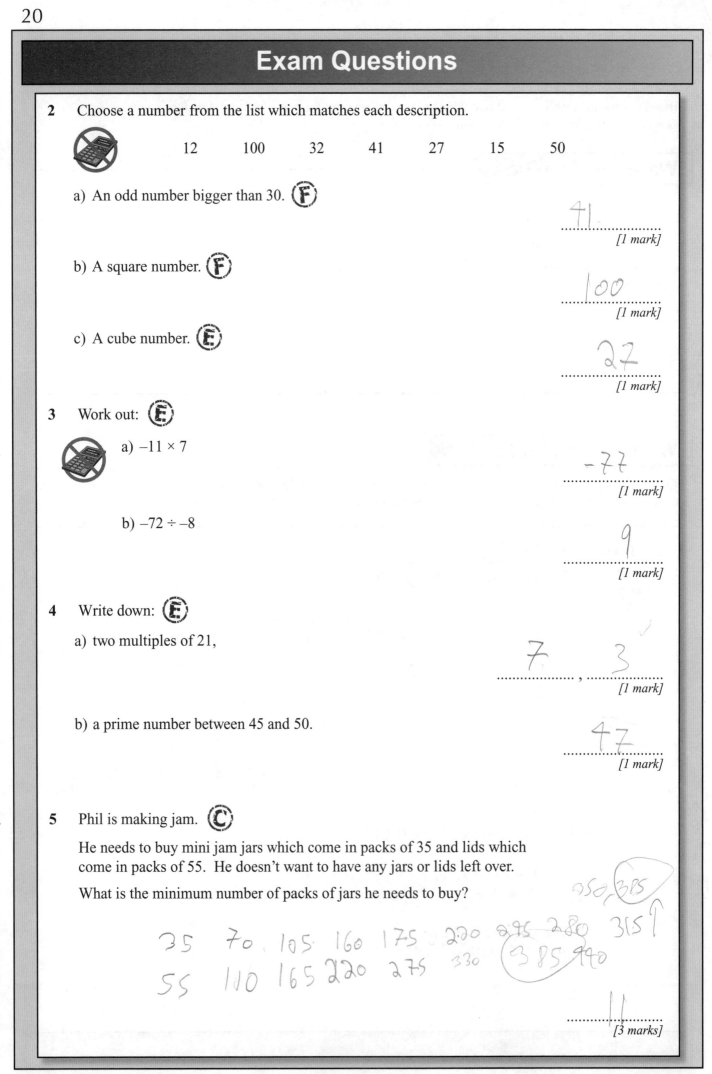

12   100   32   41   27   15   50

a) An odd number bigger than 30. Ⓕ

........................41........................
[1 mark]

b) A square number. Ⓕ

........................100........................
[1 mark]

c) A cube number. Ⓔ

........................27........................
[1 mark]

3 Work out: Ⓔ

a) -11×7

........................-77........................
[1 mark]

b) $-72 \div -8$

........................9........................
[1 mark]

4 Write down: Ⓔ

a) two multiples of 21,

........7........ , ...3........
[1 mark]

b) a prime number between 45 and 50.

........................47........................
[1 mark]

5 Phil is making jam. Ⓒ

He needs to buy mini jam jars which come in packs of 35 and lids which come in packs of 55. He doesn't want to have any jars or lids left over.

What is the minimum number of packs of jars he needs to buy?

250, 385

35 70 105 160 175 220 275 280 315

55 110 165 220 275 330 385 440

........................11........................
[3 marks]

Fractions, Decimals and Percentages

Fractions, decimals and percentages are <u>three different ways</u> of describing when you've got <u>part</u> of a <u>whole thing</u>. They're <u>closely related</u> and you can <u>convert between them</u>.

Converting Between *Fractions*, *Decimals* and *Percentages*

This table shows the really common conversions which you should know straight off without having to work them out:

Fractions with a 1 on the top (e.g. $\frac{1}{2}$, $\frac{1}{3}$, $\frac{1}{4}$, etc.) are called <u>unit fractions</u>.

Fraction	Decimal	Percentage
$\frac{1}{2}$	0.5	50%
$\frac{1}{4}$	0.25	25%
$\frac{3}{4}$	0.75	75%
$\frac{1}{3}$	0.333333...	$33\frac{1}{3}$%
$\frac{2}{3}$	0.666666...	$66\frac{2}{3}$%
$\frac{1}{10}$	0.1	10%
$\frac{2}{10}$	0.2	20%
$\frac{1}{5}$	0.2	20%
$\frac{2}{5}$	0.4	40%

0.3333... and 0.6666... are known as 'recurring' decimals — the same pattern of numbers carries on repeating itself forever. See p25.

The more of those conversions you learn, the better — but for those that you <u>don't know</u>, you must <u>also learn</u> how to <u>convert</u> between the three types. These are the methods:

$$\text{Fraction} \xrightarrow[\substack{\text{E.g. } \frac{7}{20} \text{ is } 7 \div 20}]{\text{Divide}} \text{Decimal} \xrightarrow[= 0.35]{} \xrightarrow[\text{e.g. } 0.35 \times 100]{\times \text{ by } 100} \text{Percentage}_{= 35\%}$$

$$\text{Fraction} \xleftarrow[\text{The awkward one}]{} \text{Decimal} \xleftarrow[\div \text{ by } 100]{} \text{Percentage}$$

<u>Converting decimals to fractions</u> is a bit more awkward.
The digits after the decimal point go on the top, and a <u>power of 10</u> on the bottom — with the same number of zeros as there were decimal places.

$0.6 = \frac{6}{10}$	$0.3 = \frac{3}{10}$	$0.7 = \frac{7}{10}$ etc.
$0.12 = \frac{12}{100}$	$0.78 = \frac{78}{100}$	$0.05 = \frac{5}{100}$ etc.
$0.345 = \frac{345}{1000}$	$0.908 = \frac{908}{1000}$	$0.024 = \frac{24}{1000}$ etc.

These can often be <u>cancelled down</u> — see p22.

Fractions, decimals and percentages are interchangeable

It's important you remember that a fraction, decimal or percentage can be converted into either of the other two forms. And it's even more important that you learn how to do it.

Fractions

These pages show you how to cope with fraction calculations without your <u>beloved calculator</u>.

Equivalent Fractions and *Cancelling Down*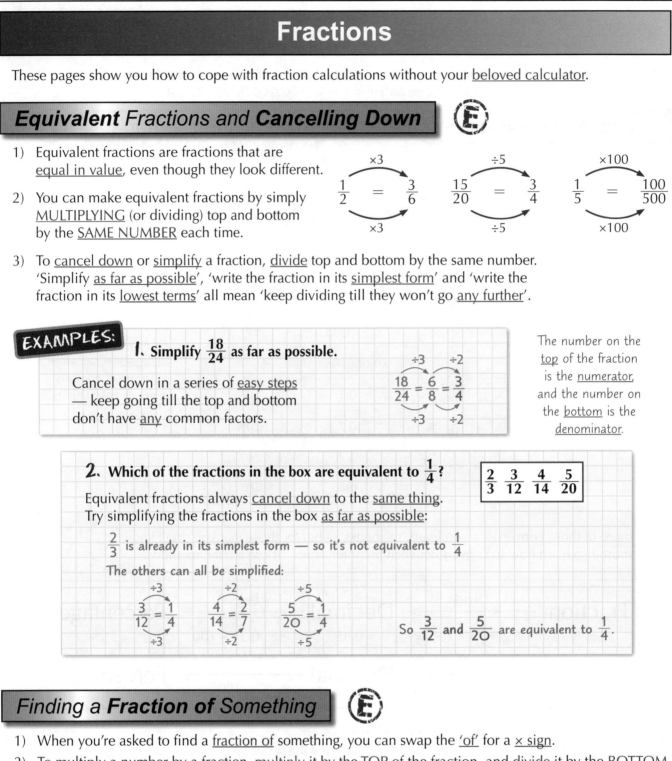

1) Equivalent fractions are fractions that are <u>equal in value</u>, even though they look different.

2) You can make equivalent fractions by simply <u>MULTIPLYING</u> (or dividing) top and bottom by the <u>SAME NUMBER</u> each time.

3) To <u>cancel down</u> or <u>simplify</u> a fraction, <u>divide</u> top and bottom by the same number. 'Simplify <u>as far as possible</u>', 'write the fraction in its <u>simplest form</u>' and 'write the fraction in its <u>lowest terms</u>' all mean 'keep dividing till they won't go <u>any further</u>'.

EXAMPLES:

1. Simplify $\frac{18}{24}$ as far as possible.

Cancel down in a series of <u>easy steps</u> — keep going till the top and bottom don't have <u>any</u> common factors.

$$\frac{18}{24} = \frac{6}{8} = \frac{3}{4}$$

The number on the top of the fraction is the <u>numerator</u>, and the number on the <u>bottom</u> is the <u>denominator</u>.

2. Which of the fractions in the box are equivalent to $\frac{1}{4}$?

$$\frac{2}{3} \quad \frac{3}{12} \quad \frac{4}{14} \quad \frac{5}{20}$$

Equivalent fractions always <u>cancel down</u> to the <u>same thing</u>. Try simplifying the fractions in the box <u>as far as possible</u>:

$\frac{2}{3}$ is already in its simplest form — so it's not equivalent to $\frac{1}{4}$

The others can all be simplified:

$$\frac{3}{12} = \frac{1}{4} \qquad \frac{4}{14} = \frac{2}{7} \qquad \frac{5}{20} = \frac{1}{4}$$

So $\frac{3}{12}$ and $\frac{5}{20}$ are equivalent to $\frac{1}{4}$.

Finding a *Fraction of* Something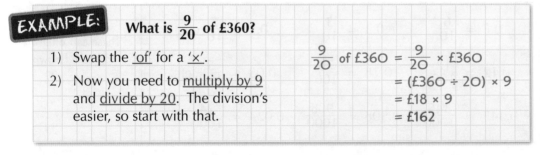

1) When you're asked to find a <u>fraction of</u> something, you can swap the <u>'of'</u> for a <u>× sign</u>.

2) To multiply a number by a fraction, <u>multiply</u> it by the <u>TOP</u> of the fraction, and <u>divide</u> it by the <u>BOTTOM</u>. It doesn't matter which order you do those two steps in — just start with whatever's easiest.

EXAMPLE: What is $\frac{9}{20}$ of £360?

1) Swap the <u>'of'</u> for a <u>'×'</u>.

2) Now you need to <u>multiply by 9</u> and <u>divide by 20</u>. The division's easier, so start with that.

$$\frac{9}{20} \text{ of } £360 = \frac{9}{20} × £360$$
$$= (£360 ÷ 20) × 9$$
$$= £18 × 9$$
$$= £162$$

Equivalent fractions are equal in value to each other

Make sure you know how to make equivalent fractions — either multiply the numerator (the top number) and the denominator (the bottom number) by the same number, or divide each of them by the same number.

Fractions

Fractions are really great — there are just so many things you can do with them. Here are a few more:

Multiplying (D)

Multiply top and bottom separately.

EXAMPLE: Find $\frac{3}{5} \times \frac{4}{7}$.

Multiply the top and bottom numbers <u>separately</u>: $\frac{3}{5} \times \frac{4}{7} = \frac{3 \times 4}{5 \times 7} = \frac{12}{35}$

Remember that multiplying by e.g. $\frac{1}{4}$ is the same as dividing by 4.

Dividing (D)

Turn the 2nd fraction <u>UPSIDE DOWN</u> and then <u>multiply</u>:

EXAMPLE: Find $\frac{3}{4} \div \frac{1}{3}$.

Turn $\frac{1}{3}$ <u>upside down</u> and <u>multiply</u>: $\frac{3}{4} \div \frac{1}{3} = \frac{3}{4} \times \frac{3}{1} = \frac{3 \times 3}{4 \times 1} = \frac{9}{4}$

Mixed Numbers (D)

<u>Mixed numbers</u> are things like $3\frac{1}{3}$, with an integer part and a fraction part. <u>Improper fractions</u> are ones where the top number is larger than the bottom number. You need to be able to convert between the two.

EXAMPLES: **1.** **Write $4\frac{2}{3}$ as an improper fraction.**

1) Think of the <u>mixed number</u> as an <u>addition</u>: $4\frac{2}{3} = 4 + \frac{2}{3}$

2) Turn the <u>whole number part</u> into a <u>fraction</u>: $4 + \frac{2}{3} = \frac{12}{3} + \frac{2}{3} = \frac{12+2}{3} = \frac{14}{3}$

2. **Write $\frac{31}{4}$ as a mixed number.**

<u>Divide</u> the top number by the bottom.

1) The <u>answer</u> gives the <u>whole number part</u>.

2) The <u>remainder</u> goes <u>on top</u> of the fraction.

$31 \div 4 = 7$ remainder 3

so $\frac{31}{4} = 7\frac{3}{4}$

*If you have to do a calculation with mixed numbers, just turn
them into improper fractions first, then carry on as normal.
(You might have to change the answer back to a mixed number at the end.)*

You have to know how to handle mixed numbers and improper fractions

Mixed numbers look difficult, but they're OK once you've converted them into normal fractions.
If you keep on practising working with fractions, you'll bag some easy marks in the exam.

Fractions

There's an awkward-looking maths word looming large on this page — but don't let it put you off.
Remember, the denominator is just the number on the bottom of the fraction.

Common Denominators

This comes in handy for <u>comparing</u> the sizes of fractions and for <u>adding</u> or <u>subtracting</u> fractions.

> You need to find a number that <u>all</u> the denominators <u>divide into</u> — this will be your <u>common denominator</u>.
> The simplest way is to <u>multiply</u> all the different denominators together.

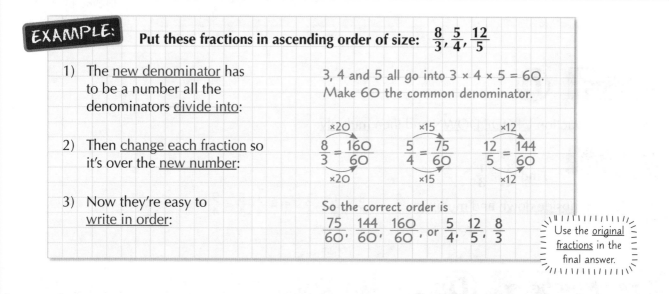

EXAMPLE: Put these fractions in ascending order of size: $\frac{8}{3}, \frac{5}{4}, \frac{12}{5}$

1) The <u>new denominator</u> has to be a number all the denominators <u>divide into</u>:

   3, 4 and 5 all go into $3 \times 4 \times 5 = 60$.
   Make 60 the common denominator.

2) Then <u>change each fraction</u> so it's over the <u>new number</u>:

   $\frac{8}{3} = \frac{160}{60}$ (×20) $\frac{5}{4} = \frac{75}{60}$ (×15) $\frac{12}{5} = \frac{144}{60}$ (×12)

3) Now they're easy to <u>write in order</u>:

   So the correct order is
   $\frac{75}{60}, \frac{144}{60}, \frac{160}{60}$, or $\frac{5}{4}, \frac{12}{5}, \frac{8}{3}$

   Use the <u>original</u> <u>fractions</u> in the final answer.

To **Add** and **Subtract** — sort the denominators first

1) Make sure the denominators are <u>the same</u> (see above).
2) Add (or subtract) the top lines (numerators) <u>only</u>.

If you're adding or subtracting <u>mixed numbers</u>, it usually helps to convert them to improper fractions first.

EXAMPLES:

1. Calculate $\frac{1}{2} - \frac{1}{5}$.

Find a <u>common denominator</u>: $\frac{1}{2} - \frac{1}{5} = \frac{5}{10} - \frac{2}{10}$

Combine the <u>top lines</u>: $= \frac{5-2}{10} = \frac{3}{10}$

2. Work out $2\frac{4}{7} + 1\frac{5}{7}$.

1) Write the mixed numbers as <u>improper fractions</u>: $2\frac{4}{7} + 1\frac{5}{7} = \frac{18}{7} + \frac{12}{7}$

2) The <u>denominators</u> are the same, so just add the <u>top lines</u>: $= \frac{18+12}{7}$

3) Turn the answer back into a <u>mixed number</u>: $= \frac{30}{7} = 4\frac{2}{7}$

Common denominators are really handy

You need to find a common denominator to order, add or subtract fractions.
Just don't forget to turn your fractions back into their original form when you give your answer.

Fractions and Recurring Decimals

Terminating and recurring decimals can always be written as fractions. You saw how to convert terminating decimals on page 21 — now it's time for recurring decimals.

Recurring or Terminating...

1) Recurring decimals have a pattern of numbers which repeats forever.

   For example, $\frac{1}{3}$ is the decimal 0.333333...

2) It doesn't have to be a single digit that repeats.
   E.g. You could have 0.143143143...

3) The repeating part is usually marked with dots on top of the number.

4) If there's one dot, only one digit is repeated. If there are two dots, then everything from the first dot to the second dot is the repeating bit.

   E.g. $0.2\dot{5}$ = 0.2555555...,
   $0.\dot{2}\dot{5}$ = 0.25252525...,
   $0.\dot{2}6\dot{5}$ = 0.265265265...

5) Terminating decimals don't go on forever.

   E.g. $\frac{1}{20}$ is the terminating decimal 0.05

6) All terminating and recurring decimals can be written as fractions.

Fraction	Recurring decimal or terminating decimal?	Decimal
$\frac{1}{2}$	Terminating	0.5
$\frac{1}{3}$	Recurring	$0.\dot{3}$
$\frac{1}{4}$	Terminating	0.25
$\frac{1}{5}$	Terminating	0.2
$\frac{1}{6}$	Recurring	$0.1\dot{6}$
$\frac{1}{7}$	Recurring	$0.\dot{1}4285\dot{7}$
$\frac{1}{8}$	Terminating	0.125
$\frac{1}{9}$	Recurring	$0.\dot{1}$
$\frac{1}{10}$	Terminating	0.1

Turning Fractions into Recurring Decimals

You might find this cropping up in your exam too — and if they're being really unpleasant, they'll stick it in a non-calculator paper.

EXAMPLE: Without using a calculator, write $\frac{5}{11}$ as a recurring decimal.

1) Remember, $\frac{5}{11}$ means 5 ÷ 11, so you can just do the division. The trick is to treat the 5 as a decimal — write it as 5.000...

   For more about division, see p8-9.

   11 into 50 goes 4 times...
   ```
        0. 4
   11 | 5 .5 0 6 0 0 0 0
   ```
   ...and carry the 6

   11 into 60 goes 5 times...
   ```
        0. 4  5
   11 | 5 .5 0 6 0 5 0 0 0
   ```
   ...and carry the 5

   ```
        0. 4  5  4  5
   11 | 5 .5 0 6 0 5 0 6 0 5 0
   ```

2) Keep going until you can see the repeating pattern. Write the recurring decimal using dots above the repeating part.

   5 ÷ 11 = 0.454545...

   so $\frac{5}{11}$ = $0.\dot{4}\dot{5}$

You need to know how to write fractions as recurring decimals

Simply divide the top of the fraction by the bottom, writing extra zeros after the decimal point as needed.

Warm-up and Worked Exam Questions

Here's a set of warm-up questions for this section. Work through them to check you've got the hang of fractions and to limber up for the exam questions that follow.

Warm-up Questions

1) What decimal is the same as $\frac{7}{10}$?

2) What percentage is the same as $\frac{2}{3}$?

3) What fraction is the same as 0.4?

4) Simplify $\frac{48}{64}$ as far as possible.

5) Which of these fractions are equivalent to $\frac{1}{3}$? $\frac{2}{6}, \frac{5}{15}, \frac{9}{36}, \frac{6}{20}$

6) Work these out, then simplify your answers where possible:

   a) $\frac{2}{5} \times \frac{2}{3}$ b) $\frac{2}{5} \div \frac{2}{3}$ c) $\frac{2}{5} + \frac{2}{3}$ d) $\frac{2}{3} - \frac{2}{5}$

7) Write $\frac{2}{7}$ as a recurring decimal.

Worked Exam Question

Make sure you understand what's going on in this question before trying the next page for yourself.

1   Work out the following. Write your answers in their simplest form. **(D)**

a) $\frac{1}{3} + \frac{2}{5}$

Common denominator is 15

$$\frac{1}{3} + \frac{2}{5} = \frac{5}{15} + \frac{6}{15}$$

$$= \frac{5+6}{15} = \frac{11}{15}$$

11 and 15 don't have any common factors (apart from 1), so this fraction is in its simplest form.

$$\frac{11}{15}$$
....
[2 marks]

b) $\frac{1}{2} - \frac{2}{7}$

$$\frac{1}{2} - \frac{2}{7} = \frac{7}{14} - \frac{4}{14}$$

$$= \frac{7-4}{14} = \frac{3}{14}$$

$$\frac{3}{14}$$
....
[2 marks]

Exam Questions

2 Convert each of the following: Ⓔ

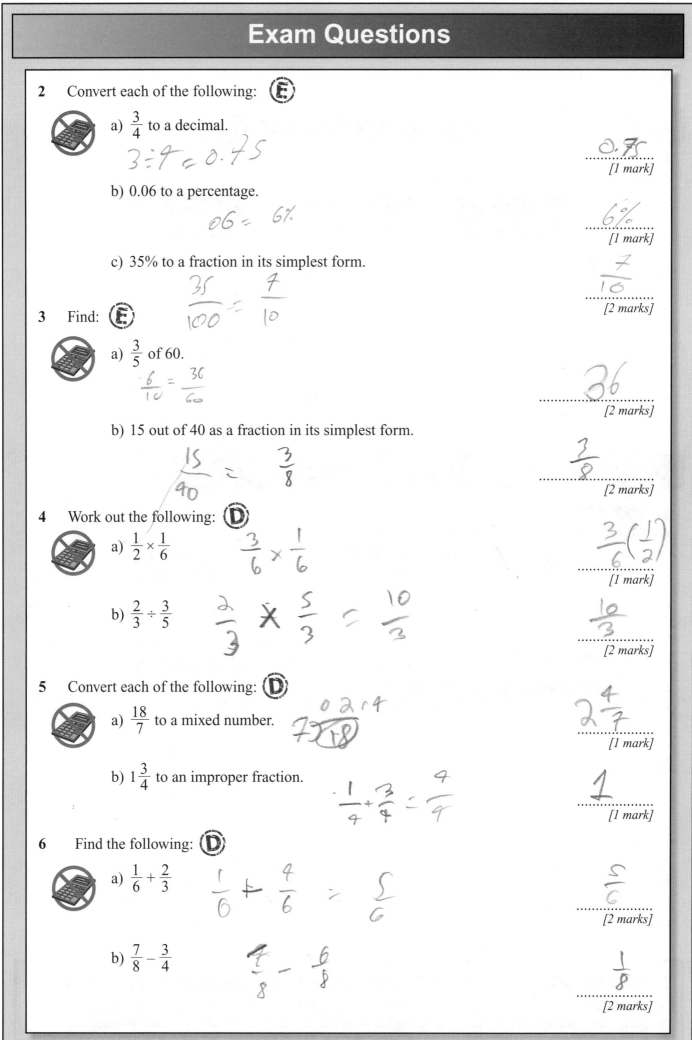

a) $\frac{3}{4}$ to a decimal.

$3 \div 4 = 0.75$

0.75
[1 mark]

b) 0.06 to a percentage.

$06 = 6\%$

6%
[1 mark]

c) 35% to a fraction in its simplest form.

$\frac{35}{100} = \frac{7}{10}$

$\frac{7}{10}$
[2 marks]

3 Find: Ⓔ

a) $\frac{3}{5}$ of 60.

$\frac{6}{10} = \frac{36}{60}$

36
[2 marks]

b) 15 out of 40 as a fraction in its simplest form.

$\frac{15}{40} = \frac{3}{8}$

$\frac{3}{8}$
[2 marks]

4 Work out the following: Ⓓ

a) $\frac{1}{2} \times \frac{1}{6}$

$\frac{3}{6} \times \frac{1}{6}$

$\frac{3}{6} \left(\frac{1}{2} \right)$
[1 mark]

b) $\frac{2}{3} \div \frac{3}{5}$

$\frac{2}{3} \times \frac{5}{3} = \frac{10}{3}$

$\frac{10}{3}$
[2 marks]

5 Convert each of the following: Ⓓ

a) $\frac{18}{7}$ to a mixed number.

$7 \overline{)18}^{\,0\,2\,r4}$

$2\frac{4}{7}$
[1 mark]

b) $1\frac{3}{4}$ to an improper fraction.

$\frac{1}{4} + \frac{3}{4} = \frac{4}{4}$

1
[1 mark]

6 Find the following: Ⓓ

a) $\frac{1}{6} + \frac{2}{3}$

$\frac{1}{6} + \frac{4}{6} = \frac{5}{6}$

$\frac{5}{6}$
[2 marks]

b) $\frac{7}{8} - \frac{3}{4}$

$\frac{7}{8} - \frac{6}{8}$

$\frac{1}{8}$
[2 marks]

Proportion Problems

Proportion problems all involve amounts that increase or decrease together.

Learn the **Golden Rule** for **Proportion** Questions ⓓ

There are lots of exam questions which at first sight seem completely
different but in fact they can all be done using the <u>GOLDEN RULE</u>...

DIVIDE FOR ONE, THEN TIMES FOR ALL

EXAMPLE: **5 pints of milk cost £1.30. How much will 3 pints cost?**

The <u>GOLDEN RULE</u> says: | DIVIDE FOR ONE, THEN TIMES FOR ALL |

which means: <u>Divide the price by 5</u> to find how much <u>FOR ONE PINT</u>,
then <u>multiply by 3</u> to find how much <u>FOR 3 PINTS</u>.

So for 1 pint: £1.30 ÷ 5 = 0.26 = 26p
For 3 pints: 26p × 3 = 78p

Use the **Golden Rule** to Scale **Recipes** Up or Down ⓓ

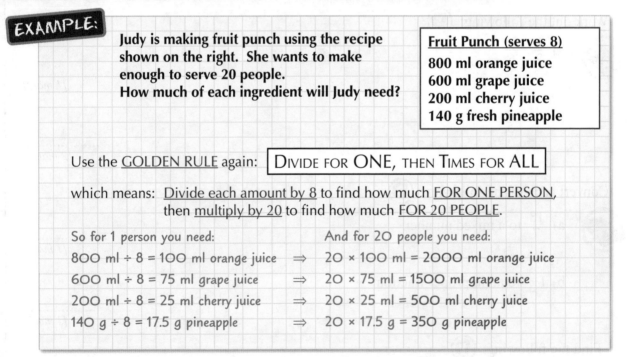

EXAMPLE:

Judy is making fruit punch using the recipe
shown on the right. She wants to make
enough to serve 20 people.
How much of each ingredient will Judy need?

Fruit Punch (serves 8)
800 ml orange juice
600 ml grape juice
200 ml cherry juice
140 g fresh pineapple

Use the <u>GOLDEN RULE</u> again: | DIVIDE FOR ONE, THEN TIMES FOR ALL |

which means: <u>Divide each amount by 8</u> to find how much <u>FOR ONE PERSON</u>,
then <u>multiply by 20</u> to find how much <u>FOR 20 PEOPLE</u>.

So for 1 person you need: And for 20 people you need:

800 ml ÷ 8 = 100 ml orange juice   ⇒   20 × 100 ml = 2000 ml orange juice

600 ml ÷ 8 = 75 ml grape juice   ⇒   20 × 75 ml = 1500 ml grape juice

200 ml ÷ 8 = 25 ml cherry juice   ⇒   20 × 25 ml = 500 ml cherry juice

140 g ÷ 8 = 17.5 g pineapple   ⇒   20 × 17.5 g = 350 g pineapple

For <u>some</u> questions like this, you can <u>just multiply</u> — e.g. in the example above, if you wanted
to know the ingredients for <u>16 servings</u> of punch, you could just times everything in the recipe by 2.

This is OK if you're <u>confident</u> you know what you're doing, but remember,
the GOLDEN RULE <u>always works</u>...

Divide for one, then times for all... divide for one, then times for all... divide for one...
Memorise this golden rule — it'll help make proportion questions a whole lot easier. But make sure you
get plenty of practice using it too — close the book and have a go at the example questions above.

Proportion Problems

Another page, another golden rule — this one's been specially handcrafted to deal with 'best buy' questions.

Best Buy Questions — Find the **Amount per Penny** Ⓓ

A slightly different type of proportion question is comparing the 'value for money' of 2 or 3 similar items. For these, follow the second GOLDEN RULE...

DIVIDE BY THE **PRICE** IN PENCE (TO GET THE AMOUNT PER PENNY)

EXAMPLE: **The local 'Supplies 'n' Vittals' stocks three sizes of Jamaican Gooseberry Jam, as shown below. Which of these represents the best value for money?**

500 g at £1.08   350 g at 80p   100 g at 42p

The GOLDEN RULE says:

DIVIDE BY THE PRICE IN PENCE TO GET THE AMOUNT PER PENNY

In the 500 g jar you get   500 g ÷ 108p = 4.63 g per penny
In the 350 g jar you get   350 g ÷ 80p = 4.38 g per penny
In the 100 g jar you get   100 g ÷ 42p = 2.38 g per penny
The 500 g jar is the best value for money, because you get more jam per penny.

With any question comparing 'value for money', DIVIDE BY THE PRICE (in pence) and it will always be the BIGGEST ANSWER that is the BEST VALUE FOR MONEY.

...or Find the **Price per Unit** Ⓓ

For some questions, the numbers mean it's easier to divide by the amount to get the cost per unit (e.g. per gram, per litre, etc.). In that case, the best buy is the smallest answer — the lowest cost per unit. Doing the example above in this way, you'd get:

The jam in the 500 g jar costs   108p ÷ 500 g = 0.216p per gram
The jam in the 350 g jar costs   80p ÷ 350 g = 0.229p per gram
The jam in the 100 g jar costs   42p ÷ 100 g = 0.42p per gram
The 500 g jar is the best value for money, because it's the cheapest per gram.

To compare prices, find the amount per penny... or the price per unit

You can use either method to solve a 'best buy' problem — look at the numbers in the question and decide which way is easier. But remember that for the 'amount per penny' method, the biggest answer is the best value. For the 'price per unit' method, the smallest answer is the best value.

Percentages

There are lots of different types of percentage questions. Read the examples on these two pages carefully and make sure you can recognise the different percentage questions you might meet.

Three **Simple** Question Types   (D)

Type 1 — "Find x% of y"

Turn the percentage into a <u>decimal</u>, then <u>multiply</u>.

EXAMPLE: **Find 15% of £46.**

divide by 100 to turn a percentage into a decimal

1) Write 15% as a <u>decimal</u>: $15\% = 15 \div 100 = 0.15$
2) <u>Multiply</u> £46 by 0.15: $0.15 \times £46 = £6.90$

Type 2 — "Find the new amount after a % increase/decrease"

1) Work out the "<u>% of original value</u>" as above — this is the actual increase or decrease.
2) <u>Add to or subtract from</u> the original value.

EXAMPLE: **A toaster is reduced in price by 40% in the sales. It originally cost £68. What is the new price of the toaster?**

1) Find <u>40% of £68</u> (using the method above): $40\% = 40 \div 100 = 0.4$
   So 40% of £68 = $0.4 \times £68 = £27.20$
2) It's a <u>decrease</u>, so <u>subtract</u> from the original: So the new price is: $£68 - £27.20 = £40.80$

Or if you prefer, you can use the <u>multiplier</u> method:

1) Write 40% as a <u>decimal</u>: $40\% = 40 \div 100 = 0.4$
2) It's a <u>decrease</u>, so find the <u>multiplier</u> by taking 0.4 from 1:   multiplier $= 1 - 0.4$
   (For an increase, you'd <u>add it to</u> 1 instead.) $= 0.6$
3) Multiply the <u>original</u> by the <u>multiplier</u>: $£68 \times 0.6 = £40.80$

Type 3 — "Express x as a percentage of y"

<u>Divide</u> x by y, then multiply by <u>100</u>.

EXAMPLE: **Give 40p as a percentage of £3.34.**

1) Make sure both amounts are in the <u>same units</u> — convert £3.34 to pence: $£3.34 = 334p$
2) <u>Divide</u> 40p by 334p, <u>then multiply</u> by 100: $(40 \div 334) \times 100 = 12.0\%$ (to 1 d.p.)

Learn how to solve these simple question types

Before you move on to the trickier examples on the next page, you need to be confident with the three simple types of percentages questions. So, cover the page and practise.

Percentages

That's right, there are more percentage questions to learn over here. Sorry about that.

Percentages *Without a Calculator*

Don't worry if you get a <u>non-calculator</u> percentage question.
You can use the trusty rules for <u>dividing by 10 and 100</u> (see p6) to help you work things out.

EXAMPLE: **Calculate 23% of 250 g. Show your working.**

1) You know that <u>250 g is 100%</u>,
   so it's easy to find <u>10%</u> and <u>1%</u>:

   $$100\% = 250\ g$$
   ÷10 ⟶ $10\% = 25\ g$ ⟵ ÷10
   ÷10 ⟶ $1\% = 2.5\ g$ ⟵ ÷10

2) Now use those values to <u>make 23%</u>:

   $$20\% = 2 \times 10\%$$
   $$= 2 \times 25\ g = 50\ g$$
   $$3\% = 3 \times 1\%$$
   $$= 3 \times 2.5\ g = 7.5\ g$$
   So $23\% = 20\% + 3\%$
   $$= 50\ g + 7.5\ g$$
   $$= 57.5\ g$$

Simple Interest

1) <u>Interest</u> is money that's usually paid when you borrow or save some money.
   It gets added to the <u>original amount</u> you saved or borrowed.

2) <u>Simple interest</u> means a certain percentage of the <u>original amount</u> is paid at regular
   intervals (usually once a year). So the amount of interest is <u>the same every time</u> it's paid.

3) To work out the <u>total amount</u> of simple interest paid over a certain amount of time, just find the
   size of <u>one interest payment</u>, then multiply it by the <u>number of payments</u> in the time period.

EXAMPLE: **Regina invests £380 in an account which pays
3% simple interest per annum.
How much interest will she earn in 4 years?**

*'Per annum' just
means 'each year'.*

1) Work out the amount of interest earned
   <u>in one year</u>:

   $$3\% = 3 \div 100 = 0.03$$
   $$3\%\ of\ £380 = 0.03 \times £380$$
   $$= £11.40$$

2) Multiply by 4 to get the <u>total interest</u> for
   <u>4 years</u>:

   $$4 \times £11.40 = £45.60$$

Percentages are one of the most useful things you'll ever learn

Whenever you open a newspaper, see an advert, watch TV or do a maths exam paper you will see
percentages. So it's really important you get confident with using them — so practise.

Ratios

Ratios can be a murky topic — but work through these examples, and it should all become crystal clear...

Reducing **Ratios** to their **Simplest Form**

To reduce a ratio to a <u>simpler form</u>, divide <u>all the numbers</u> in the ratio by the <u>same thing</u> (a bit like simplifying a fraction). It's in its <u>simplest form</u> when there's nothing left you can divide by.

> **EXAMPLE:** **Write the ratio 15:18 in its simplest form.**
>
> For the ratio 15:18, both numbers have a <u>factor</u> of 3, so <u>divide them by 3</u>.
>
> $$\div 3 \left(\frac{15:18}{5:6} \right) \div 3$$
>
> You can't reduce this any further. So the simplest form of 15:18 is <u>5:6</u>.

A <u>handy trick</u> for the calculator paper — use the <u>fraction button</u>

If you enter a fraction with the [⊟] or [a b/c] button, the calculator will cancel it down when you press [=].

So for 8:12, enter $\frac{8}{12}$ as a fraction and it'll get reduced to $\frac{2}{3}$.

Now just change it back to a ratio, i.e. <u>2:3</u>.

Proportional **Division**

If you know the <u>ratio</u> and the <u>TOTAL AMOUNT</u> you can split the total into <u>separate amounts</u>.

The key is to think about the <u>PARTS</u> that make up each amount — just follow these three steps:

> 1) <u>ADD UP THE PARTS</u>
> 2) <u>DIVIDE TO FIND ONE "PART"</u>
> 3) <u>MULTIPLY TO FIND THE AMOUNTS</u>

> **EXAMPLE:** **Jess, Mo and Greg share £9100 in the ratio 2:4:7. How much does Mo get?**
>
> 1) The ratio 2:4:7 means there will be a total of 13 <u>parts</u>:   2 + 4 + 7 = 13 parts
> 2) Divide the <u>total amount</u> by the number of <u>parts</u>:   £9100 ÷ 13 = £700   (= 1 part)
> 3) We want to know <u>Mo's share</u>, which is <u>4 parts</u>:   4 parts = 4 × 700 = £2800

Scaling Up **Ratios**

If you know the <u>ratio</u> and the actual size of <u>ONE AMOUNT</u>, you can <u>scale the ratio up</u> to find the other amounts.

> **EXAMPLE:** **Mortar is made from sand and cement in the ratio 7:2.**
> **If 21 buckets of sand are used, how much cement is needed?**
>
> You need to <u>multiply by 3</u> to go from 7 to 21 on the left-hand side — do that to <u>both sides</u>:
>
> sand:cement
> $$= \; \times 3 \left(\frac{7:2}{21:6} \right) \times 3$$
>
> So 6 buckets of cement are needed.

In some ways, ratios are kind of like fractions... in others, they're not

You can reduce a ratio to a simpler form like a fraction, or scale a ratio up by multiplying both sides.
You have to treat them a whole lot differently when doing proportional division questions though.

Warm-up and Worked Exam Questions

Here is another set of warm-up questions for this section. Give them a go before you tackle the exam questions over the page.

Warm-up Questions

1) If three chocolate bars cost 96p, how much will four of the bars cost?
2) Marmalade can be bought in 3 different sizes: 250 g (£1.25), 350 g (£2.10) or 525 g (£2.50). Which size is best value for money?
3) Calculate 34% of £50.
4) A suit costs £120 during a sale. Once the sale is over, the price of the suit rises by 15%. What is the new price of the suit?
5) What is 37 out of 50 as a percentage?
6) Find 15% of 60 (without using a calculator).
7) You pay £200 into a bank account that pays 2.5% interest per year. How much money will be in the account after one year?
8) Write these ratios in their simplest forms:
   a) 4:8   b) 20:15
9) A nursery group has 5 girls and 9 boys. Write the ratio of girls to boys.
10) Divide £2400 in the ratio 5:7.

Worked Exam Question

With the answers written in, it's very easy to just skim over this worked exam question. But that's not really going to help you, so take the time to make sure you've really understood it.

1   Aled wants to buy a suitcase to take on holiday. (D)

He sees a suitcase which was £18, but today it has 35% off.

a) How much money would he get off the suitcase if he bought it today?

35% = 35 ÷ 100 = 0.35 ———— First write 35% as a decimal.

0.35 × £18 = £6.30

£6.30
....
[2 marks]

He sees another suitcase which was £24, but the ticket says today the price is reduced by £6.

b) What is £6 as a percentage of £24?

£6 ÷ £24 = 0.25

0.25 × 100 = 25%

Multiply by 100 to convert to a percentage.

25%
....
[2 marks]

Exam Questions

2 Brown sauce can be bought in three different sizes. **(D)**
The price of each is shown on the right.
Which size of bottle is the best value for money?

Size	Price
250 ml	£2.30
330 ml	£2.97
500 ml	£4.10

[handwritten: 10, 11, 12p]

[handwritten working: 92 × 730 =]

[handwritten: 0. / 250)2.30]

[handwritten: 92p = 10ml / 250]

.... **250** ml
[2 marks]

3 Bill collects elephant ornaments. **(D)**

60% of the ornaments are made from wood.
30% of the ornaments are made from stone.

Bill has 3 other elephant ornaments which are made from other materials.
How many elephant ornaments does Bill have in total?

[handwritten: 60 + 30 = 90]

[handwritten: 30 ornaments]

[handwritten: 10% = 3]

.... **30**
[4 marks]

4 In a class of 26 children, 12 are boys and 14 are girls. **(D)**

a) What is the ratio of boys to girls? Give your answer in its simplest form.

[handwritten: 12 ÷ 2 = 6   6 : 7]
[handwritten: 2 ÷ 14 = 7]

.... **6 : 7**
[1 mark]

b) In another class, the ratio of boys to girls is 2 : 3. There are 25 children in the class.
How many girls are there?

[handwritten: 3, 6 9, 12, 18]
[handwritten: 2, 6, 8, 10, 12]
[handwritten: 15 ~ 6]

.... **15**
[2 marks]

5 Andy, Louise and Christine have £160. They share it in the ratio 3 : 6 : 7. **(D)**
How much money did Christine get?

[handwritten: 3 + 6 + 7 = 16   10 = 1 part]
[handwritten: 7 × 10 = 70]

£ **70**
[2 marks]

Rounding Off

You need to be able to use 3 different rounding methods.
We'll do decimal places first, but there's the same basic idea behind all three.

Decimal Places (d.p.)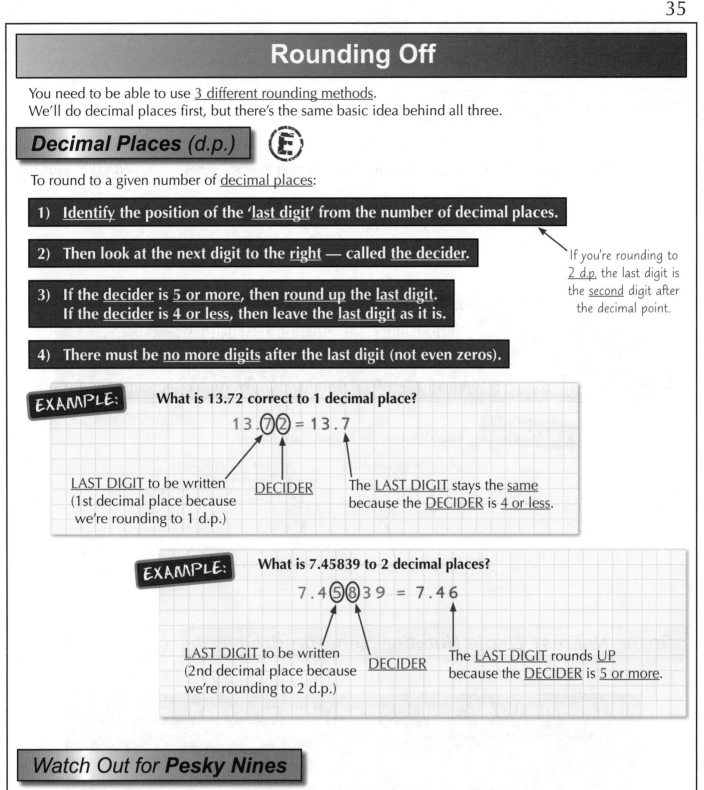

To round to a given number of decimal places:

1) **Identify** the position of the 'last digit' from the number of decimal places.

2) Then look at the next digit to the **right** — called **the decider**.

3) If the **decider** is **5 or more**, then **round up** the **last digit**.
   If the **decider** is **4 or less**, then leave the **last digit** as it is.

4) There must be **no more digits** after the last digit (not even zeros).

If you're rounding to 2 d.p. the last digit is the second digit after the decimal point.

EXAMPLE: **What is 13.72 correct to 1 decimal place?**

13.72 = 13.7

LAST DIGIT to be written (1st decimal place because we're rounding to 1 d.p.)   DECIDER   The LAST DIGIT stays the same because the DECIDER is 4 or less.

EXAMPLE: **What is 7.45839 to 2 decimal places?**

7.45839 = 7.46

LAST DIGIT to be written (2nd decimal place because we're rounding to 2 d.p.)   DECIDER   The LAST DIGIT rounds UP because the DECIDER is 5 or more.

Watch Out for Pesky Nines

If you have to round up a 9 (to 10), replace the 9 with 0, and add 1 to digit on the left.

EXAMPLE: **Round 45.698 to 2 d.p.**

decider

45.698 → 45.69 → 45.70 to 2 d.p.

last digit — round up   The question asks for 2 d.p. so you must put 45.70 not 45.7.

You'll need to round off a lot of your answers in the exam

Rounding is a really important skill, and you'll be throwing easy marks away if you get it wrong.
Make sure you're completely happy with the basic method, then get plenty of practice.

Rounding Off

Significant Figures (s.f.) (D)

The 1st significant figure of any number is the first digit which isn't a zero.

The 2nd, 3rd, 4th, etc. significant figures follow immediately after the 1st — they're allowed to be zeros.

$$0.002309 \qquad\qquad 506.07$$

SIG. FIGS: 1st   2nd   3rd   4th   1st   2nd   3rd   4th

To round to a given number of significant figures:

1) **Find the last digit — if you're rounding to, say 3 s.f., then the 3rd significant figure is the last digit.**

2) **Use the digit to the right of it as the decider, just like for d.p.**

3) **Once you've rounded, fill up with zeros, up to but not beyond the decimal point.**

EXAMPLE:

Round 506.07 to 2 significant figures.

Last digit is the 2nd sig. fig.   Need one zero to fill up to decimal point.

$$5\,\widehat{0\,6}.07 = 510$$

DECIDER is 5 or more ⟶ Last digit rounds up

To the *Nearest Whole Number*, *Ten*, *Hundred* etc. (E)

You might be asked to round to the nearest whole number, ten, hundred, thousand or million:

1) <u>Identify the last digit</u>, e.g. for the nearest <u>whole number</u> it's the <u>units</u> position, and for the '<u>nearest ten</u>' it's the <u>tens</u> position, etc.

2) <u>Round the last digit</u> and <u>fill in with zeros</u> up to the decimal point, just like for significant figures.

EXAMPLE:

Round 6751 to the nearest hundred.

Last digit is in the 'hundreds' position   Fill in 2 zeros up to decimal point.

$$6\,\widehat{7\,5}1 = 6800$$

DECIDER is 5 or more ⟶ Last digit rounds up.

Significant figures can be a bit tricky to get your head round

Decimal places are easy, but significant figures take a bit more thinking about. Learn the method on this page for identifying significant figures, and make sure you really understand the example.

Estimating Calculations

"Estimate" doesn't mean "take a wild guess", so don't just make something up...

Estimating

> 1) **Round everything off** to **1 significant figure.**
> 2) Then **work out the answer** using these nice easy numbers.
> 3) **Show all your working** or you won't get the marks.

Have a look at the previous page to remind yourself how to round to 1 s.f.

It's okay to keep more than 1 significant figure if it makes the calculation easier.

EXAMPLE:

Estimate the value of $\frac{42.6 \times 12.1}{7.9}$.

1) <u>Round</u> each number to <u>1 s.f.</u>

$$\frac{42.6 \times 12.1}{7.9} \approx \frac{40 \times 10}{8}$$

2) Do the <u>calculation</u> with the rounded numbers.

$$= \frac{400}{8}$$

$$= 50$$

$\approx$ means 'approximately equal to'

EXAMPLE:

If the number on the <u>bottom</u> is still <u>smaller than 1</u> after you've rounded, <u>multiply the top and bottom by 10</u>. Repeat until you've <u>got rid of the decimal point</u> (see p5).

Find an estimate for the answer to the calculation $\frac{3.2 \times 98.6}{0.485}$.

1) <u>Round</u> each number to <u>1 s.f.</u>

$$\frac{3.2 \times 98.6}{0.485} \approx \frac{3 \times 100}{0.5}$$

2) <u>Multiplying</u> top and bottom by <u>10</u> gets rid of the decimal point.

$$= \frac{300 \times 10}{0.5 \times 10}$$

$$= \frac{3000}{5}$$

$$= 3000 \div 5 = 600$$

EXAMPLE:

Jo has a cake-making business. She spent £984.69 on flour last year. A bag of flour costs £1.89, and she makes an average of 5 cakes from each bag of flour. Work out an estimate of how many cakes she made last year.

Don't panic if you get a 'real-life' estimating question — just round everything to 1 s.f. as before.

1) Estimate the number of bags of flour — <u>round</u> numbers to <u>1 s.f.</u>

Number of bags of flour $= \frac{984.69}{1.89}$

$$\approx \frac{1000}{2} = 500$$

2) Multiply to find the number of cakes.

Number of cakes $\approx 500 \times 5 = 2500$

Make sure you show all your working when estimating calculations

There's nothing too tricky here — once you've rounded everything to 1 s.f. the calculations are really easy.

Square Roots and Cube Roots

Square roots and cube roots might look a bit odd, but once you know what they mean they suddenly don't seem all that scary.

Square Roots (F)

'Squared' means 'multiplied by itself': $8^2 = 8 \times 8 = 64$

SQUARE ROOT $\sqrt{\ }$ is the reverse process: $\sqrt{64} = 8$

The best way to think of it is: **'Square Root' means 'What Number Times by Itself gives...'**

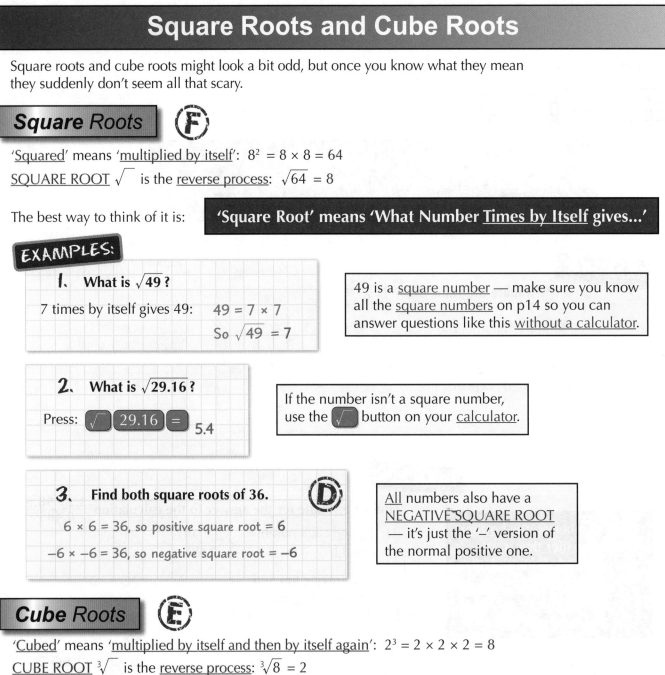

EXAMPLES:

1. What is $\sqrt{49}$?

7 times by itself gives 49: $49 = 7 \times 7$

So $\sqrt{49} = 7$

> 49 is a square number — make sure you know all the square numbers on p14 so you can answer questions like this without a calculator.

2. What is $\sqrt{29.16}$?

Press: $\sqrt{\ }$ [29.16] [=] 5.4

> If the number isn't a square number, use the $\sqrt{\ }$ button on your calculator.

3. Find both square roots of 36. (D)

$6 \times 6 = 36$, so positive square root = 6

$-6 \times -6 = 36$, so negative square root = -6

> All numbers also have a NEGATIVE SQUARE ROOT — it's just the '−' version of the normal positive one.

Cube Roots (E)

'Cubed' means 'multiplied by itself and then by itself again': $2^3 = 2 \times 2 \times 2 = 8$

CUBE ROOT $\sqrt[3]{\ }$ is the reverse process: $\sqrt[3]{8} = 2$

'Cube Root' means 'What Number Times by Itself and then by Itself Again gives...'

You need to be able to write down the cube roots of the cube numbers given on p14 without a calculator. To find the cube root of any other number you can use your calculator — press $\sqrt[3]{\ }$.

EXAMPLES:

1. What is $\sqrt[3]{27}$? 27 is a cube number.

3 times by itself and then by itself again gives 27: $27 = 3 \times 3 \times 3$

So $\sqrt[3]{27} = 3$

2. What is $\sqrt[3]{4913}$?

Press: $\sqrt[3]{\ }$ [4913] [=] 17

Make sure you learn the square numbers and cube numbers

If you can't see the cube root button on your calculator, it might be lurking behind the square root one. Hit shift and then the square root button and it should bring up something like this on the display: $\sqrt[3]{\Box}$

Powers

You've already seen 'to the power 2' and 'to the power 3' — they're just 'squared' and 'cubed'.

Powers are a very *Useful Shorthand* 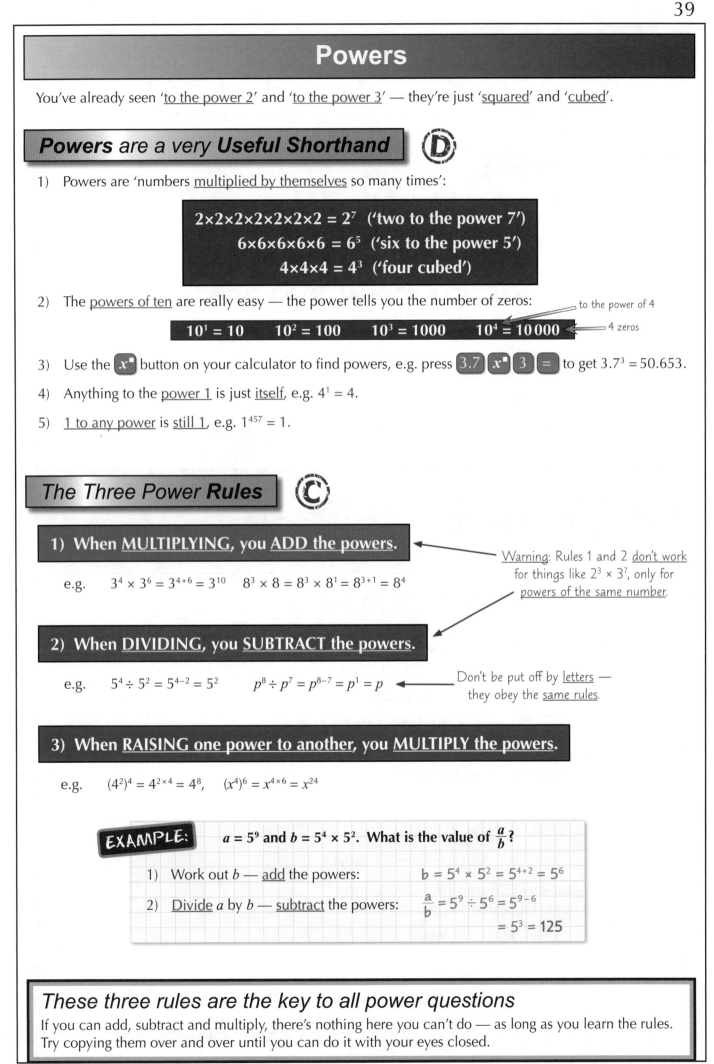 **D**

1) Powers are 'numbers <u>multiplied by themselves</u> so many times':

$$2\times2\times2\times2\times2\times2\times2 = 2^7 \text{ ('two to the power 7')}$$
$$6\times6\times6\times6\times6 = 6^5 \text{ ('six to the power 5')}$$
$$4\times4\times4 = 4^3 \text{ ('four cubed')}$$

2) The <u>powers of ten</u> are really easy — the power tells you the number of zeros:

$$10^1 = 10 \qquad 10^2 = 100 \qquad 10^3 = 1000 \qquad 10^4 = 10\,000$$

to the power of 4

4 zeros

3) Use the $x^{\blacksquare}$ button on your calculator to find powers, e.g. press 3.7 $x^{\blacksquare}$ 3 $=$ to get $3.7^3 = 50.653$.

4) Anything to the <u>power 1</u> is just <u>itself</u>, e.g. $4^1 = 4$.

5) <u>1 to any power</u> is <u>still 1</u>, e.g. $1^{457} = 1$.

The Three Power **Rules** **C**

1) When <u>MULTIPLYING</u>, you <u>ADD the powers.</u>

e.g. $\quad 3^4 \times 3^6 = 3^{4+6} = 3^{10} \quad 8^3 \times 8 = 8^3 \times 8^1 = 8^{3+1} = 8^4$

<u>Warning</u>: Rules 1 and 2 <u>don't work</u>
for things like $2^3 \times 3^7$, only for
<u>powers of the same number.</u>

2) When <u>DIVIDING</u>, you <u>SUBTRACT the powers.</u>

e.g. $\quad 5^4 \div 5^2 = 5^{4-2} = 5^2 \qquad p^8 \div p^7 = p^{8-7} = p^1 = p$

Don't be put off by <u>letters</u> —
they obey the <u>same rules</u>.

3) When <u>RAISING</u> one power to another, you <u>MULTIPLY the powers.</u>

e.g. $\quad (4^2)^4 = 4^{2\times4} = 4^8, \quad (x^4)^6 = x^{4\times6} = x^{24}$

EXAMPLE: $\quad a = 5^9$ and $b = 5^4 \times 5^2$. **What is the value of $\frac{a}{b}$?**

1) Work out b — <u>add</u> the powers: $\qquad b = 5^4 \times 5^2 = 5^{4+2} = 5^6$

2) <u>Divide</u> a by b — <u>subtract</u> the powers: $\qquad \frac{a}{b} = 5^9 \div 5^6 = 5^{9-6}$

$$= 5^3 = 125$$

These three rules are the key to all power questions

If you can add, subtract and multiply, there's nothing here you can't do — as long as you learn the rules.
Try copying them over and over until you can do it with your eyes closed.

Warm-up and Worked Exam Questions

Before you dive into the exam questions on the next page, have a paddle in these friendly-looking warm-up questions. If you're not sure about any of them, go back and look at the topic again.

Warm-up Questions

1) Round these numbers off to 1 decimal place:
   a) 3.24   b) 1.78   c) 2.31   d) 0.46   e) 9.76

2) Round these off to the nearest whole number:
   a) 3.4   b) 5.2   c) 1.84   d) 6.9   e) 3.26

3) Round these numbers to the stated number of significant figures:
   a) 352 to 2 s.f.   b) 465 to 1 s.f.   c) 12.38 to 3 s.f.   d) 0.03567 to 2 s.f.

4) Round these numbers off to the nearest hundred:
   a) 2865   b) 450   c) 123

5) Use your calculator to find the answers to 2 decimal places: a) $\sqrt{200}$ b) $\sqrt[3]{8000}$
   For a), what is the other value that your calculator didn't give?

6) Estimate the following: $\dfrac{29.5 - 9.6}{4.87}$

7) Simplify: a) $3^2 \times 3^6$ b) $4^3 \div 4^2$ c) $(8^3)^4$ d) $(3^2 \times 3^3 \times 1^6) / 3^5$ e) $7^3 \times 7 \times 7^2$

8) Simplify: a) $5^2 \times 5^7 \times 5^3$ b) $1^3 \times 5^0 \times 6^2$ c) $(2^5 \times 2 \times 2^6) \div (2^3 \times 2^4)$

Worked Exam Question

Look at that — an exam question with all the answers filled in. How unexpected.

1   Round the following to the given degree of accuracy.

   a) Josh has 123 people coming to his party. **(F)**
      Write this number to the nearest 10.

      1②3 ———— Last digit is in the 'tens' position
      12③ ———— Decider is less than 5
      120 ———— 1 space to fill before the decimal point

      ....120..........
      [1 mark]

   b) The attendance at a football match was 2568 people. **(F)**
      What is this to the nearest hundred?

      Last digit   Decider

      2⑤68
      2600 ———— 2 spaces to fill

      ....2600..........
      [1 mark]

   c) The population of Ulverpool is 452 529. **(E)**
      Round this to the nearest 100 000.

      Last digit   Decider

      ④52 529
      500 000 ———— 5 spaces to fill

      ....500 000..........
      [1 mark]

Exam Questions

2 Use your calculator to find the following: (F)

a) 8.7^3

658.503
....
[1 mark]

b) $\sqrt{2025}$

45
....
[1 mark]

3 Estimate the value of $\sqrt{42}$. (F)

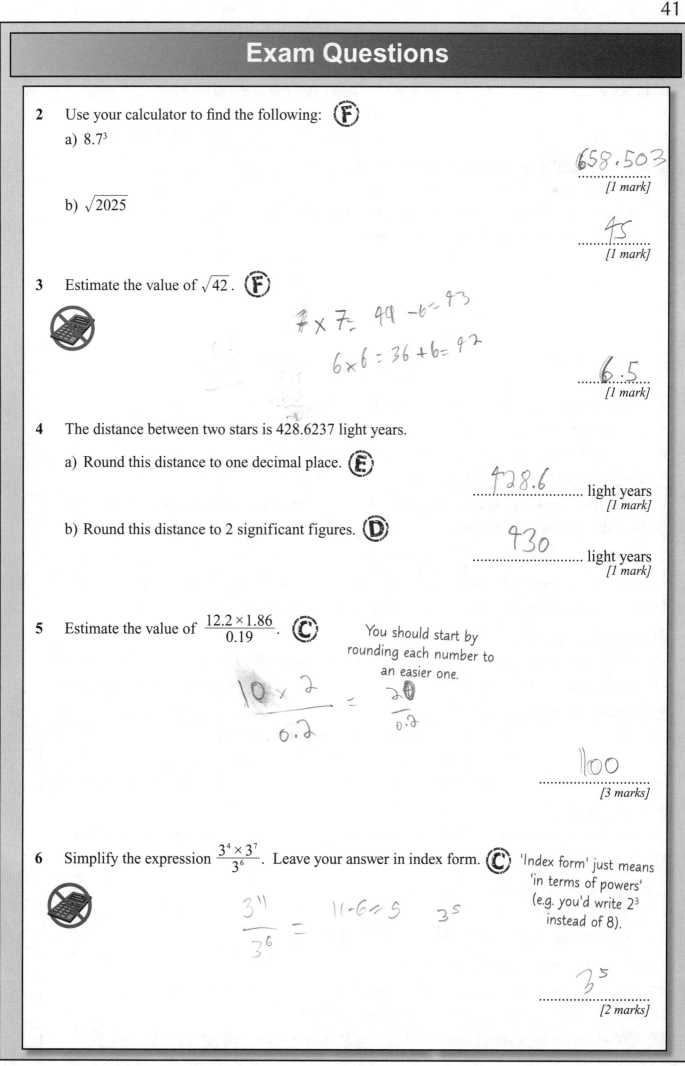

$7 \times 7 = 49 - 6 = 43$

$6 \times 6 = 36 + 6 = 42$

6.5
....
[1 mark]

4 The distance between two stars is 428.6237 light years.

a) Round this distance to one decimal place. (E)

428.6 .... light years
[1 mark]

b) Round this distance to 2 significant figures. (D)

430
.... light years
[1 mark]

5 Estimate the value of $\frac{12.2 \times 1.86}{0.19}$. (C)

You should start by rounding each number to an easier one.

$\frac{10 \times 2}{0.2} = \frac{20}{0.2}$

100
....
[3 marks]

6 Simplify the expression $\frac{3^4 \times 3^7}{3^6}$. Leave your answer in index form. (C)

'Index form' just means 'in terms of powers' (e.g. you'd write 2^3 instead of 8).

$\frac{3^{11}}{3^6} = 11-6=5 \quad 3^5$

3^5
....
[2 marks]

Revision Questions for Section One

Well, that wraps up <u>Section One</u> — time to put yourself to the test and find out <u>how much you really know</u>.
- Try these questions and <u>tick off each one</u> when you <u>get it right</u>.
- When you've done <u>all the questions</u> for a topic and are <u>completely happy</u> with it, tick off the topic.

Ordering Numbers and Arithmetic (p1-9) ☑

1) Write this number out in words: 21 306 515 ☑
2) Put these numbers in order of size: 2.2, 4.7, 3.8, 3.91, 2.09, 3.51 ☑
   <u>Don't</u> use your calculator for questions 3 and 4.
3) Calculate: a) 258 + 624　　b) 533 – 87　　c) £2.30 + £1.12 + 75p ☑
4) Work out: a) £1.20 × 100　　b) £150 ÷ 300　　c) 338 ÷ 13　d) 3.3 × 19　e) 4.2 ÷ 12 ☑

Types of Number, Factors and Multiples (p13-18) ☑

5) Find: a) –10 – 6　b) –35 ÷ –5　c) –4 + –5 + 22 – –7 ☑
6) What are square numbers? Write down the first ten of them. ☑
7) Find all the prime numbers between 40 and 60 (there are 5 of them). ☑
8) What are multiples? Find the first six multiples of: a) 10　　b) 4 ☑
9) Express each of these as a product of prime factors: a) 210　b) 1050 ☑
10) Find: a) the HCF of 42 and 28　b) the LCM of 8 and 10 ☑

Fractions and Decimals (p21-25) ☑

11) Write: a) 0.04 as: (i) a fraction (ii) a percentage　b) 65% as: (i) a fraction (ii) a decimal ☑
12) How do you simplify a fraction? ☑
13) Calculate a) $\frac{4}{7}$ of 560　b) $\frac{2}{5}$ of £150 ☑
14) Work out without a calculator: a) $\frac{5}{8}+\frac{9}{4}$　b) $\frac{2}{3}-\frac{1}{7}$　c) $\frac{25}{6}\div\frac{8}{3}$　d) $\frac{2}{3}\times4\frac{2}{5}$ ☑
15) What is a recurring decimal? How do you show that a decimal is recurring? ☑

Proportions (p28-29) ☑

16) Rick ordered 5 pints of milk from the milkman. His bill was £2.35. How much would 3 pints cost? ☑
17) Tins of Froggatt's Ham come in two sizes. Which is the best buy, 100 g for 24p or 250 g for 52p? ☑

Percentages (p30-31) ☑

18) What's the method for finding one amount as a percentage of another? ☑
19) A DVD player costs £50 plus VAT. If VAT is 20%, how much does the DVD player cost? ☑
20) A top that should cost £45 has been reduced by 15%. Carl has £35. Can he afford the top? ☑

Ratios (p32) ☑

21) Sarah has 150 carrots and 240 turnips. Write the ratio of carrots to turnips in its simplest form. ☑
22) Divide 3000 in the ratio 5:8:12. ☑

Rounding and Estimating (p35-37) ☑

23) Round a) 17.65 to 1 d.p.　b) 6743 to 2 s.f.　c) 3 643 510 to the nearest million. ☑
24) Estimate the value of a) $\frac{17.8\times32.3}{6.4}$　b) $\frac{96.2\times7.3}{0.463}$ ☑

Roots and Powers (p38-39) ☑

25) Find without using a calculator: a) $\sqrt{121}$　b) $\sqrt[3]{64}$　c) 8^2-2^3　d) Ten thousand as a power of ten. ☑
26) Use a calculator to find: a) 7.5^3　b) $\sqrt{23.04}$　c) $\sqrt[3]{512}$ ☑
27) a) What are the three power rules? b) If $f = 7^6 \times 7^4$ and $g = 7^5$, what is $f \div g$? ☑

Algebra — Simplifying Terms

Algebra really terrifies so many people. But honestly, it's not that bad. You just have to make sure you understand and learn these basic rules for dealing with algebraic expressions.

Terms (E)

Before you can do anything else with algebra, you must understand what a term is:

> **A TERM IS A COLLECTION OF NUMBERS, LETTERS AND BRACKETS, ALL MULTIPLIED/DIVIDED TOGETHER**

There's more on multiplying letters together on the next page.

Terms are separated by + and – signs. Every term has a + or – attached to the front of it.

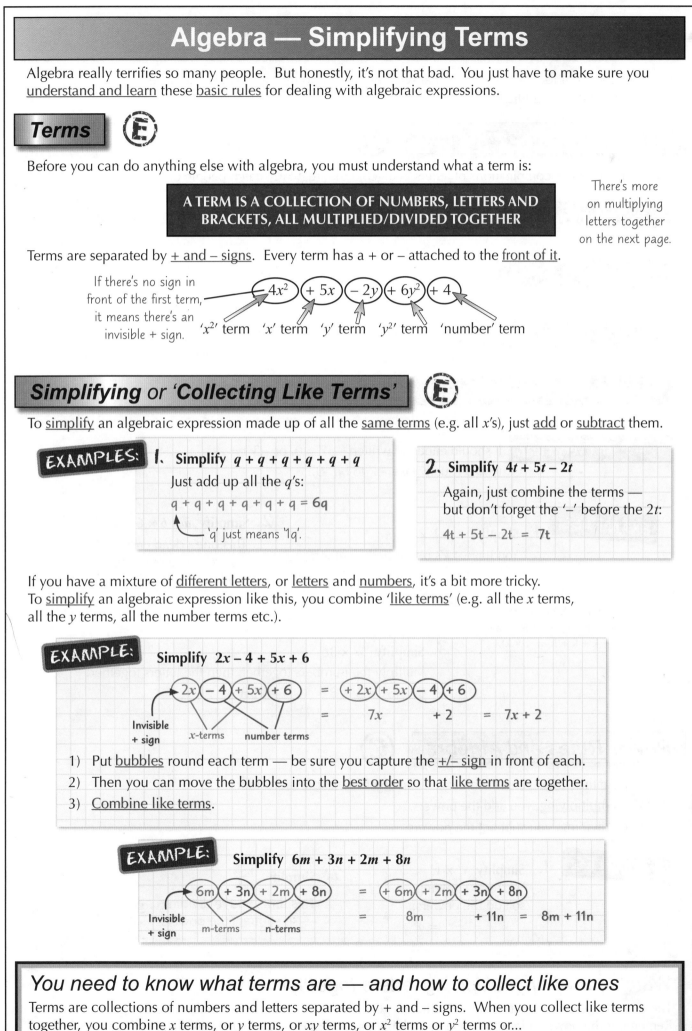

If there's no sign in front of the first term, it means there's an invisible + sign.

$-4x^2$ $+ 5x$ $- 2y$ $+ 6y^2$ $+ 4$

'x^2' term   'x' term   'y' term   'y^2' term   'number' term

Simplifying or 'Collecting Like Terms' (E)

To simplify an algebraic expression made up of all the same terms (e.g. all x's), just add or subtract them.

EXAMPLES:

1. Simplify $q + q + q + q + q + q$

Just add up all the q's:

$q + q + q + q + q + q = 6q$

'q' just means '$1q$'.

2. Simplify $4t + 5t – 2t$

Again, just combine the terms — but don't forget the '–' before the $2t$:

$4t + 5t – 2t = 7t$

If you have a mixture of different letters, or letters and numbers, it's a bit more tricky.
To simplify an algebraic expression like this, you combine 'like terms' (e.g. all the x terms, all the y terms, all the number terms etc.).

EXAMPLE: Simplify $2x – 4 + 5x + 6$

$2x$ $- 4$ $+ 5x$ $+ 6$ = $+ 2x$ $+ 5x$ $- 4$ $+ 6$

= $7x$ $+ 2$ = $7x + 2$

Invisible + sign x-terms   number terms

1) Put bubbles round each term — be sure you capture the +/– sign in front of each.
2) Then you can move the bubbles into the best order so that like terms are together.
3) Combine like terms.

EXAMPLE: Simplify $6m + 3n + 2m + 8n$

$6m$ $+ 3n$ $+ 2m$ $+ 8n$ = $+ 6m$ $+ 2m$ $+ 3n$ $+ 8n$

= $8m$ $+ 11n$ = $8m + 11n$

Invisible + sign m-terms n-terms

You need to know what terms are — and how to collect like ones

Terms are collections of numbers and letters separated by + and – signs. When you collect like terms together, you combine x terms, or y terms, or xy terms, or x^2 terms or y^2 terms or...

Algebra — Simplifying Terms

On this page we'll look at some <u>rules</u> that will help you <u>simplify</u> expressions that have <u>letters and numbers multiplied together</u>.

Letters **Multiplied** Together ⓓ

Watch out for these combinations of letters in algebra that regularly catch people out:

1) *abc* means $a \times b \times c$ and $3a$ means $3 \times a$. The ×'s are often left out to make it clearer.

2) gn^2 means $g \times n \times n$. Note that only the *n* is squared, not the *g* as well.

> There's more on powers on p39.

3) $(gn)^2$ means $g \times g \times n \times n$. The brackets mean that <u>BOTH</u> letters are squared.

4) <u>Powers</u> tell you <u>how many</u> letters are multiplied together. So $r^6 = r \times r \times r \times r \times r \times r$.

5) -3^2 isn't very clear. It should either be written $(-3)^2 = 9$, or $-(3^2) = -9$ (you'd usually take -3^2 to be -9).

EXAMPLES:

1. Simplify $k \times k \times k \times k$
You have 4 *k*'s multiplied together:
$k \times k \times k \times k = k^4$

Careful — k times itself 4 times is k^4, <u>not</u> 4k (4k means $k + k + k + k$ or $4 \times k$).

2. Simplify $a \times b \times 6$
This one's dead easy —
just combine into one term
(and put the number at the front):
$a \times b \times 6 = 6ab$

3. Simplify $5s \times 3t$
Multiply the numbers together, then the letters together:
$5s \times 3t = 5 \times 3 \times s \times t = 15st$

Power Rules and Algebra ©

You can use the <u>power rules</u> from page 39 on <u>algebraic expressions</u> too:
1) When <u>multiplying</u>, you <u>add</u> the powers.
2) When <u>dividing</u>, you <u>subtract</u> the powers.

EXAMPLES:

1. Simplify $v^2 \times v^3$
You're multiplying,
so <u>add</u> the powers:
$v^2 \times v^3 = v^{2+3} = v^5$

2. Simplify $\dfrac{w^{11}}{w^8}$
This time, you're dividing
— so <u>subtract</u> the powers: $\dfrac{w^{11}}{w^8} = w^{11-8} = w^3$

Watch out when multiplying letters together

There are some cases when multiplying letters gets a bit tricky, but just remember the five points above. Remember the power rules too — when multiplying, add powers and when dividing, subtract powers.

Algebra — Multiplying Out Brackets

If you have an algebraic expression with <u>brackets</u> in, you might be asked to get rid of them by <u>multiplying out the brackets</u>. Your multiplying skills from the previous page will come in handy here.

Multiplying Out Brackets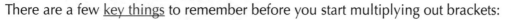

There are a few <u>key things</u> to remember before you start multiplying out brackets:

1) The thing <u>outside</u> the brackets multiplies <u>each separate term</u> inside the brackets.

2) When <u>letters</u> are multiplied together, they are just written next to each other, e.g. pq.

3) Remember from the previous page that $r \times r = r^2$, and when you multiply terms with numbers <u>and</u> letters in, you multiply the <u>numbers</u> together then the <u>letters</u>.

4) Be very careful with <u>MINUS SIGNS</u> — remember the rules for multiplying them from p13.

EXAMPLES:

1. **Expand $3(2x + 5)$**
Multiply the $2x$ and 5 inside by the 3 outside:
$$3(2x + 5) = (3 \times 2x) + (3 \times 5)$$
$$= 6x + 15$$

2. **Expand $4a(3b - 2)$**
Multiply the $3b$ and -2 inside by the $4a$ outside:
$$4a(3b - 2) = (4a \times 3b) + (4a \times -2)$$
$$= 12ab - 8a \qquad 4 \times -2 = -8$$

3. **Expand $-4(3p^2 - 7q^3)$**
Be very careful with the minus signs here:
$$-4(3p^2 - 7q^3) = (-4 \times 3p^2) + (-4 \times -7q^3)$$
Note that the minus sign outside the brackets <u>reverses</u> all the signs when you multiply.
$$= -12p^2 + 28q^3$$

4. **Expand $2e(e - 3)$**
This time, be careful when you multiply $2e$ by e — you'll end up with a $2e^2$:
$$2e(e - 3) = (2e \times e) + (2e \times -3)$$
$$= 2e^2 - 6e$$

Collecting Like Terms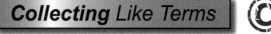

If you're given <u>more than one</u> set of brackets to expand like $2(x + 2) + 3(x - 4)$, you'll have to <u>simplify</u> at the end by <u>collecting like terms</u> (see p43).

EXAMPLE: **Expand and simplify $3(x + 2) + 4(4 - x)$**

First, <u>expand</u> each of the brackets separately (as you did above):
$$3(x + 2) + 4(4 - x) = (3 \times x) + (3 \times 2) + (4 \times 4) + (4 \times -x)$$
$$= 3x + 6 + 16 - 4x$$

Careful with the negatives here — the 4 and $-x$ multiply to give $-4x$.

Then <u>collect like terms</u> to simplify the expression:
$$= 3x - 4x + 6 + 16 = -x + 22 \text{ OR } 22 - x$$

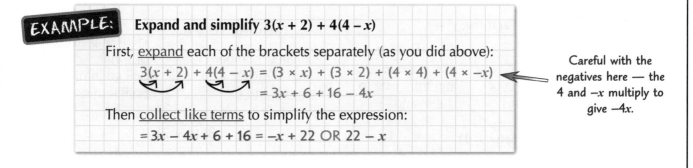

Take your time when multiplying out brackets

When multiplying out brackets, don't rush — you'll make mistakes and throw away easy marks. Remember to keep an eye out for minus signs too, and give your answer in its simplest form.

Algebra — Taking Out Common Factors

Right, now you know how to expand brackets, it's time to put them back in. This is known as <u>factorising</u>.

Factorising — Putting Brackets In

This is the <u>exact reverse</u> of multiplying out brackets. You have to look for <u>common factors</u> — numbers or letters that go into <u>every term</u>. Here's the method to follow:

1) Take out the <u>biggest number</u> that goes into all the terms.

2) <u>For each letter in turn</u>, take out the <u>highest power</u> (e.g. x, x^2 etc.) that will go into <u>EVERY</u> term.

3) Open the brackets and fill in all the bits needed to <u>reproduce each term</u>.

4) Check your answer by multiplying out the brackets again.

REMEMBER: The bits <u>taken out</u> and put at the front of the brackets are the <u>common factors</u>. The bits <u>inside</u> are what get you back to the <u>original terms</u> when you multiply out again.

Taking Out a Number

If <u>both</u> terms of the expression you're trying to factorise have a <u>number part</u>, you can look for a <u>common factor</u> of both numbers and take it <u>outside</u> the brackets. The common factor is the <u>biggest number</u> that the numbers in <u>both terms</u> divide by.

EXAMPLES:

1. **Factorise $3x - 9$**

3 and 9 both <u>divide by 3</u>.

Decide what you need to <u>multiply</u> 3 by to get to $3x$ and 9.

$$3(x - 3)$$

Check: $3(x - 3) = 3x - 9$ ✓

2. **Factorise $16x + 20y$**

The biggest number that 16 and 20 both <u>divide by is 4</u>.

The letters are <u>different</u>, so they can't be a common factor.

$$4(4x + 5y)$$

Check: $4(4x + 5y) = 16x + 20y$ ✓

Taking Out a Letter

If the <u>same letter</u> appears in <u>all</u> the terms (but to <u>different powers</u>), you can take out some <u>power</u> of the <u>letter</u> as a <u>common factor</u>. You might be able to take out a <u>number</u> as well.

EXAMPLES:

1. **Factorise $y^3 - y$**

<u>Highest power</u> of y in both terms.

Decide what you need to <u>multiply</u> y by to get y^3 and $-y$.

$$y(y^2 - 1)$$

Check: $y(y^2 - 1) = y^3 - y$ ✓

2. **Factorise $3x^2 + 6x$**

Biggest number that'll <u>divide</u> into 3 and 6.

Highest <u>power of x</u> that will go into both terms.

$$3x(x + 2)$$

Check: $3x(x + 2) = 3x^2 + 6x$ ✓

Factorising is the opposite of multiplying out brackets

There's no excuse for making mistakes when factorising — you can check your answer by multiplying the brackets out again. Do it right, and you'll get back to the original expression.

Warm-up and Worked Exam Questions

It's easy to think you've learnt everything in the section until you try the warm-up questions.
Don't panic if you've forgotten bits. Just go back over them until they're fixed in your brain.

Warm-up Questions

1) Simplify: a) $5x + 3y - 4 - 2y - x$ b) $3x + 2 + 5xy + 6x - 7$
               c) $2x + 3x^2 + 5y^2 + 3x$ d) $3y - 6xy + 3y + 2yx$

2) Simplify: a) $e \times e \times e \times e \times e$ b) $3f \times 6g$

3) Simplify: a) $h^4 \times h^5$ b) $\dfrac{s^9}{s^6}$

4) Multiply out: a) $2(x - 2)$ b) $x(5 + x)$ c) $y(y + x)$ d) $3y(2x - 6)$

5) Factorise: a) $5xy + 15x$ b) $5a - 7ab$ c) $12xy + 6y - 36y^2$

Worked Exam Question

I've gone through this question and written in answers, just like you'll do in the exam.
It should really help with the questions which follow, so don't say I never do anything for you.

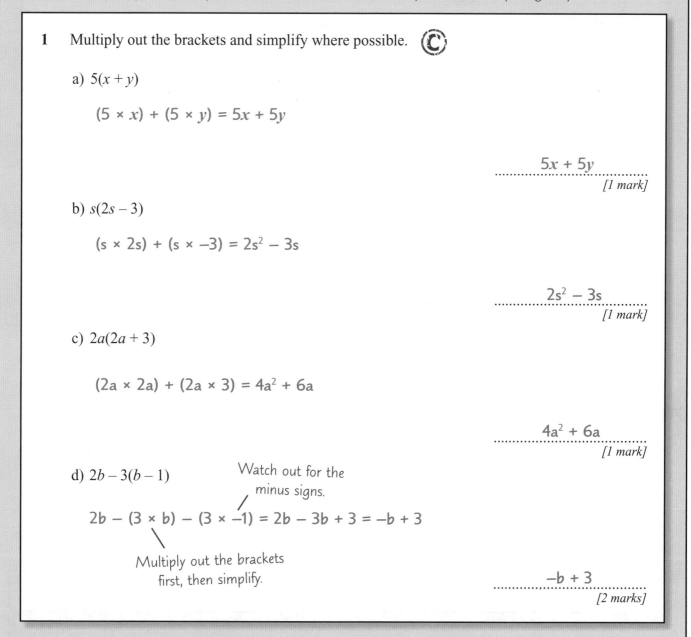

1   Multiply out the brackets and simplify where possible. Ⓒ

a) $5(x + y)$

$(5 \times x) + (5 \times y) = 5x + 5y$

.............$5x + 5y$.............
[1 mark]

b) $s(2s - 3)$

$(s \times 2s) + (s \times -3) = 2s^2 - 3s$

.............$2s^2 - 3s$.............
[1 mark]

c) $2a(2a + 3)$

$(2a \times 2a) + (2a \times 3) = 4a^2 + 6a$

.............$4a^2 + 6a$.............
[1 mark]

d) $2b - 3(b - 1)$ Watch out for the minus signs.

$2b - (3 \times b) - (3 \times -1) = 2b - 3b + 3 = -b + 3$

Multiply out the brackets first, then simplify.

.............$-b + 3$.............
[2 marks]

Exam Questions

2 Write the following in their simplest form. (D)

a) $2a \times 5b$

....
[1 mark]

b) $5pq + pq - 2pq$

....
[1 mark]

c) $2x^2 + 8x - 4x - x^2$

....
[2 marks]

3 Expand the following. (C)

a) $3(x - 2)$

....
[1 mark]

b) $x(x + 4)$

....
[1 mark]

4 Factorise the following expressions. (C)

a) $6x + 3$

....
[1 mark]

b) $x^2 + 7x$

....
[1 mark]

5 Factorise these expressions fully. (C)

a) $4x^2 + 6xy$

....
[2 marks]

b) $2vw + 8v^2$

....
[2 marks]

Solving Equations

'Solving equations' basically means 'find the <u>value of x</u> (or whatever letter is used) that makes the equation <u>true</u>'. To do this, you usually have to <u>rearrange</u> the equation to get x <u>on its own</u>.

The 'Common Sense' Approach (E)

The trick here is to realise that the <u>unknown quantity</u> 'x' is just a <u>number</u> and the 'equation' is a <u>cryptic clue</u> to help you find it.

EXAMPLE: **Solve the equation $3x + 4 = 46$.** ← This just means 'find the value of x'.

This is what you should say to yourself:

'Something + 4 = 46', so that 'something' must be 42.
So that means $3x = 42$, which means '3 × something = 42'.
So it must be $42 ÷ 3 = 14$, so $x = 14$.

If you were writing this down in an exam question, just write down the bits in blue.

In other words don't think of it as algebra, but as '<u>find the mystery number</u>'.

The 'Proper' Way (E)

The 'proper' way to solve equations is to keep <u>rearranging</u> them until you end up with '<u>x = </u>' on one side. There are a few <u>important points</u> to remember when rearranging.

Golden Rules
1) Always do the <u>SAME thing</u> to <u>both sides of the equation</u>.
2) To get rid of something, do the <u>opposite</u>.
   The opposite of + is – and the opposite of – is +.
   The opposite of × is ÷ and the opposite of ÷ is ×.
3) Keep going until you have a letter <u>on its own</u>.

EXAMPLES:

1. Solve $x + 7 = 11$ — The opposite of +7 is –7
(−7) $x + 7 - 7 = 11 - 7$
$x = 4$
This means 'take away 7 from both sides'.

2. Solve $x - 3 = 7$ — The opposite of –3 is +3
(+3) $x - 3 + 3 = 7 + 3$
$x = 10$

3. Solve $5x = 15$ — $5x$ means $5 × x$, so do the opposite — divide both sides by 5
(÷5) $5x ÷ 5 = 15 ÷ 5$
$x = 3$

4. Solve $\frac{x}{3} = 2$ — $\frac{x}{3}$ means $x ÷ 3$, so do the opposite — multiply both sides by 3
(×3) $\frac{x}{3} × 3 = 2 × 3$
$x = 6$

That's it — all the steps you need to solve any of these equations

It's always good to know the proper way to solve equations, just in case you get thrown a curveball in the exam and they give you a real nightmare of an equation to solve.

Solving Equations

You're not done with solving equations yet — not by a long shot.

Two-Step Equations (E)

If you come across an equation like $4x + 3 = 19$ (where there's an _x-term_ and a _number_ on the _same side_), use the methods from the previous page to solve it — just do it in _two steps_:

1) _Add or subtract_ the number first.   2) _Multiply or divide_ to get '$x = $'.

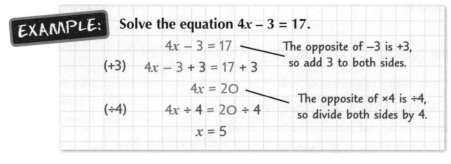

EXAMPLE: **Solve the equation $4x - 3 = 17$.**

$$4x - 3 = 17$$ — The opposite of -3 is $+3$, so add 3 to both sides.

(+3) $4x - 3 + 3 = 17 + 3$

$$4x = 20$$ — The opposite of $\times 4$ is $\div 4$, so divide both sides by 4.

(÷4) $4x \div 4 = 20 \div 4$

$$x = 5$$

Equations with an 'x' on Both Sides (D)

For equations like $2x + 3 = x + 7$ (where there's an _x-term_ on _each side_), you have to:

1) Get all the _x's_ on one side and all the _numbers_ on the other.
2) _Multiply or divide_ to get '$x = $'.

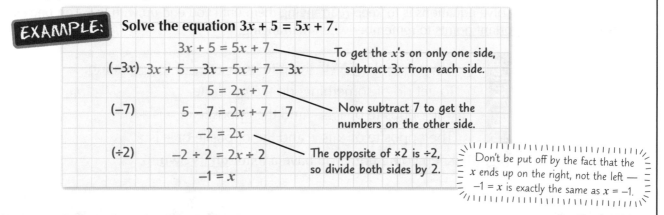

EXAMPLE: **Solve the equation $3x + 5 = 5x + 7$.**

$$3x + 5 = 5x + 7$$ — To get the x's on only one side, subtract $3x$ from each side.

(−3x) $3x + 5 - 3x = 5x + 7 - 3x$

$$5 = 2x + 7$$ — Now subtract 7 to get the numbers on the other side.

(−7) $5 - 7 = 2x + 7 - 7$

$$-2 = 2x$$ — The opposite of $\times 2$ is $\div 2$, so divide both sides by 2.

(÷2) $-2 \div 2 = 2x \div 2$

$$-1 = x$$

Don't be put off by the fact that the x ends up on the right, not the left — $-1 = x$ is exactly the same as $x = -1$.

Equations with Brackets (C)

If the equation has _brackets_ in, you have to _multiply out_ the brackets (see p45) before solving it as above.

EXAMPLE: **Solve the equation $5x + 3 = 4(x + 2)$.**

$$5x + 3 = 4(x + 2)$$ — Multiply out the brackets.

$$5x + 3 = 4x + 8$$ — To get the x's on only one side, subtract $4x$ from each side.

(−4x) $5x + 3 - 4x = 4x + 8 - 4x$

$$x + 3 = 8$$ — The opposite of $+3$ is -3, so subtract 3 from each side.

(−3) $x + 3 - 3 = 8 - 3$

$$x = 5$$

You can use the same methods to solve all sorts of equations

Even if your equation has an 'x' on both sides or has brackets, you can still use the 'common sense' or 'proper' way to solve it. Just make sure you always end up with the x's all on one side.

Using Formulas

Formulas come up again and again in GCSE Maths, so make sure you're happy using them.
The first thing you need to be able to do is pretty easy — it's just putting <u>numbers</u> into them.

Putting **Numbers** into **Formulas**

You might be given a <u>formula</u> and asked to work out its <u>value</u> when you put in <u>certain numbers</u>.
All you have to do here is follow this <u>method</u>.

1) Write out the <u>formula</u>.

2) Write it <u>again</u>, directly underneath, but <u>substituting numbers for letters</u> on the <u>RHS</u> (right-hand side).

3) Work it out <u>in stages</u>. Use <u>BODMAS</u> (see p1) to work things out in the <u>right order</u>.
<u>Write down</u> values for each bit as you go along.

4) <u>DO NOT</u> attempt to do it <u>all in one go</u> on your calculator — you're more likely to make <u>mistakes</u>.

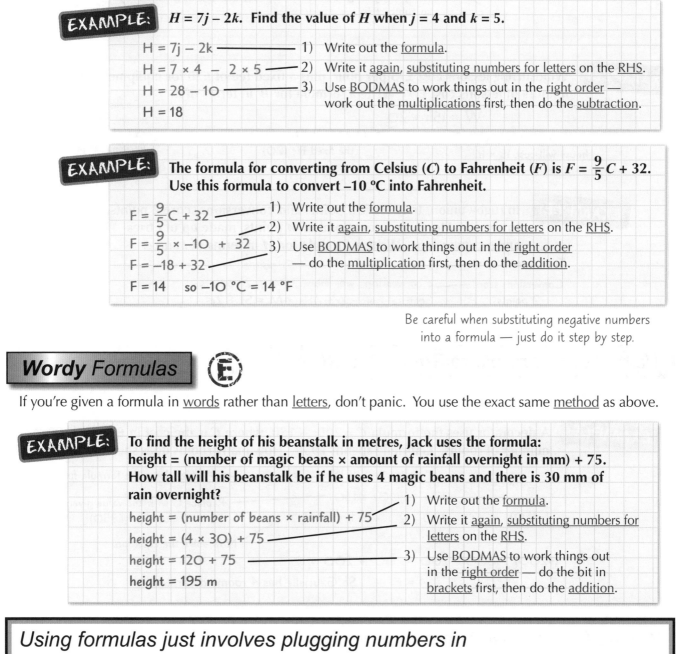

EXAMPLE: $H = 7j - 2k$. **Find the value of H when $j = 4$ and $k = 5$.**

$H = 7j - 2k$ ———————— 1) Write out the <u>formula</u>.
$H = 7 \times 4 - 2 \times 5$ ———— 2) Write it <u>again</u>, substituting numbers for letters on the <u>RHS</u>.
$H = 28 - 10$ ———————— 3) Use <u>BODMAS</u> to work things out in the <u>right order</u> —
$H = 18$ work out the <u>multiplications</u> first, then do the <u>subtraction</u>.

EXAMPLE: **The formula for converting from Celsius (C) to Fahrenheit (F) is $F = \frac{9}{5}C + 32$.**
Use this formula to convert –10 ºC into Fahrenheit.

$F = \frac{9}{5}C + 32$ ———— 1) Write out the <u>formula</u>.
                       2) Write it <u>again</u>, substituting numbers for letters on the <u>RHS</u>.
$F = \frac{9}{5} \times -10 + 32$ 3) Use <u>BODMAS</u> to work things out in the <u>right order</u>
$F = -18 + 32$ — do the <u>multiplication</u> first, then do the <u>addition</u>.
$F = 14$ so –10 °C = 14 °F

Be careful when substituting negative numbers
into a formula — just do it step by step.

Wordy Formulas

If you're given a formula in <u>words</u> rather than <u>letters</u>, don't panic. You use the exact same <u>method</u> as above.

EXAMPLE: **To find the height of his beanstalk in metres, Jack uses the formula:**
height = (number of magic beans × amount of rainfall overnight in mm) + 75.
How tall will his beanstalk be if he uses 4 magic beans and there is 30 mm of
rain overnight?

height = (number of beans × rainfall) + 75
height = (4 × 30) + 75
height = 120 + 75
height = 195 m

1) Write out the <u>formula</u>.
2) Write it <u>again</u>, substituting numbers for letters on the <u>RHS</u>.
3) Use <u>BODMAS</u> to work things out in the <u>right order</u> — do the bit in brackets first, then do the <u>addition</u>.

Using formulas just involves plugging numbers in

If you have more than one number to put into a formula, make sure you put them in the right places
in the formula — don't get them mixed up.

Making Formulas from Words

Before we get started, there are a few <u>definitions</u> you need to know:

> 1) **EXPRESSION** — a <u>collection</u> of <u>terms</u> (see p43). Expressions <u>DON'T</u> have an = sign in them.
> 2) **EQUATION** — an expression with an = sign in it (so you can solve it).
> 3) **FORMULA** — a <u>rule</u> that helps you work something out (it will also have an = sign in it).

Making a **Formula** from **Given Information** Ⓔ

Making <u>formulas</u> from <u>words</u> can be a bit confusing as you're given a lot of <u>information</u> in one go. You just have to go through it slowly and carefully and <u>extract the maths</u> from it.

EXAMPLE: Tiana is x years old. Leah is 5 years younger than Tiana. Martin is 4 times as old as Tiana.

a) **Write an expression for Leah's age in terms of x.**

Tiana's age is x Leah is 5 years younger, so subtract 5
So Leah's age is $x - 5$

b) **Write an expression for Martin's age in terms of x.**

         4 times older
Tiana's age is x
So Martin's age is $4 \times x = 4x$

EXAMPLE: Windsurfing lessons cost £15 per hour, plus a fixed fee of £20 for equipment hire. h hours of lessons cost £W. Write a formula for W in terms of h.

$$W = 15h + 20$$

One hour costs 15, so h hours will cost 15 × h

Don't forget to add on the fixed fee (20)

Because you're asked for a formula, you must include the 'W =' bit to get full marks (i.e. don't just put 15h + 20).

EXAMPLE: In rugby union, tries score 5 points and conversions score 2 points. In a game, Morgan scores a total of M points, made up of t tries and c conversions. Write a formula for M in terms of t and c.

Tries score 5 points —— t tries will score 5 × t = 5t points
Conversions score —— c conversions will score 2 × c = 2c points
2 points   So total points scored are $M = 5t + 2c$

Using Your **Formula** to **Solve Equations** Ⓓ

Sometimes, you might be asked to <u>use</u> a formula to <u>solve an equation</u>.

EXAMPLE: A decorator uses the formula $C = 200r + 150$, where C is the cost in £ and r is the number of rooms. Gabrielle spends £950. How many rooms does she have decorated?

$$C = 200r + 150$$ —— Write down the formula first.
$$950 = 200r + 150$$ —— Replace C with the value given in the question (£950).
(-150) $950 - 150 = 200r + 150 - 150$
$$800 = 200r$$ —— Now solve the equation.
$(\div 200)$ $800 \div 200 = 200r \div 200$
$$4 = r$$ So Gabrielle has 4 rooms decorated

Writing formulas isn't hard once you get the idea

Once you've written down the formula it's worth checking it by putting in some numbers.
E.g. two hours of Windsurfing lessons must cost £50, so you'd check that when h is 2, W comes out as 50.

Rearranging Formulas

The <u>subject</u> of a formula is the letter <u>on its own</u> before the = (so x is the subject of $x = 2y + 3z$).

Changing the Subject of a Formula ©

<u>Rearranging formulas</u> means making a different letter the <u>subject</u>, e.g. getting '$y =$ ' from '$x = 3y + 2$' — you have to get the subject <u>on its own</u>. Fortunately, you can use the <u>same methods</u> that you used for <u>solving equations</u> (see p49-50) — here's a quick reminder:

Golden Rules
1) Always do the <u>SAME thing</u> to <u>both sides of the formula</u>.
2) To get rid of something, do the <u>opposite</u>.
   The opposite of + is – and the opposite of – is +.
   The opposite of × is ÷ and the opposite of ÷ is ×.
3) Keep going until you have the letter you want <u>on its own</u>.

EXAMPLE: **Rearrange $p = q + 12$ to make q the subject of the formula.**

$p = q + 12$

(–12) $p - 12 = q + 12 - 12$ The opposite of +12 is –12, so take away 12 from both sides.

$p - 12 = q$ OR $q = p - 12$

EXAMPLE: **Rearrange $a = 3b + 4$ to make b the subject of the formula.**

$a = 3b + 4$

(–4) $a - 4 = 3b + 4 - 4$ The opposite of +4 is –4, so take away 4 from both sides.

$a - 4 = 3b$

(÷3) $(a - 4) \div 3 = 3b \div 3$ The opposite of ×3 is ÷3, so divide both sides by 3.

Careful here — you divide the <u>whole side</u> by 3, not just one term.

$\dfrac{a - 4}{3} = b$ OR $b = \dfrac{a - 4}{3}$

EXAMPLE: **Rearrange $m = \dfrac{n}{4} - 7$ to make n the subject of the formula.**

$m = \dfrac{n}{4} - 7$ The opposite of –7 is +7, so add 7 to both sides.

(+7) $m + 7 = \dfrac{n}{4} - 7 + 7$

$m + 7 = \dfrac{n}{4}$ The opposite of ÷4 is ×4, so multiply both sides by 4.

(×4) $(m + 7) \times 4 = \dfrac{n}{4} \times 4$

$4(m + 7) = n$ OR $n = 4m + 28$

Just remember — the subject is the letter on its own

Rearranging formulas is really just like solving equations — so if you learn the method for one, you know the method for the other. All you have to do is remember the golden rules and you'll be set.

Warm-up and Worked Exam Questions

I know that you'll be champing at the bit to get into the exam questions, but these warm-up questions are invaluable to get the basic facts straight first.

Warm-up Questions

1) Solve: $4x - 12 = 20$
2) Solve: $3x + 5 = 5x - 9$
3) Solve these equations: a) $3x + 1 = 13$ b) $\frac{q}{4} = 8$ c) $5y + 4 = 2y - 2$
4) Use the formula '$a = 2c^2 - 6$' to find a, when $c = 9$.
5) The value of y is found by taking x, multiplying it by five and then subtracting three. Write down a formula for y in terms of x.
6) "Froggatt's Foods" produce some kebabs which cost 95p each. Write a formula for the total cost, C pence, of buying n kebabs.
7) Rearrange this formula to make b the subject: $2(b - 3) = a$

Worked Exam Question

I'd like an exam question, and the answers written in — and a surprise.
Two out of three's not bad.

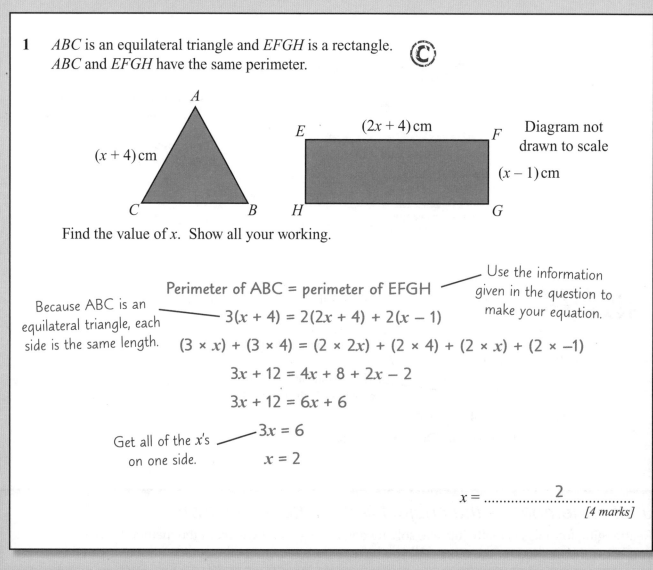

1   ABC is an equilateral triangle and $EFGH$ is a rectangle.
    ABC and $EFGH$ have the same perimeter.

Find the value of x. Show all your working.

Perimeter of ABC = perimeter of EFGH — Use the information given in the question to make your equation.

Because ABC is an equilateral triangle, each side is the same length. — $3(x + 4) = 2(2x + 4) + 2(x - 1)$

$(3 \times x) + (3 \times 4) = (2 \times 2x) + (2 \times 4) + (2 \times x) + (2 \times -1)$

$3x + 12 = 4x + 8 + 2x - 2$

$3x + 12 = 6x + 6$

Get all of the x's on one side. — $3x = 6$

$x = 2$

$x =$ ....2.............
[4 marks]

Exam Questions

2 Solve these equations for x. (E)

a) $x + 3 = 12$

$x =$ ....
[1 mark]

b) $6x = 24$

$x =$ ....
[1 mark]

c) $\frac{x}{5} = 4$

$x =$ ....
[1 mark]

3 $Q = 7x - 3y$ (E)

Find the value of Q when $x = 8$ and $y = 7$.

....
[2 marks]

4 $S = 4m^2 + 2.5n$ (E)

Calculate the value of S when $m = 6.5$ and $n = 4$.

....
[2 marks]

5 The formula for converting a temperature in Celsius (C) to (E)
a temperature in Fahrenheit (F) is:

$$F = \frac{9C}{5} + 32$$

Convert 35 °C to Fahrenheit.

.... °F
[2 marks]

6 To hire a cement mixer, Alex pays £50 per day plus a flat fee of £300. (E)

a) Write a formula to show the total cost C (in £) of hiring the cement mixer for d days.

....
[2 marks]

b) How much would Alex have to pay to hire the cement mixer for 3 days?

£..................................
[2 marks]

Exam Questions

7 Solve the following equations. 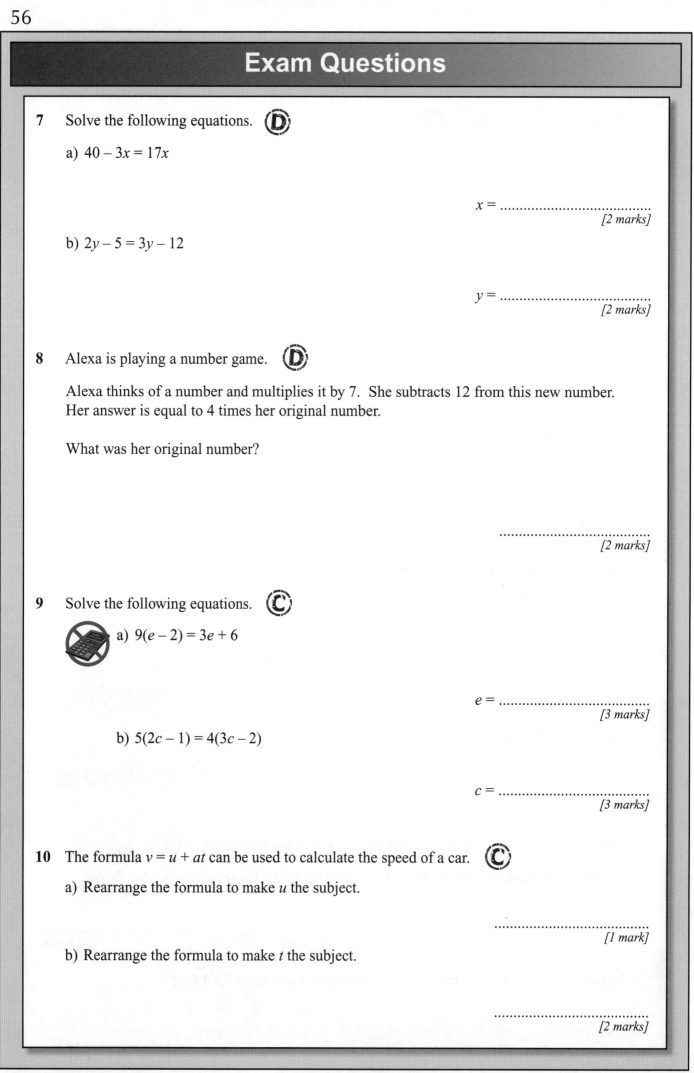(D)

a) $40 - 3x = 17x$

$x = $ ....
[2 marks]

b) $2y - 5 = 3y - 12$

$y = $ ....
[2 marks]

8 Alexa is playing a number game. (D)

Alexa thinks of a number and multiplies it by 7. She subtracts 12 from this new number.
Her answer is equal to 4 times her original number.

What was her original number?

....
[2 marks]

9 Solve the following equations. (C)

a) $9(e - 2) = 3e + 6$

$e = $ ....
[3 marks]

b) $5(2c - 1) = 4(3c - 2)$

$c = $ ....
[3 marks]

10 The formula $v = u + at$ can be used to calculate the speed of a car. (C)

a) Rearrange the formula to make u the subject.

....
[1 mark]

b) Rearrange the formula to make t the subject.

....
[2 marks]

Number Patterns and Sequences

Sequences are just patterns of numbers or shapes that follow a rule.
You need to be able to spot what the rule is.

Finding Number Patterns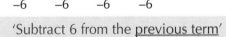

The trick to finding the rule for number patterns is to write down what you have to do to get from one number to the next in the gaps between the numbers. There are 2 main types to look out for:

1) Add or subtract the same number

E.g. 2   5   8   11   14   ...   30   24   18   12   ...
+3 +3 +3 +3 +3   −6   −6   −6   −6

The RULE: 'Add 3 to the previous term'   'Subtract 6 from the previous term'

2) Multiply or divide by the same number each time

E.g. 2   6   18   54   ...   40 000   4000   400   40   ...
×3 ×3 ×3 ×3   ÷10   ÷10   ÷10   ÷10

The RULE: 'Multiply the previous term by 3'   'Divide the previous term by 10'

You might sometimes get patterns that follow a different rule — for example, you might have to add or subtract a changing number each time, or add together the two previous terms. You probably don't need to worry about this, but if it comes up, just describe the pattern and use your rule to find the next term.

Shape Patterns

If you have a pattern of shapes, you need to be able to continue the pattern. You might also have to find the rule for the pattern to work out how many shapes there'll be in a later pattern.

EXAMPLE: **On the right, there are some patterns made of circles.**
a) Draw the next pattern in the sequence.
b) Work out how many circles there will be in the 10th pattern.

a) Just continue the pattern — each 'leg' increases by one circle.

In an exam question, you might be given a table and asked to complete it.

b) Set up a table to find the rule:

Pattern number	1	2	3	4	5	6	7	8	9	10
Number of circles	1	3	5	7	9	11	13	15	17	19

The rule is 'add 2 to the previous term'.
So just keep on adding 2 to extend the table until you get to the 10th term — which is **19**.

Always write the change in the gaps between the numbers

It's the most straightforward way to spot the pattern — you'll see straight away if the difference is the same, changing by a certain amount, or multiplying. Read on for more about sequences.

Number Patterns and Sequences

You might get asked to "find an <u>expression</u> for the <u>nth term</u> of a sequence" —
this is a rule with *n*, like 5*n* – 3. It gives <u>every term in a sequence</u> when you put in different values for *n*.

Finding the **nth Term** of a **Sequence**

This method works for sequences with a <u>common difference</u> — where you <u>add</u> or <u>subtract</u> the <u>same number</u> each time (i.e. the difference between each pair of terms is the <u>same</u>).

EXAMPLE:

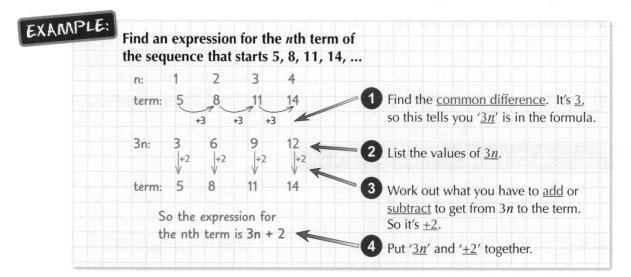

Find an expression for the *n*th term of the sequence that starts 5, 8, 11, 14, ...

① Find the <u>common difference</u>. It's <u>3</u>, so this tells you '<u>3*n*</u>' is in the formula.

② List the values of <u>3*n*</u>.

③ Work out what you have to <u>add</u> or <u>subtract</u> to get from 3*n* to the term. So it's <u>+2</u>.

④ Put '<u>3*n*</u>' and '<u>+2</u>' together.

So the expression for the nth term is 3n + 2

<u>Check</u> your formula by putting the first few values of *n* back in:

n = 1 gives 3*n* + 2 = 3 + 2 = 5 ✓
n = 2 gives 3*n* + 2 = 6 + 2 = 8 ✓

Deciding if a Term is in a Sequence

You might be given the *n*th term and asked if a <u>certain value</u> is in the sequence. The trick here is to <u>set the expression equal to that value</u> and solve to find *n*. If *n* is a <u>whole number</u>, the value is <u>in</u> the sequence.

EXAMPLE:

A sequence is given by the rule 6*n* – 2.

a) Find the 6th term in the sequence.

Just put *n* = 6 into the expression:
(6 × 6) – 2 = 36 – 2
= 34

b) Is 45 a term in this sequence?

Set it equal to 45... 6n – 2 = 45
6n = 47 ...and solve for *n*.
n = 47 ÷ 6 = 7.8333...

n is not a whole number, so 45 is <u>not</u> in the sequence 6n – 2.

It might be even <u>easier</u> to decide if a number is in a sequence or not — for example, if the sequence was all <u>odd numbers</u>, there's <u>no way</u> that an <u>even number</u> could be in the sequence. You just have to use your common sense — e.g. if all the terms in the sequence ended in <u>3</u> or <u>8</u>, 44 would <u>not</u> be in the sequence.

Follow the steps above to find the nth term of a sequence

Learn the steps above for finding the *n*th term. With it, you can easily find the 500th term, or the 500 000th — but without it, you'll need a dozen extra sheets of paper and a spare pen.

Trial and Improvement

Trial and improvement is a way of finding an approximate solution to an equation that's too hard to be solved using normal methods. You'll always be told WHEN to use trial and improvement.

Keep Trying Different Values in the Equation

The basic idea of trial and improvement is to keep trying different values of x that are getting closer and closer to the solution.

Here's the method to follow:

> STEP 1: Put 2 values into the equation that give opposite cases (one too big, one too small).
>
> STEP 2: Choose the next value between the two opposite cases.
>
> STEP 3: Repeat STEP 2 until you have two numbers:
> - both to 1 d.p.
> - differing by 1 in the last digit (e.g. 4.3 and 4.4).
> These are the two possible answers.
>
> STEP 4: Take the exact middle value to decide which one it is.

If you had to find a solution to 2 d.p., the method's just the same — except you'll end up with 2 numbers to 2 d.p. instead of 1 d.p.

Put Your Working in a Table ©

It's a good idea to keep track of your working in a table — see the example below.

EXAMPLE:

The solution to the equation $x^3 + 9x = 40$ lies between 2 and 3. Use trial and improvement to find the solution to this equation to 1 d.p.

STEP 1: Put in **2 and 3** first (given in question)

STEP 2: Try **2.5**

STEP 3: Try **2.7**

Try **2.6**

x	$x^3 + 9x$	
2	26	Too small
3	54	Too big
2.5	38.125	Too small
2.7	43.983	Too big
2.6	40.976	Too big
2.55	39.531375	Too small

... so solution is between **2 and 3**
... so solution is between **2.5 and 3**
... so solution is between **2.5 and 2.7**
... so solution is between **2.5 and 2.6**
— so the answer to 1 d.p. has to be either **2.5 or 2.6**...

STEP 4: Take the exact middle value

... so solution is between 2.55 and 2.6, so $x = 2.6$ to 1 d.p.

Make sure you show all your working — otherwise the examiner won't be able to tell what method you've used and you'll lose marks.

It's like playing a game of higher and lower

You need to commit this method to memory, otherwise you've wasted your time reading it. Luckily it's simple — like a guessing game where you guess a number too high, and then too low and get gradually closer until you get to the right answer. That's really all you're doing.

Inequalities

Inequalities are a bit tricky, but once you've learned the tricks involved, most of the algebra for them is identical to ordinary equations (have a look back at pages 49-50 if you need a reminder).

The **Inequality Symbols** C

> means 'Greater than'   ≥ means 'Greater than or equal to'
< means 'Less than'   ≤ means 'Less than or equal to'

REMEMBER — the one at the BIG end is BIGGEST so $x > 4$ and $4 < x$ both say: 'x is greater than 4'.

EXAMPLE: **x is an integer such that $-4 < x \leq 3$. Write down all possible values of x.**

Work out what each bit of the inequality is telling you:

$-4 < x$ means 'x is greater than -4',
and $x \leq 3$ means 'x is less than or equal to 3'.

Now just write down all the values that x can take:

→ $-3, -2, -1, 0, 1, 2, 3$

Remember, integers are just whole numbers (+ve and −ve, including 0).

-4 isn't included because of the <, but 3 is included because of the ≤.

You Can Show Inequalities on **Number Lines** C

Drawing inequalities on a number line is really easy — all you have to remember is that you use an open circle (○) for > or < and a coloured-in circle (●) for ≥ or ≤.

EXAMPLE: **Show the inequality $-4 < x \leq 3$ on a number line.**

Closed circle because 3 is included.

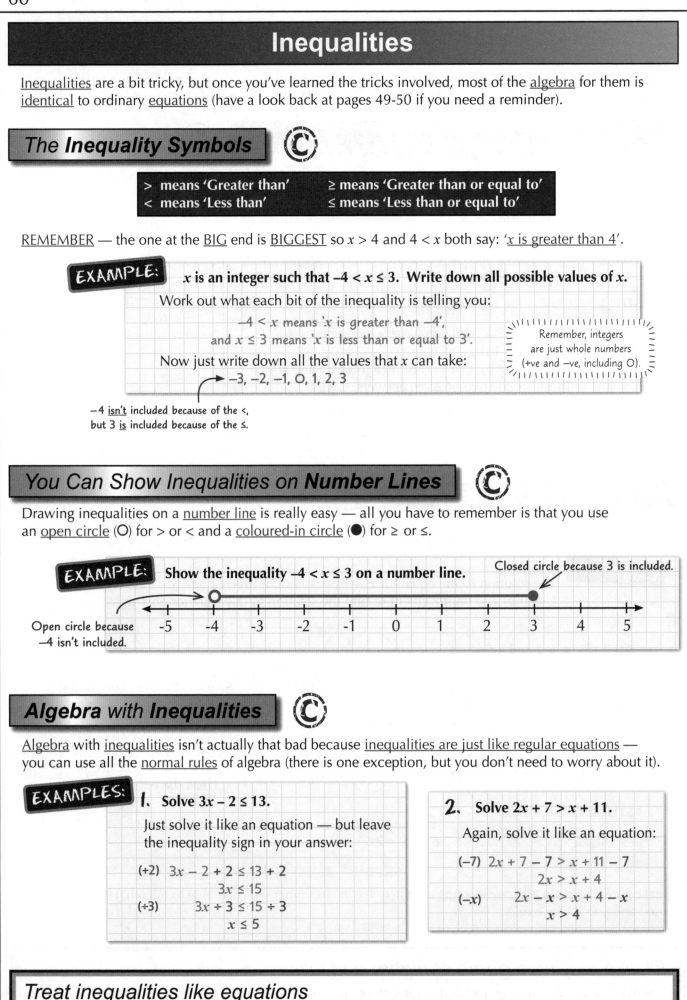

Open circle because -4 isn't included.

Algebra with **Inequalities** C

Algebra with inequalities isn't actually that bad because inequalities are just like regular equations — you can use all the normal rules of algebra (there is one exception, but you don't need to worry about it).

EXAMPLES:

1. Solve $3x - 2 \leq 13$.

Just solve it like an equation — but leave the inequality sign in your answer:

(+2) $3x - 2 + 2 \leq 13 + 2$
          $3x \leq 15$
(÷3) $3x \div 3 \leq 15 \div 3$
             $x \leq 5$

2. Solve $2x + 7 > x + 11$.

Again, solve it like an equation:

(−7) $2x + 7 - 7 > x + 11 - 7$
            $2x > x + 4$
(−x) $2x - x > x + 4 - x$
              $x > 4$

Treat inequalities like equations

There's just one exception though — if you multiply or divide by a negative number, you have to flip the inequality sign round (so < becomes > and ≤ becomes ≥). But you shouldn't need to do this in the exam.

Warm-up and Worked Exam Questions

Sequences, trial and improvement and inequalities — lovely topics, if you ask me...
Anyway, love them or hate them, you have to do them — it's just a case of learning the method and
practising lots of questions... So let's start with some warm-up questions...

Warm-up Questions

1) Find the next two numbers in each of these sequences, and say in words what the rule is for extending each one: a) 2, 5, 9, 14 ...   b) 2, 20, 200 ...   c) 64, 32, 16, 8 ...

2) Find the expression for the nth number in this sequence: 7, 9, 11, 13

3) The equation $x^3 - 2x = 1$ has a solution between 1 and 2.
Use trial and improvement to find it to 1 d.p.

4) Show the inequality $-1 < n \leq 5$ on a number line.

5) Solve the inequalities and find the integer values of x which satisfy both:
$2x + 9 \geq 1$ and $4x < 6 + x$

Worked Exam Question

Work through the question below and give all the questions on the next page a good go.
Then you'll be well and truly prepared for the exam.

1   The first four terms in a sequence are 2, 9, 16, 23, ... Ⓒ

a) Find the nth term of the sequence.

$$2 \quad 9 \quad 16 \quad 23$$
$$+7 \quad +7 \quad +7$$

The common difference is 7,
so 7n is in the formula.

$$n = \quad 1 \quad\quad 2 \quad\quad 3 \quad\quad 4$$
$$7n = \quad 7 \quad\quad 14 \quad\quad 21 \quad\quad 28$$
$$-5 \quad -5 \quad -5 \quad -5$$
$$\text{term} = \quad 2 \quad\quad 9 \quad\quad 16 \quad\quad 23$$

You have to subtract
5 to get to the term.

So the expression for the nth term is 7n − 5

.... $7n - 5$ ....
[2 marks]

b) What is the 30th term of the sequence?

30th term = (7 × 30) − 5 = 205

.... 205 ....
[1 mark]

Exam Questions

2 A sequence is made from patterns of triangles. The first three patterns are shown below.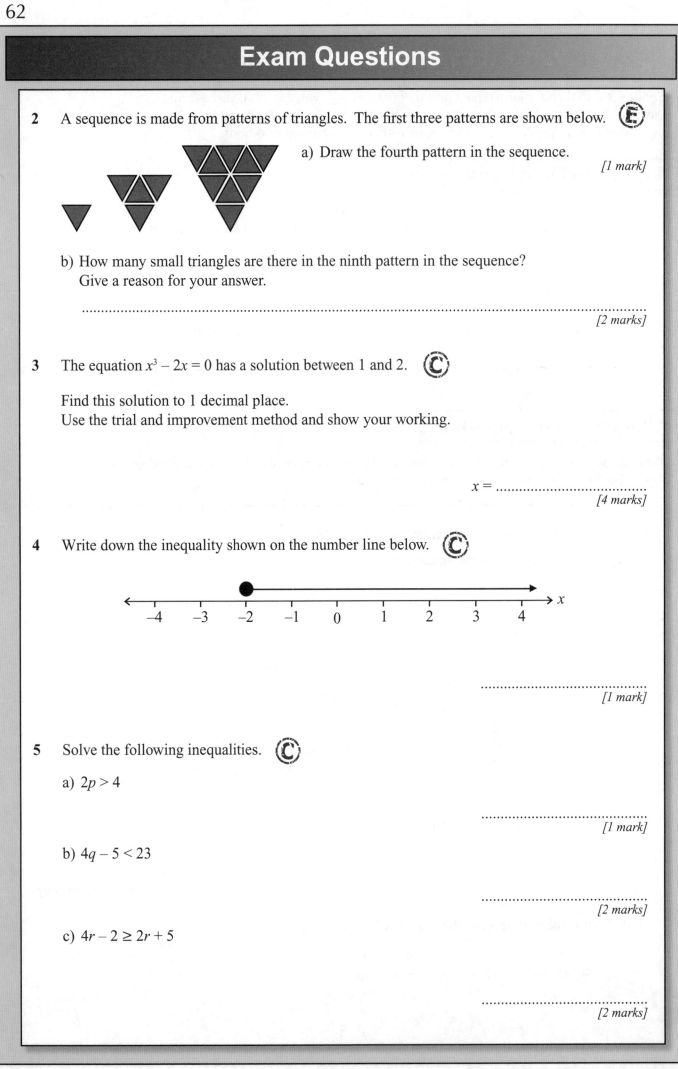

  a) Draw the fourth pattern in the sequence.

  [1 mark]

  b) How many small triangles are there in the ninth pattern in the sequence?
     Give a reason for your answer.

  ....
  [2 marks]

3 The equation $x^3 - 2x = 0$ has a solution between 1 and 2. ©

  Find this solution to 1 decimal place.
  Use the trial and improvement method and show your working.

  $x =$ ....
  [4 marks]

4 Write down the inequality shown on the number line below. ©

  ....
  [1 mark]

5 Solve the following inequalities. ©

  a) $2p > 4$

  ....
  [1 mark]

  b) $4q - 5 < 23$

  ....
  [2 marks]

  c) $4r - 2 \geq 2r + 5$

  ....
  [2 marks]

Revision Questions for Section Two

There was a lot of <u>nasty algebra</u> in that section — let's see how much you remember.

- Try these questions and <u>tick off each one</u> when you <u>get it right</u>.
- When you've done <u>all the questions</u> for a topic and are <u>completely happy</u> with it, tick off the topic.

Algebra (p43-46) ☑

1) Simplify: a) $e + e + e$ b) $4f + 5f - f$

2) Simplify: a) $2x + 3y + 5x - 4y$ b) $11a + 2 - 8a + 7$

3) Simplify: a) $m \times m \times m$ b) $p \times q \times 7$ c) $2x \times 9y$

4) Simplify: a) $g^5 \times g^6$ b) $c^{15} \div c^{12}$

5) Expand: a) $6(x + 3)$ b) $-3(3x - 4)$ c) $x(5 - x)$

6) Expand and simplify $4(3 + 5x) - 2(7x + 6)$

7) What is factorising?

8) Factorise: a) $8x + 24$ b) $18x + 27y$ c) $5x^2 + 15x$

Solving Equations (p49-50) ☑

9) Solve: a) $x + 9 = 16$ b) $x - 4 = 12$ c) $6x = 18$

10) Solve: a) $4x + 3 = 19$ b) $3x + 6 = x + 10$ c) $3(x + 2) = 5x$

Formulas (p51-53) ☑

11) $Q = 5r + 6s$. Work out the value of Q when $r = -2$ and $s = 3$.

12) Imran buys d DVDs and c CDs. DVDs cost £7 each and CDs cost £5 each. He spends £P in total. Write a formula for P in terms of d and c.

13) Hiring ice skates costs £3 per hour plus a deposit of £5. Lily paid £11. How long did she hire the skates for?

14) Rearrange the formula $W = 4v + 5$ to make v the subject.

Number Patterns and Sequences (p57-58) ☑

15) For each of the following sequences, find the next term and write down the rule you used.
    a) 3, 10, 17, 24, ...   b) 1, 4, 16, 64, ...   c) 2, 5, 7, 12, ...

16) Find an expression for the nth term of the sequence that starts 4, 10, 16, 22, ...

17) Is 34 a term in the sequence given by the expression $7n - 1$?

Trial and Improvement (p59) ☑

18) What is trial and improvement?

19) Given that $x^3 + 8x = 103$ has a solution between 4 and 5, use trial and improvement to find this solution to 1 d.p.

20) Given that $x^3 - 3x = 41$ has a solution between 3 and 4, use trial and improvement to find this solution to 1 d.p.

Inequalities (p60) ☑

21) Write the following inequalities out in words: a) $x > -7$ b) $x \leq 6$

22) $0 < k \leq 7$. Find all the possible integer values of k.

23) Solve the following inequalities: a) $x + 4 < 14$ b) $x - 11 > 3$ c) $7x \geq 21$

24) Solve the inequality $3x + 5 \leq 26$.

Coordinates and Midpoints

To get on well with graph questions you need to get to grips with the basics.

The Four Quadrants (F)

A graph has <u>four different quadrants</u> (regions).

The top-right region is the easiest because
<u>ALL THE COORDINATES IN IT ARE POSITIVE</u>.

You have to be careful in the <u>other regions</u> though,
because the x- and y- coordinates could be <u>negative</u>,
and that makes life much more difficult.

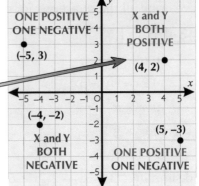

THREE IMPORTANT POINTS ABOUT COORDINATES:

1) The coordinates are always in <u>ALPHABETICAL ORDER, x then y</u>. (x, y)

2) x is always the flat axis going <u>ACROSS</u> the page.
   In other words '<u>x is a..cross</u>' (x is a '×').

3) Remember it's always <u>IN THE HOUSE</u> (→) and then <u>UP THE STAIRS</u> (↑)
   so it's <u>ALONG first</u> and <u>then UP</u>, i.e. x-coordinate first, and then y-coordinate.

The Midpoint of a Line (C)

The '<u>MIDPOINT OF A LINE SEGMENT</u>' is the <u>POINT THAT'S RIGHT IN THE MIDDLE</u> of it.

Finding the coordinates of a midpoint is pretty easy.

<u>LEARN THESE THREE STEPS...</u>

> 1) Find the <u>average</u> of the <u>x-coordinates</u>.
> 2) Find the <u>average</u> of the <u>y-coordinates</u>.
> 3) Put them in <u>brackets</u>.

Use the coordinates of the <u>two end points</u>.

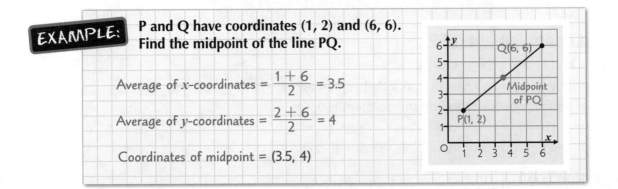

EXAMPLE: **P and Q have coordinates (1, 2) and (6, 6).**
Find the midpoint of the line PQ.

Average of x-coordinates $= \dfrac{1+6}{2} = 3.5$

Average of y-coordinates $= \dfrac{2+6}{2} = 4$

Coordinates of midpoint $= (3.5, 4)$

Coordinates should always be written as (x, y)

Learn the three points for getting x and y the right way round and the three easy steps for finding the
midpoint of a line segment. Then close the book and write them all down.

Straight-Line Graphs

You ought to know these simple graphs straight off with no hesitation.

Horizontal and *Vertical* lines: 'x = a' and 'y = a'

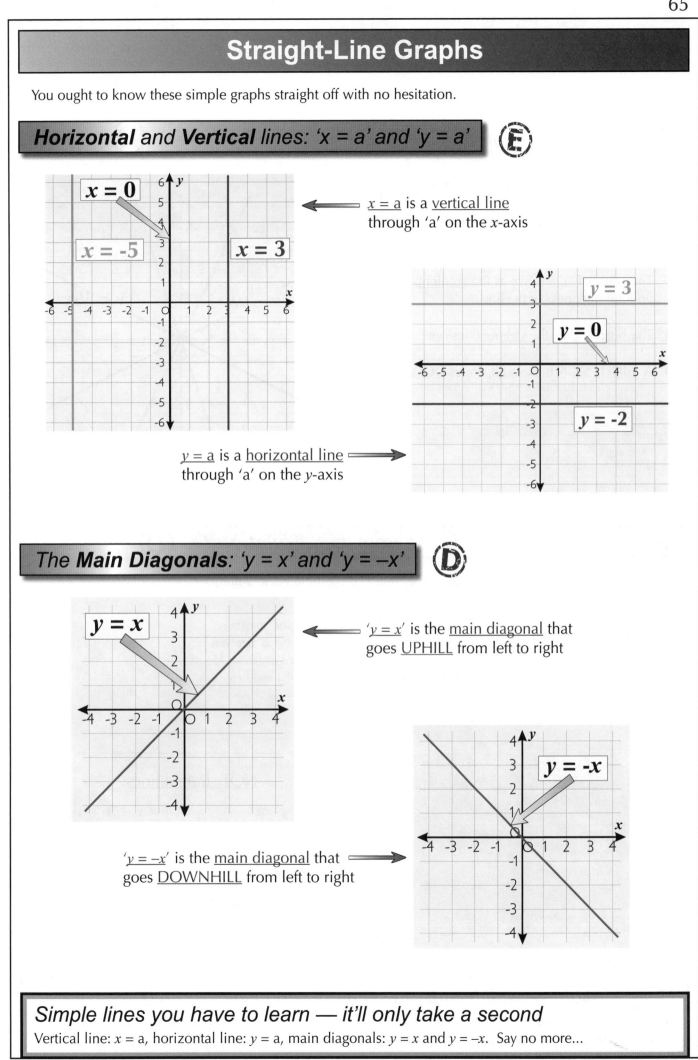

$x = a$ is a vertical line through 'a' on the x-axis

$y = a$ is a horizontal line through 'a' on the y-axis

The *Main Diagonals*: 'y = x' and 'y = –x'

'$y = x$' is the main diagonal that goes UPHILL from left to right

'$y = –x$' is the main diagonal that goes DOWNHILL from left to right

Simple lines you have to learn — it'll only take a second
Vertical line: $x = a$, horizontal line: $y = a$, main diagonals: $y = x$ and $y = –x$. Say no more...

Straight-Line Graphs

Here are some more of the basic straight-line graphs that you really need to know.

Other **Lines** *Through the Origin:* 'y = ax' *and* 'y = –ax' (D)

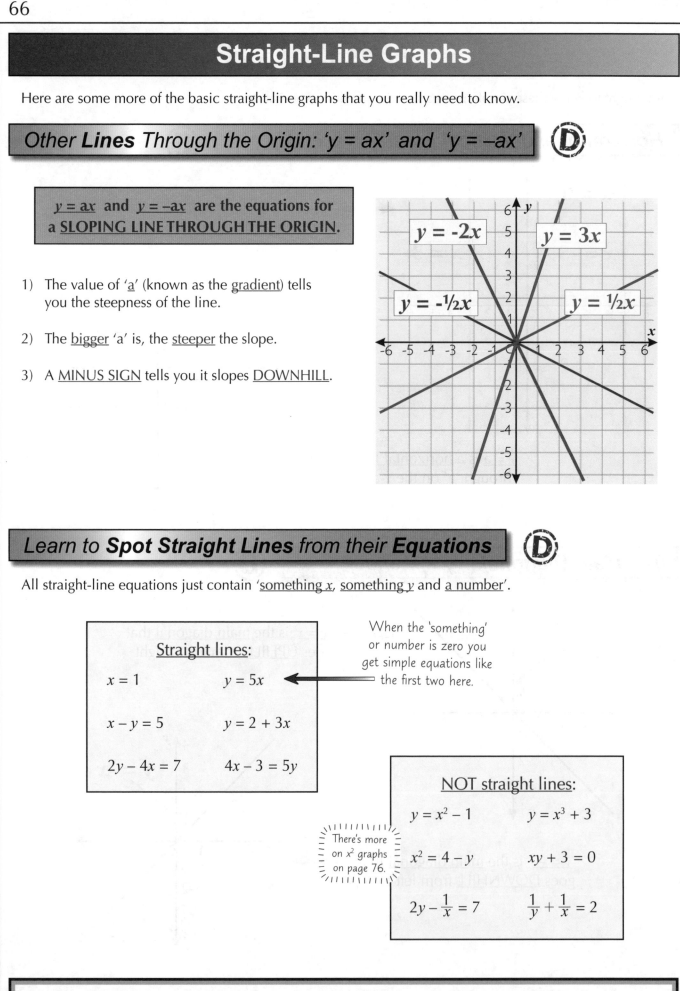

$y = ax$ and $y = -ax$ are the equations for a SLOPING LINE THROUGH THE ORIGIN.

1) The value of 'a' (known as the gradient) tells you the steepness of the line.

2) The bigger 'a' is, the steeper the slope.

3) A MINUS SIGN tells you it slopes DOWNHILL.

Learn to **Spot Straight Lines** *from their* **Equations** (D)

All straight-line equations just contain 'something x, something y and a number'.

Straight lines:

$x = 1$ $y = 5x$

$x - y = 5$ $y = 2 + 3x$

$2y - 4x = 7$ $4x - 3 = 5y$

When the 'something' or number is zero you get simple equations like the first two here.

There's more on x^2 graphs on page 76.

NOT straight lines:

$y = x^2 - 1$ $y = x^3 + 3$

$x^2 = 4 - y$ $xy + 3 = 0$

$2y - \frac{1}{x} = 7$ $\frac{1}{y} + \frac{1}{x} = 2$

Get it straight — which lines are straight (and which aren't)

The graphs $y = ax$ and $y = -ax$ are diagonals like $y = x$ and $y = -x$ on the previous page. They're steeper or flatter depending on the value of 'a'. Make sure you can spot when an equation will be a straight line.

Plotting Straight-Line Graphs

You're likely to be asked to <u>draw the graph</u> of an equation in the exam.
This <u>EASY METHOD</u> will net you the marks every time:

> 1) Choose **3 values of x** and <u>**draw up a table**</u>,
> 2) <u>**Work out the corresponding y-values**</u>,
> 3) <u>**Plot the coordinates, and draw the line**</u>.

You might get lucky and be given a table in an exam question. Don't worry if it contains 5 or 6 values.

Doing the 'Table of Values' (D)

EXAMPLE: **Draw the graph of $y = 2x - 3$ for values of x from –2 to 4.**

1) <u>Choose 3 easy x-values for your table:</u>
Use x-values from the grid you're given. Avoid negative ones if you can.

x	O	2	4
y			

2) <u>Find the y-values</u> by putting each x-value into the equation:

x	O	2	4
y	–3	1	5

When $x = $ O,
$y = 2x - 3$
$= (2 \times O) - 3 = -3$

When $x = 4$,
$y = 2x - 3$
$= (2 \times 4) - 3 = 5$

Plotting the Points and *Drawing the Graph* (D)

EXAMPLE: **...continued from above.**

3) <u>Plot each pair</u> of x- and y- values from your table.

The table gives the coordinates (O, –3), (2, 1) and (4, 5).

Now draw a <u>straight line</u> through your points.

> If one point looks a bit odd, check 2 things:
> – the <u>y-value</u> you worked out in the table
> – that you've <u>plotted</u> it properly.

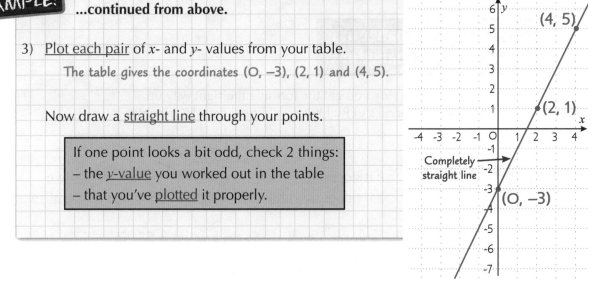

Spot and plot a straight line — then check it looks right

In the exam you might get an equation like $3x + y = 5$ to plot, making finding the y-values a bit trickier.
Just substitute the x-value and find the y-value that makes the equation true.
E.g. when $x = 1$, $3x + y = 5$ → $(3 \times 1) + y = 5$ → $3 + y = 5$ → $y = 2$.
Or you can rearrange the equation to get y on its own if you find that easier.

Straight-Line Graphs — Gradients

Time to find some accurate gradients. I'm afraid "quite steep" won't do.

Finding the Gradient C

The <u>gradient</u> of a line is a measure of its <u>slope</u>. The <u>bigger</u> the number, the <u>steeper</u> the line.

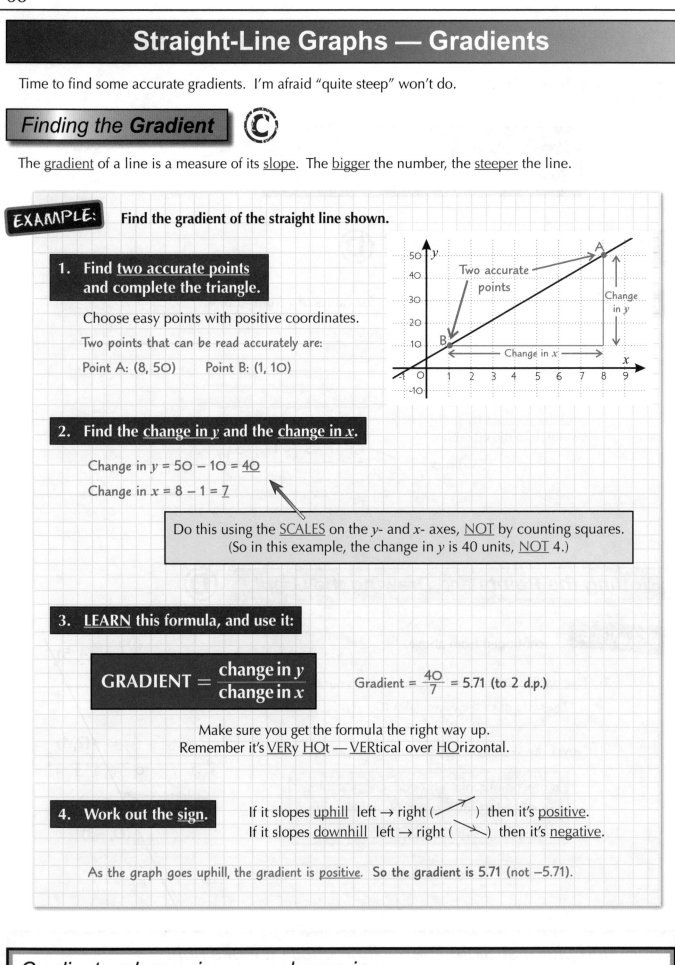

EXAMPLE: **Find the gradient of the straight line shown.**

1. **Find <u>two accurate points</u> and complete the triangle.**

Choose easy points with positive coordinates.

Two points that can be read accurately are:

Point A: (8, 50)   Point B: (1, 10)

2. **Find the <u>change in *y*</u> and the <u>change in *x*</u>.**

Change in *y* = 50 – 10 = <u>40</u>

Change in *x* = 8 – 1 = <u>7</u>

Do this using the <u>SCALES</u> on the *y*- and *x*- axes, <u>NOT</u> by counting squares. (So in this example, the change in *y* is 40 units, <u>NOT</u> 4.)

3. **<u>LEARN</u> this formula, and use it:**

$$\text{GRADIENT} = \frac{\text{change in } y}{\text{change in } x}$$

$\text{Gradient} = \frac{40}{7} = 5.71 \text{ (to 2 d.p.)}$

Make sure you get the formula the right way up. Remember it's <u>VER</u>y <u>HO</u>t — <u>VER</u>tical over <u>HO</u>rizontal.

4. **Work out the <u>sign</u>.**

If it slopes <u>uphill</u> left → right (⟋) then it's <u>positive</u>.

If it slopes <u>downhill</u> left → right (⟍) then it's <u>negative</u>.

As the graph goes uphill, the gradient is <u>positive</u>. So the gradient is 5.71 (not –5.71).

Gradient = change in y over change in x

It's really important that you get to grips with this method for finding gradients. Learn the four steps, then test yourself by writing them down. Gradients crop up in a few different types of graph question — for example, when you're asked to find the equation of a straight line (coming up on the next page).

Straight-Line Graphs — "y = mx + c"

Here is your last page on straight lines for now. There's lots of information here, so read carefully.

y = mx + c is the Equation of a Straight Line ©

$y = mx + c$ is the general equation for a straight-line graph, and you need to remember:

'm' is equal to the GRADIENT of the graph

'c' is the value WHERE IT CROSSES THE Y-AXIS and is called the Y-INTERCEPT.

'm' and 'c' are always just numbers — so $y = 3x - 1$ and $y = -x + 2$ are equations of straight lines.

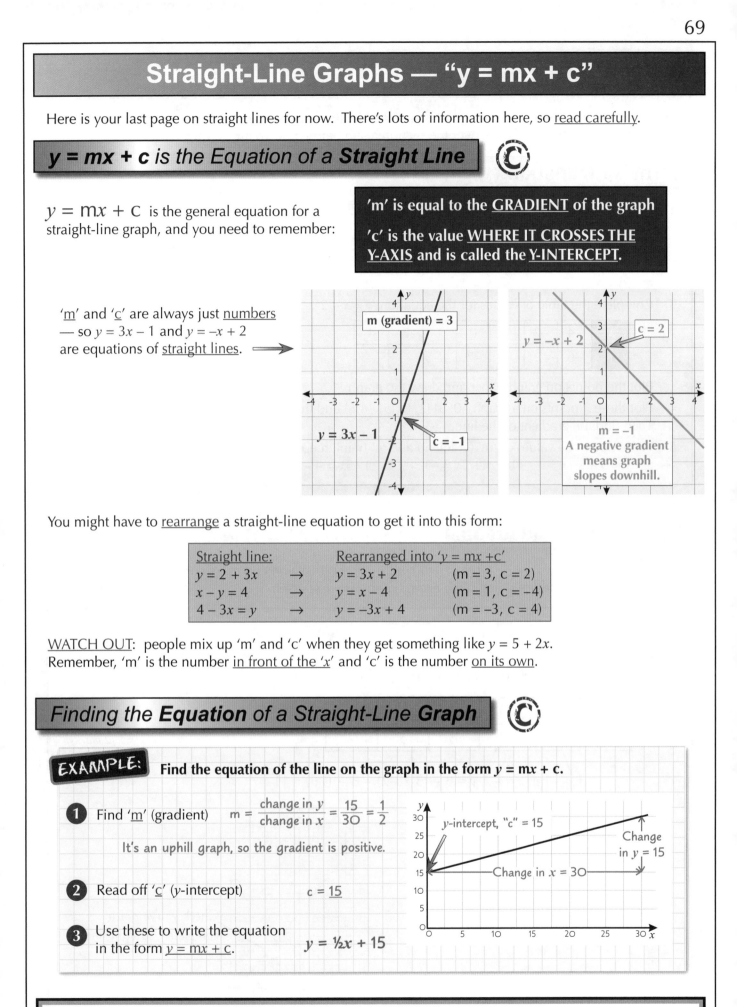

You might have to rearrange a straight-line equation to get it into this form:

Straight line:		Rearranged into 'y = mx + c'	
$y = 2 + 3x$	→	$y = 3x + 2$	(m = 3, c = 2)
$x - y = 4$	→	$y = x - 4$	(m = 1, c = -4)
$4 - 3x = y$	→	$y = -3x + 4$	(m = -3, c = 4)

WATCH OUT: people mix up 'm' and 'c' when they get something like $y = 5 + 2x$. Remember, 'm' is the number in front of the 'x' and 'c' is the number on its own.

Finding the Equation of a Straight-Line Graph ©

EXAMPLE: Find the equation of the line on the graph in the form y = mx + c.

1. Find 'm' (gradient) $m = \dfrac{\text{change in } y}{\text{change in } x} = \dfrac{15}{30} = \dfrac{1}{2}$

   It's an uphill graph, so the gradient is positive.

2. Read off 'c' (y-intercept) $c = \underline{15}$

3. Use these to write the equation in the form $y = mx + c$. $y = ½x + 15$

'm' is the gradient and 'c' is the y-intercept

The key thing to remember is that 'm' is the number in front of the x, and 'c' is the number on its own. Remember that and you'll be able to find the equation of any straight line they throw at you.

Warm-up and Worked Exam Questions

In the exam, you'll have to know straight-line graphs like the back of your hand. If you struggle with any of the warm-up questions, go back over the section again before you go any further.

Warm-up Questions

1) Plot these points on some graph paper: A(1, 4), B(5, 6), C(3, 2), D(7, 0).
   a) Draw a line between points A and B and find the midpoint of the line AB.
   b) Draw a line between points C and D and find the midpoint of line CD.

2) Say whether the graphs of the following equations will be a straight line.
   a) $2y = -x + 7$ b) $y = 4x^2 - 1$ c) $x = 3y$

3) Complete the table of values for the equation '$y = x - 2$'.

x	−4	−2	−1	0	1	2	4
y	−6		−2				

   Plot the points on graph paper and draw the graph.

4) Plot these 3 points on a graph: (0, 3) (2, 0) (5, −4.5) and then join them up with a straight line. Then find the gradient of the line.

5) Write down the values of 'm' (gradient) and 'c' (y-intercept) for the following equations.
   a) $y = 5 + x$ b) $y + 2x = -3$ c) $5x = 4 - y$

Worked Exam Question

You know the routine by now — work carefully through this question and make sure you understand it. Then it's on to the real test of doing some exam questions for yourself.

1 This is a question about the equation $y = 8 - 3x$. **(D)**

a) Complete this table of values for the equation $y = 8 - 3x$.

x	−2	−1	0	1	2
y	14	11	8	5	2

[2 marks]

E.g. when $x = -2$, $y = 8 - (3 \times -2) = 8 - (-6) = 14$

b) Using the table, draw the graph of $y = 8 - 3x$ on the grid to the right.

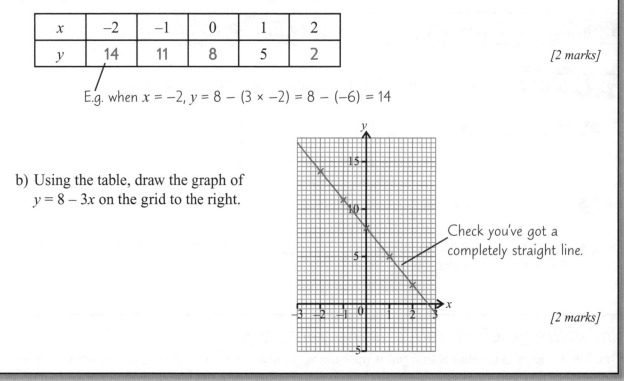

Check you've got a completely straight line.

[2 marks]

Exam Questions

2 Points **Q** and **R** have been plotted on the grid below.

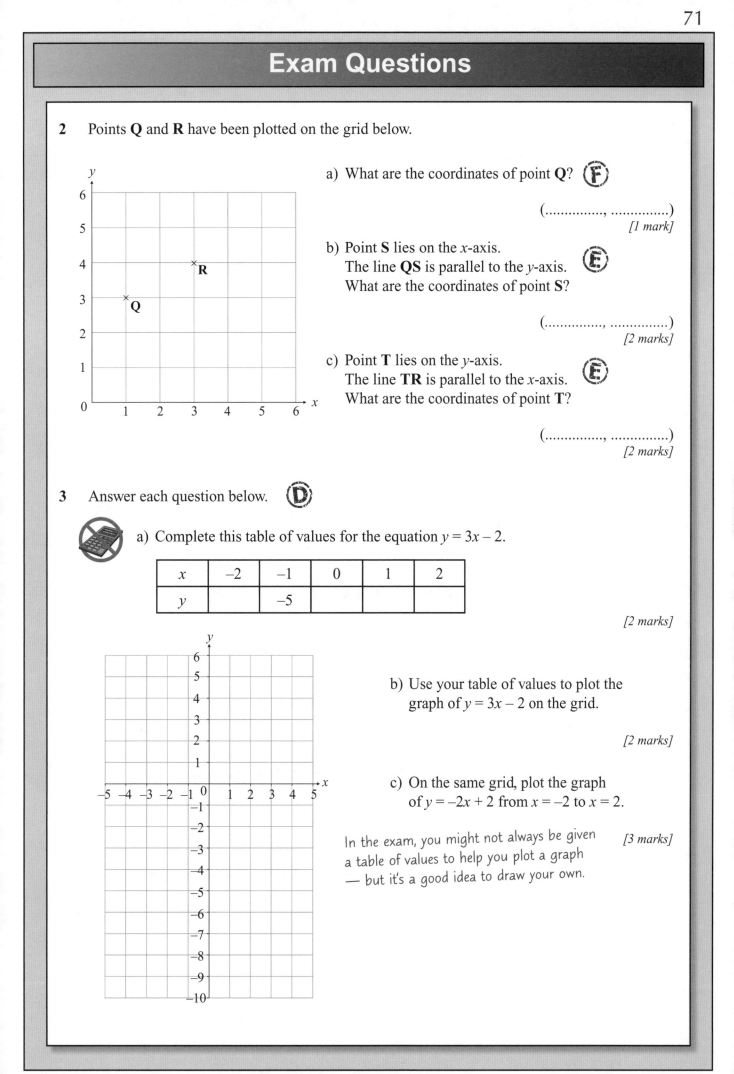

a) What are the coordinates of point **Q**? (F)

(............., ...)
[1 mark]

b) Point **S** lies on the *x*-axis.
The line **QS** is parallel to the *y*-axis. (E)
What are the coordinates of point **S**?

(............., ...)
[2 marks]

c) Point **T** lies on the *y*-axis.
The line **TR** is parallel to the *x*-axis. (E)
What are the coordinates of point **T**?

(............., ...)
[2 marks]

3 Answer each question below. (D)

a) Complete this table of values for the equation $y = 3x - 2$.

x	-2	-1	0	1	2
y		-5			

[2 marks]

b) Use your table of values to plot the graph of $y = 3x - 2$ on the grid.

[2 marks]

c) On the same grid, plot the graph of $y = -2x + 2$ from $x = -2$ to $x = 2$.

[3 marks]

In the exam, you might not always be given a table of values to help you plot a graph — but it's a good idea to draw your own.

72

Exam Questions

4 Use the grid to answer the questions below.

a) Draw the graph of $2x + y = 6$ for values of x from –4 to 4. Ⓓ

[3 marks]

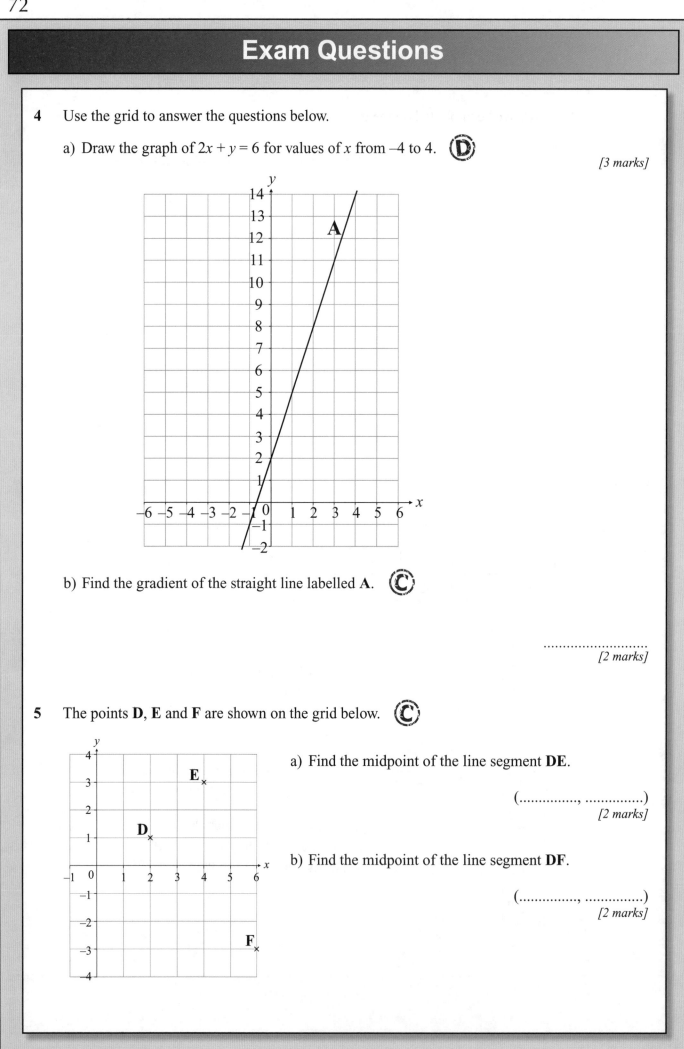

b) Find the gradient of the straight line labelled **A**. Ⓒ

....
[2 marks]

5 The points **D**, **E** and **F** are shown on the grid below. Ⓒ

a) Find the midpoint of the line segment **DE**.

(..............., ...)
[2 marks]

b) Find the midpoint of the line segment **DF**.

(..............., ...)
[2 marks]

Travel Graphs

If <u>travel graphs</u> come up in the exam, you'll need to know all the <u>vital details</u> about them. Read on...

Distance-Time Graphs Ⓓ

1) The graph <u>GOING UP</u> means it's travelling <u>AWAY</u>. The graph <u>COMING DOWN</u> means it's <u>COMING BACK AGAIN</u>.

2) At any point, <u>GRADIENT = SPEED</u>, but watch out for the <u>UNITS</u>.

3) The <u>STEEPER</u> the graph, the <u>FASTER</u> it's going

4) <u>FLAT SECTIONS</u> are where it is <u>STOPPED</u>.

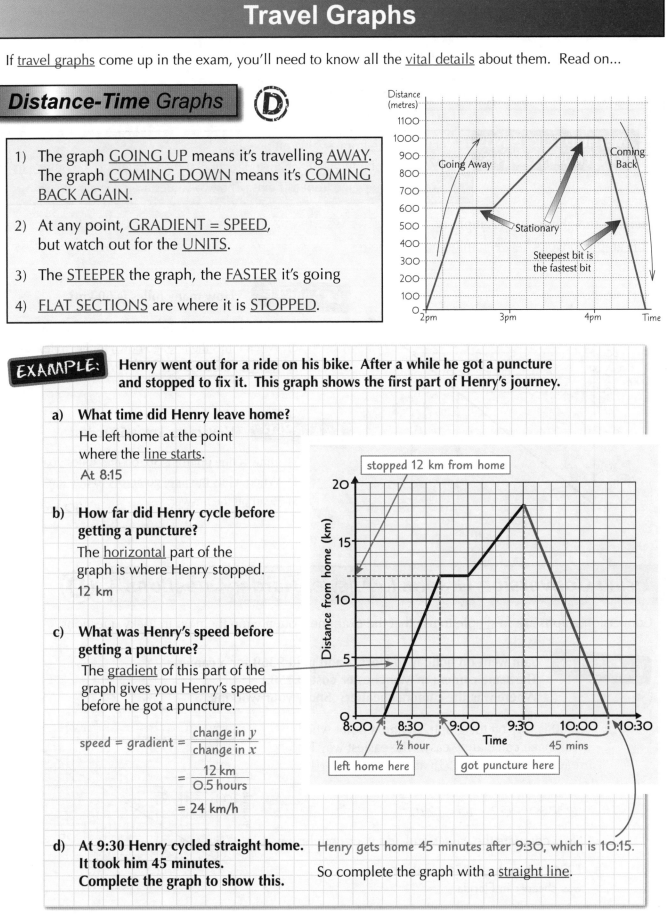

EXAMPLE: Henry went out for a ride on his bike. After a while he got a puncture and stopped to fix it. This graph shows the first part of Henry's journey.

a) **What time did Henry leave home?**

He left home at the point where the <u>line starts</u>.

At 8:15

b) **How far did Henry cycle before getting a puncture?**

The <u>horizontal</u> part of the graph is where Henry stopped.

12 km

c) **What was Henry's speed before getting a puncture?**

The <u>gradient</u> of this part of the graph gives you Henry's speed before he got a puncture.

$$\text{speed} = \text{gradient} = \frac{\text{change in } y}{\text{change in } x}$$

$$= \frac{12\,\text{km}}{0.5\,\text{hours}}$$

$$= 24\,\text{km/h}$$

d) **At 9:30 Henry cycled straight home. It took him 45 minutes. Complete the graph to show this.**

Henry gets home 45 minutes after 9:30, which is 10:15. So complete the graph with a <u>straight line</u>.

The gradient of a distance-time graph equals the speed

Learn the four important details about distance-time graphs, then cover the page and write them down. Exam questions on this topic can look a bit daunting, but spot the key details and you'll be fine.

Conversion Graphs

In the exam you're likely to get a graph which converts something like <u>£ to dollars</u> or <u>mph to km/h</u>.

Conversion Graphs are Easy to Use Ⓔ

METHOD FOR USING CONVERSION GRAPHS:	1) <u>Draw a line</u> from a value on <u>one axis</u>.
	2) When you hit the LINE, <u>change direction</u> and go straight to <u>the other axis</u>.
	3) <u>Read off the value</u> from this axis. The two values are <u>equivalent</u>.

Here's a straightforward example:

<u>This graph converts between miles and kilometres</u>

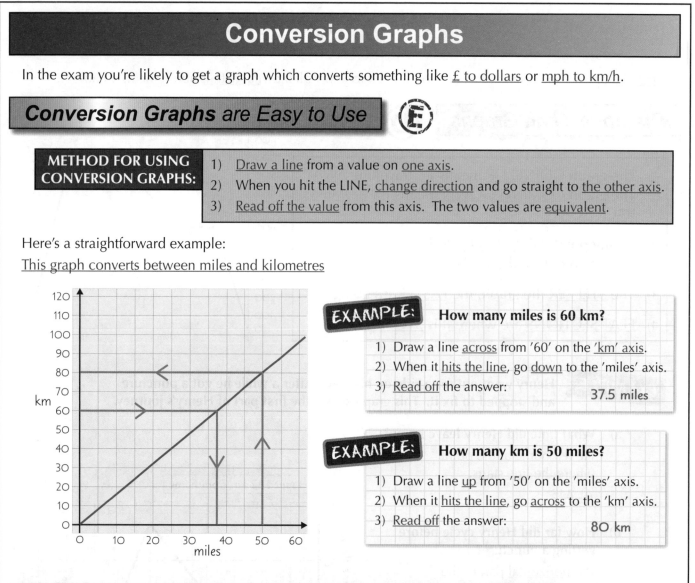

EXAMPLE: **How many miles is 60 km?**

1) Draw a line <u>across</u> from '60' on the <u>'km' axis</u>.
2) When it <u>hits the line</u>, go <u>down</u> to the 'miles' axis.
3) <u>Read off</u> the answer: 37.5 miles

EXAMPLE: **How many km is 50 miles?**

1) Draw a line <u>up</u> from '50' on the 'miles' axis.
2) When it <u>hits the line</u>, go <u>across</u> to the 'km' axis.
3) <u>Read off</u> the answer: 80 km

Using *Conversion Graphs* to Answer **Harder Questions** Ⓓ

Conversion graphs are so <u>simple</u> to use that the examiners often wrap them up in tricky questions.

EXAMPLE: **Sam went on holiday to Florida and paid $360 for a camera. The same camera in Manchester costs £250. Where was the camera cheaper? Show your working.**

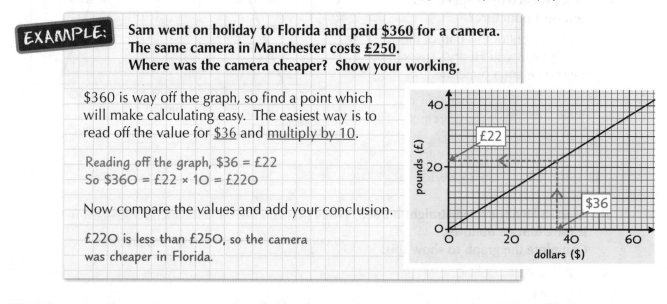

$360 is way off the graph, so find a point which will make calculating easy. The easiest way is to read off the value for <u>$36</u> and <u>multiply by 10</u>.

Reading off the graph, $36 = £22
So $360 = £22 × 10 = £220

Now compare the values and add your conclusion.

£220 is less than £250, so the camera was cheaper in Florida.

Learn how to convert graph questions into marks

Questions on conversion graphs aren't too bad, as long as you learn the three-step method above. Always leave your conversion lines showing on the graph — they might get you a mark for your working.

Real-Life Graphs

Graphs can be drawn for just about anything. Here are some useful examples.

Graphs Can Show **How Much** You'll **Pay**

Graphs are great for showing how much you'll be charged for using a service or buying multiple items.

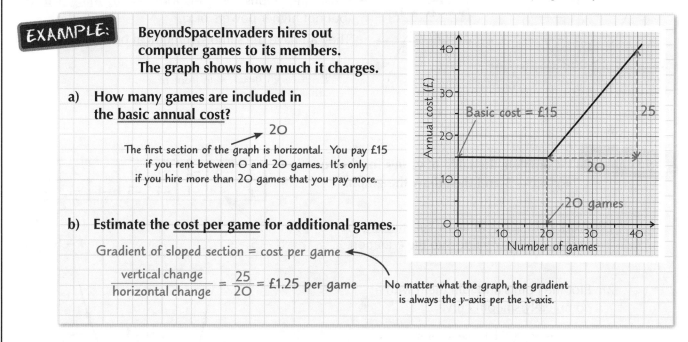

EXAMPLE: BeyondSpaceInvaders hires out computer games to its members. The graph shows how much it charges.

a) **How many games are included in the basic annual cost?**

20

The first section of the graph is horizontal. You pay £15 if you rent between 0 and 20 games. It's only if you hire more than 20 games that you pay more.

b) **Estimate the cost per game for additional games.**

Gradient of sloped section = cost per game

$\frac{\text{vertical change}}{\text{horizontal change}} = \frac{25}{20} = £1.25$ per game

No matter what the graph, the gradient is always the y-axis per the x-axis.

Graphs Can Show **Other Changes** Too

Graphs aren't always about money. They can also show things like water depth or temperature.

EXAMPLE: At a lemonade factory, a cylindrical tank is filled up before being transferred into a tanker. This graph shows how the height of lemonade in the tank changes.

a) **During the morning, one of the juicing machines broke. This slowed down the rate of lemonade production.**

What height was the lemonade when this happened?

The gradient decreases when the rate of lemonade production slowed down. This happened when the lemonade height was 1.75 m.

b) **How long did it take to empty the tank?**

The "downhill" part of the graph shows the tank being emptied.

This starts at 11:30 and ends at 13:00, which is 1.5 hours.

The gradient tells you the number of y-axis units per x-axis unit

If the question talks about 'rate' or 'something per something else', it's a big clue that the gradient's involved. Make sure you're happy using the gradient formula and reading values from graphs.

Quadratic Graphs

That's it for straight lines. Now you're on to <u>quadratic graphs</u>, which are always drawn as <u>smooth curves</u>.

Drawing a **Quadratic Graph** Ⓒ

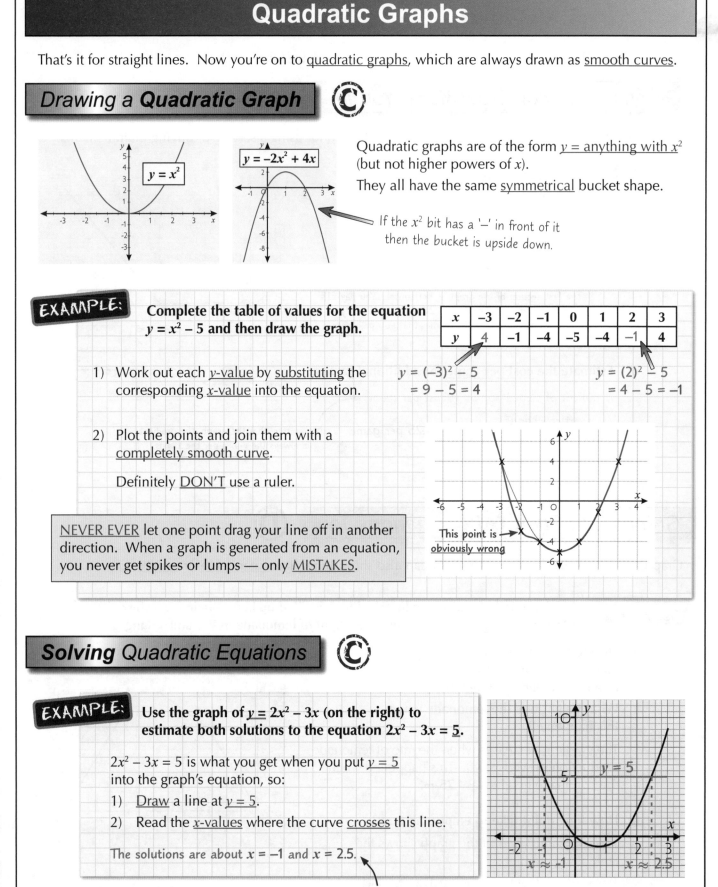

Quadratic graphs are of the form <u>y = anything with x^2</u> (but not higher powers of x).

They all have the same <u>symmetrical</u> bucket shape.

If the x^2 bit has a '−' in front of it then the bucket is upside down.

EXAMPLE: **Complete the table of values for the equation $y = x^2 - 5$ and then draw the graph.**

x	−3	−2	−1	0	1	2	3
y	4	−1	−4	−5	−4	−1	4

1) Work out each <u>y-value</u> by <u>substituting</u> the corresponding <u>x-value</u> into the equation.

$$y = (-3)^2 - 5$$
$$= 9 - 5 = 4$$

$$y = (2)^2 - 5$$
$$= 4 - 5 = -1$$

2) Plot the points and join them with a <u>completely smooth curve</u>.

Definitely <u>DON'T</u> use a ruler.

<u>NEVER EVER</u> let one point drag your line off in another direction. When a graph is generated from an equation, you never get spikes or lumps — only <u>MISTAKES</u>.

This point is obviously wrong

Solving Quadratic Equations Ⓒ

EXAMPLE: **Use the graph of $y = 2x^2 - 3x$ (on the right) to estimate both solutions to the equation $2x^2 - 3x = \underline{5}$.**

$2x^2 - 3x = 5$ is what you get when you put <u>$y = 5$</u> into the graph's equation, so:

1) <u>Draw</u> a line at <u>$y = 5$</u>.
2) Read the <u>x-values</u> where the curve <u>crosses</u> this line.

The solutions are about $x = -1$ and $x = 2.5$.

$y = 5$

$x \approx -1$ $x \approx 2.5$

Quadratic equations usually have 2 solutions.

Quadratic graphs have an x^2 term, but no higher powers of x

Learn the details of the method for drawing quadratic graphs and practise drawing some smooth curves. To solve equations, work out where to draw your line and read off the points where it crosses the curve.

Warm-up and Worked Exam Questions

The warm-up questions run quickly over the basic skills you'll need in the exam. The exam questions come later — but unless you've learnt the basics first, you'll find the exams tricky.

Warm-up Questions

1) Use the graph opposite to answer the questions below.
   a) A petrol tank holds 8 gallons. How many litres is this?
   b) Approximately how many gallons of water would fit into a 20 litre container?

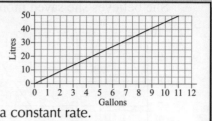

2) A barrel containing 10 litres of water is filled with more water at a constant rate. This graph shows the amount of water in the barrel against time.
   a) How much water is in the barrel after one minute?
   b) Work out the rate of flow of water in litres per second.

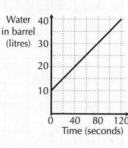

3) a) Complete the table of values for $y = x^2 - 2x - 1$.

x	-2	-1	0	1	2	3	4	5
x^2	4							
$-2x$	4							
-1	-1							
$y = x^2 - 2x - 1$	7							

   b) On a piece of graph paper, draw a set of axes using a scale of 1 unit to 1 cm on the x-axis and 2 units to 1 cm on the y-axis. Plot the x and y values from the table and join the points up to form a smooth curve.
   c) Use your curve to find the value of y when $x = 3.5$.
   d) Find the two values of x when $y = 5$.

Worked Exam Question

Wow, an exam question with the answer written in. Make the most of it... you know what's coming.

1 Katherine is going to the cinema. **(D)**

She leaves her house at 1 pm and walks at a constant speed of 3.5 mph.
Katherine arrives at the cinema 2 hours later.
She stays for 3 hours and 45 minutes, and then travels straight home by bus.

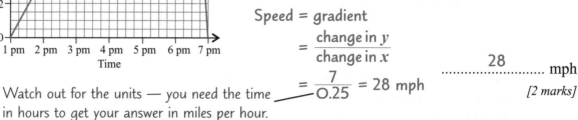

a) Given that Katherine arrives back home at 7 pm, use the grid on the left to draw a distance-time graph showing her journey.

1 home → cinema
2 at cinema
3 cinema → home

[3 marks]

b) What is Katherine's average speed on the bus home?

Speed = gradient

$= \dfrac{\text{change in } y}{\text{change in } x}$

$= \dfrac{7}{0.25} = 28$ mph

.... 28 .... mph

[2 marks]

She walks 2 × 3.5 = 7 miles to the cinema.

Watch out for the units — you need the time in hours to get your answer in miles per hour.

Exam Questions

*2 Edwige has just returned from a holiday in France.
She compares how much she spent on holiday to what she would normally spend at home.
She believes she saved money while on holiday. Explain why she is correct. Ⓓ

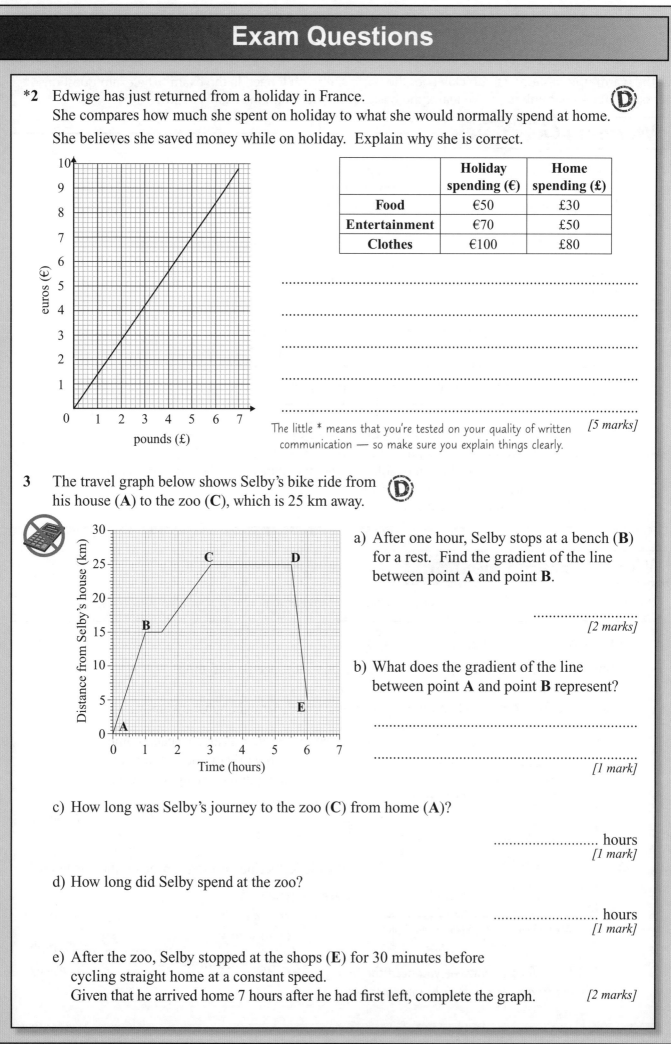

	Holiday spending (€)	Home spending (£)
Food	€50	£30
Entertainment	€70	£50
Clothes	€100	£80

....

....

....

....

....

The little * means that you're tested on your quality of written communication — so make sure you explain things clearly.

[5 marks]

3 The travel graph below shows Selby's bike ride from his house (**A**) to the zoo (**C**), which is 25 km away. Ⓓ

a) After one hour, Selby stops at a bench (**B**) for a rest. Find the gradient of the line between point **A** and point **B**.

....
[2 marks]

b) What does the gradient of the line between point **A** and point **B** represent?

....

....
[1 mark]

c) How long was Selby's journey to the zoo (**C**) from home (**A**)?

.... hours
[1 mark]

d) How long did Selby spend at the zoo?

.... hours
[1 mark]

e) After the zoo, Selby stopped at the shops (**E**) for 30 minutes before cycling straight home at a constant speed.
Given that he arrived home 7 hours after he had first left, complete the graph. *[2 marks]*

Exam Questions

4 An electricity company offers its customers two different price plans. 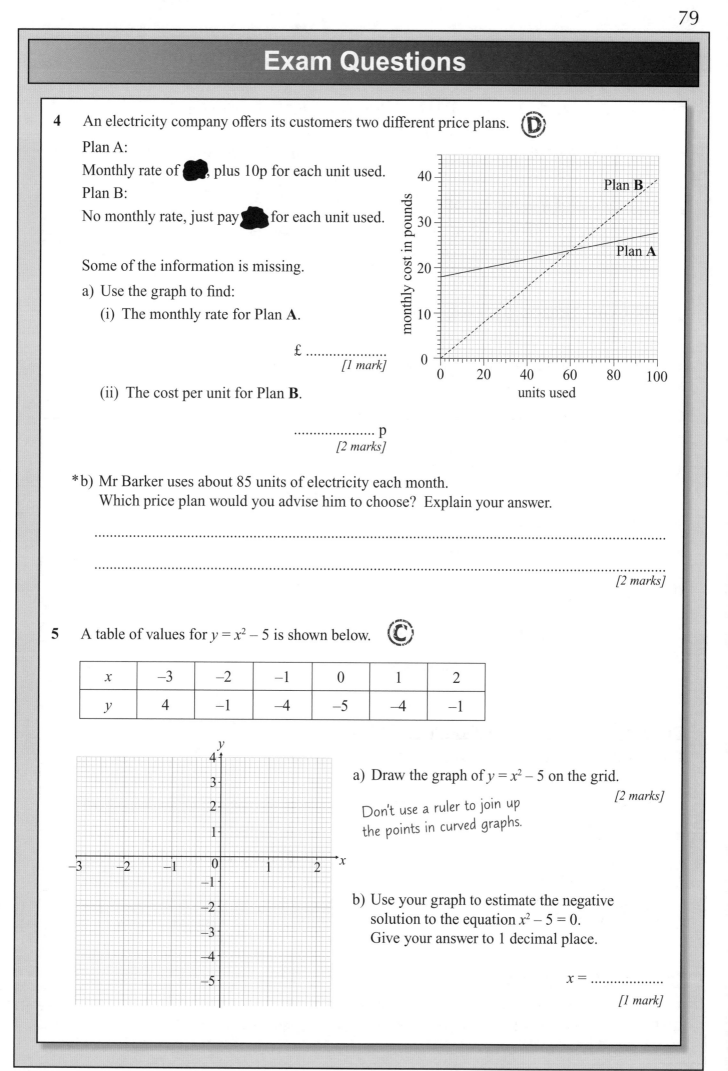 Ⓓ

Plan A:

Monthly rate of ●, plus 10p for each unit used.

Plan B:

No monthly rate, just pay ● for each unit used.

Some of the information is missing.

a) Use the graph to find:

  (i) The monthly rate for Plan **A**.

  £ ....
  [1 mark]

  (ii) The cost per unit for Plan **B**.

  .... p
  [2 marks]

*b) Mr Barker uses about 85 units of electricity each month.
   Which price plan would you advise him to choose? Explain your answer.

   ....

   ....
   [2 marks]

5 A table of values for $y = x^2 - 5$ is shown below. Ⓒ

x	-3	-2	-1	0	1	2
y	4	-1	-4	-5	-4	-1

a) Draw the graph of $y = x^2 - 5$ on the grid.
   [2 marks]

   Don't use a ruler to join up
   the points in curved graphs.

b) Use your graph to estimate the negative
   solution to the equation $x^2 - 5 = 0$.
   Give your answer to 1 decimal place.

   $x =$ ....
   [1 mark]

Revision Questions for Section Three

That wraps up <u>Section Three</u> — time to put yourself to the test and find out <u>how much you really know</u>.
* Try these questions and <u>tick off each one</u> when you <u>get it right</u>.
* When you've done <u>all the questions</u> for a topic and are <u>completely happy</u> with it, tick off the topic.

Coordinates and Midpoints (p64) ✓

1) Give the coordinates of points A to E in the diagram on the right.
2) Find the midpoint of the line segment with endpoints B and C.

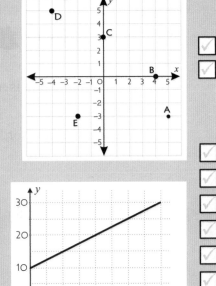

Straight-Line Graphs and their Gradients (p65-69) ✓

3) Sketch the lines a) $y = -x$, b) $y = -4$, c) $x = 2$
4) What does a straight-line equation look like?
5) Use the table of three values method to draw the graph $y = 2x + 3$.
6) What does a line with a negative gradient look like?
7) Find the gradient of the line on the right.
8) What do 'm' and 'c' represent in $y = mx + c$?

Travel Graphs (p73) ✓

9) What does a horizontal line mean on a distance-time graph?
10) The graph on the right shows Ben's car journey to the supermarket and home again.
   a) Did he drive faster on his way to the supermarket or on his way home?
   b) How long did he spend at the supermarket?

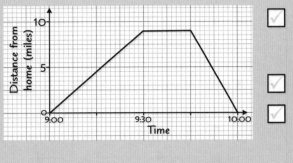

Conversion and Real-Life Graphs (p74-75) ✓

11) This graph shows the monthly cost of a mobile phone contract.
   a) What is the basic monthly fee?
   b) How many minutes does the monthly fee include?
   c) Mary uses her phone for 35 minutes one month. What will her bill be?
   d) Stuart is charged £13.50 one month. How long did he use his phone for?
   e) Estimate the cost per minute for additional minutes. Give your answer to the nearest 1p.

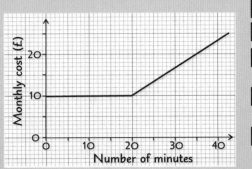

Quadratic Graphs (p76) ✓

12) Describe the shapes of the graphs $y = x^2 - 8$ and $y = -x^2 + 2$.
13) Plot the graph $y = x^2 + 2x$ for values of x between –3 and 3, and use it to solve $2 = x^2 + 2x$.

Symmetry

There are two types of <u>symmetry</u> that you need to know about — <u>line symmetry</u> and <u>rotational symmetry</u>.

Line Symmetry (F)

This is where you draw one or more <u>MIRROR LINES</u> across a shape and both sides will <u>fold exactly</u> together.

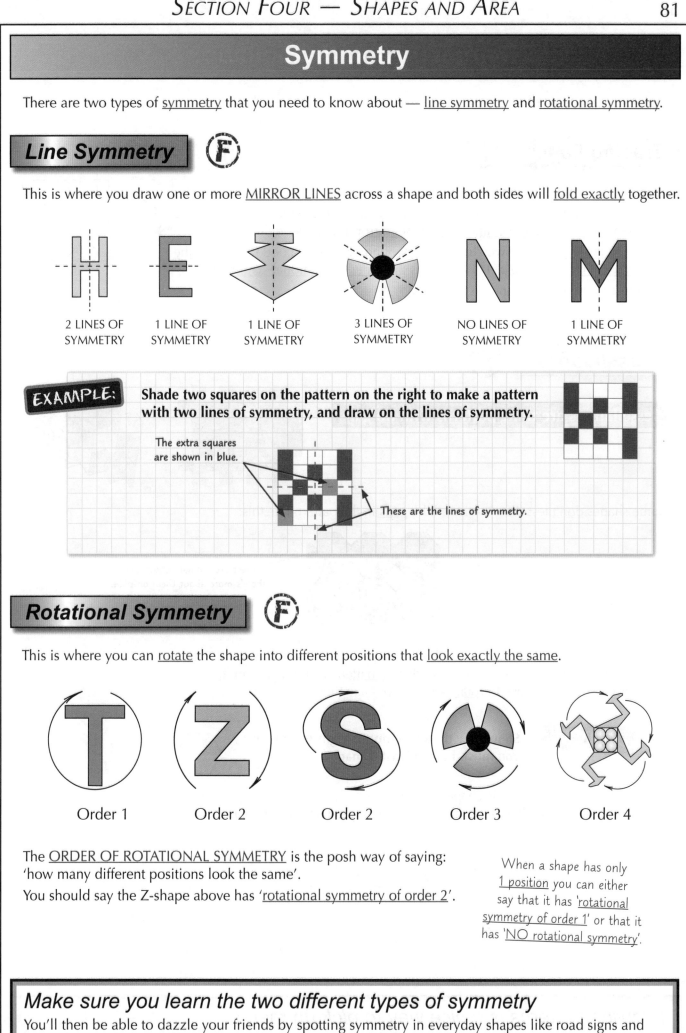

2 LINES OF SYMMETRY   1 LINE OF SYMMETRY   1 LINE OF SYMMETRY   3 LINES OF SYMMETRY   NO LINES OF SYMMETRY   1 LINE OF SYMMETRY

EXAMPLE: **Shade two squares on the pattern on the right to make a pattern with two lines of symmetry, and draw on the lines of symmetry.**

The extra squares are shown in blue.

These are the lines of symmetry.

Rotational Symmetry (F)

This is where you can <u>rotate</u> the shape into different positions that <u>look exactly the same</u>.

Order 1   Order 2   Order 2   Order 3   Order 4

The <u>ORDER OF ROTATIONAL SYMMETRY</u> is the posh way of saying: 'how many different positions look the same'.

You should say the Z-shape above has '<u>rotational symmetry of order 2</u>'.

When a shape has only <u>1 position</u> you can either say that it has '<u>rotational symmetry of order 1</u>' or that it has '<u>NO rotational symmetry</u>'.

Make sure you learn the two different types of symmetry

You'll then be able to dazzle your friends by spotting symmetry in everyday shapes like road signs and letters. More importantly, you'll get lots of nice, juicy marks in the exam.

Symmetry and Tessellations

One more handy bit of advice about <u>symmetry</u>, then it's on to <u>tessellations</u>.

Tracing Paper

<u>Tracing paper</u> makes symmetry questions a lot easier.

1) For <u>REFLECTIONS</u>, trace one side of the drawing and the mirror line too. Then <u>turn the paper over and line up the mirror line</u> in its original position.

2) For <u>ROTATIONS</u>, just swizzle the paper round. It's really good for <u>finding the centre of rotation</u> (see p119) as well as the <u>order of rotational symmetry</u>.

3) You can use tracing paper in the <u>EXAM</u>.

Tessellations

<u>Tessellations are tiling patterns with no gaps</u>

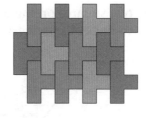

Some shapes <u>don't</u> tessellate — there'll be <u>gaps</u> in the pattern like this:

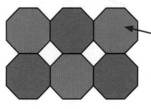

These are regular octagons — there's more about them on p114.

In the <u>exam</u>, you might have to <u>show</u> how a shape tessellates — this just means that you have to use the shape to <u>draw a pattern</u> with <u>no gaps</u> in it. Sometimes you might have to <u>rotate</u> a shape to make them fit together.

EXAMPLE: **Show how the shape on the right tessellates. You must draw at least 6 shapes.**

Just fit the shapes together, making sure you don't leave any gaps between them:

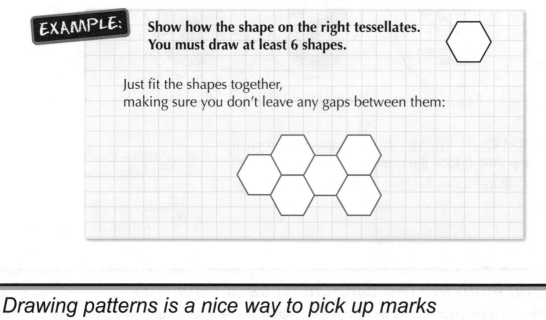

Drawing patterns is a nice way to pick up marks
Don't get carried away and make up your own shapes though (the examiners don't like it).

Properties of 2D Shapes

These two pages are jam-packed with details about <u>triangles</u> and <u>quadrilaterals</u> — and you need to learn them all.

Triangles (Three-Sided Shapes)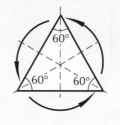

1) Equilateral Triangles

<u>3 equal sides</u> and
<u>3 equal angles</u> of <u>60°</u>.
<u>3 lines</u> of symmetry,
rotational symmetry of <u>order 3</u> (see p81).

2) Right-angled Triangles

1 <u>right angle</u> (<u>90°</u>).

<u>No</u> lines of symmetry unless
the other angles are 45°
(in which case, there's 1).

<u>No</u> rotational symmetry.

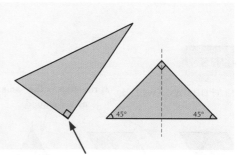

The little square means it's a right angle.

3) Isosceles Triangles

<u>2 sides</u> the same.
<u>2 angles</u> the same.
<u>1 line</u> of symmetry.
<u>No</u> rotational symmetry.

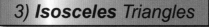

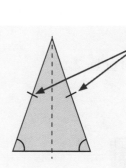

These dashes mean
that the two sides
are the same length.

4) Scalene Triangles

All three sides <u>different</u>.
All three angles <u>different</u>.
No symmetry (pretty obviously).

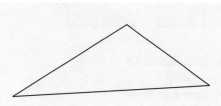

Triangles have three sides

Learn the names (and how to spell them) and the properties of all the triangles on this page.
These are easy marks in the exam — make sure you know them all.

Properties of 2D Shapes

Quadrilaterals *(Four-Sided Shapes)*

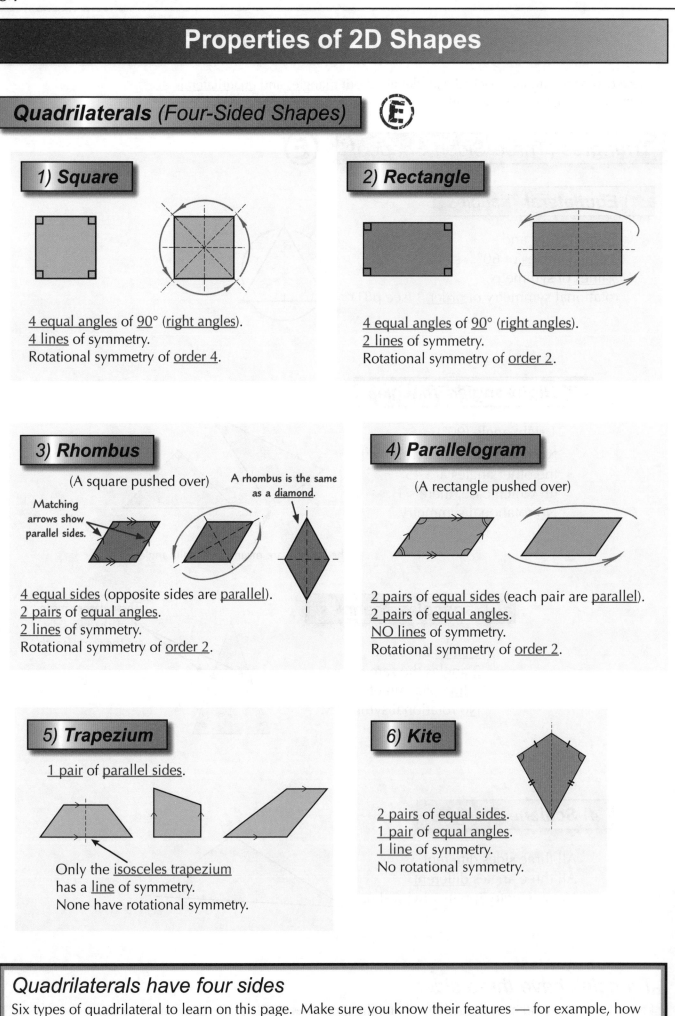

1) *Square*

4 equal angles of 90° (right angles).
4 lines of symmetry.
Rotational symmetry of order 4.

2) *Rectangle*

4 equal angles of 90° (right angles).
2 lines of symmetry.
Rotational symmetry of order 2.

3) *Rhombus*

(A square pushed over)

A rhombus is the same as a diamond.

Matching arrows show parallel sides.

4 equal sides (opposite sides are parallel).
2 pairs of equal angles.
2 lines of symmetry.
Rotational symmetry of order 2.

4) *Parallelogram*

(A rectangle pushed over)

2 pairs of equal sides (each pair are parallel).
2 pairs of equal angles.
NO lines of symmetry.
Rotational symmetry of order 2.

5) *Trapezium*

1 pair of parallel sides.

Only the isosceles trapezium has a line of symmetry.
None have rotational symmetry.

6) *Kite*

2 pairs of equal sides.
1 pair of equal angles.
1 line of symmetry.
No rotational symmetry.

Quadrilaterals have four sides

Six types of quadrilateral to learn on this page. Make sure you know their features — for example, how many lines of symmetry they have and whether they have rotational symmetry.

Congruence and Similarity

You need to learn the difference between shapes that are <u>congruent</u> and shapes that are <u>similar</u>.

Congruent — Same Shape, Same Size (F)

<u>Congruence</u> is another long maths word which sounds really complicated when it's not:

> If two shapes are <u>CONGRUENT</u>, they are <u>EXACTLY THE SAME</u>
> — the <u>SAME SIZE</u> and the <u>SAME SHAPE</u>.

These shapes are all <u>congruent</u>:

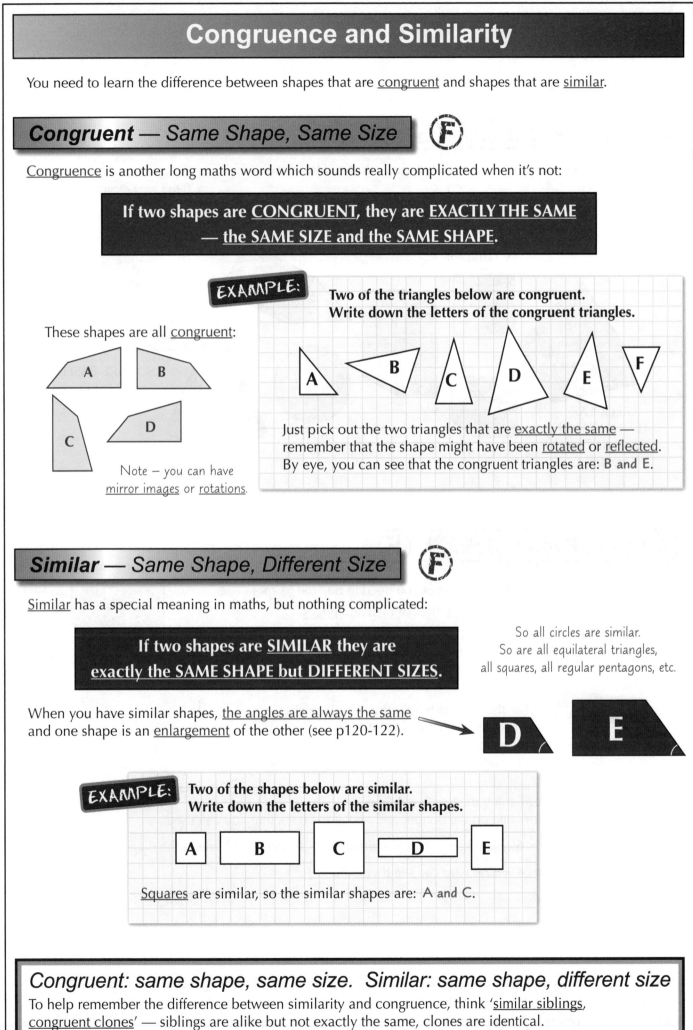

Note — you can have <u>mirror images</u> or <u>rotations</u>.

EXAMPLE: Two of the triangles below are congruent. Write down the letters of the congruent triangles.

Just pick out the two triangles that are <u>exactly the same</u> — remember that the shape might have been <u>rotated</u> or <u>reflected</u>. By eye, you can see that the congruent triangles are: **B and E**.

Similar — Same Shape, Different Size (F)

<u>Similar</u> has a special meaning in maths, but nothing complicated:

> If two shapes are <u>SIMILAR</u> they are
> exactly the <u>SAME SHAPE</u> but <u>DIFFERENT SIZES</u>.

So all circles are similar. So are all equilateral triangles, all squares, all regular pentagons, etc.

When you have similar shapes, <u>the angles are always the same</u> and one shape is an <u>enlargement</u> of the other (see p120-122).

EXAMPLE: Two of the shapes below are similar. Write down the letters of the similar shapes.

<u>Squares</u> are similar, so the similar shapes are: A and C.

Congruent: same shape, same size. Similar: same shape, different size

To help remember the difference between similarity and congruence, think '<u>similar siblings</u>, <u>congruent clones</u>' — siblings are alike but not exactly the same, clones are identical.

3D Shapes

First up are some <u>3D shapes</u> for you to learn, closely followed by a look at the <u>different parts of solids</u>.

Eight **Solids** to Learn Ⓖ

<u>3D shapes</u> are <u>solid shapes</u>. These are the ones you need to know:

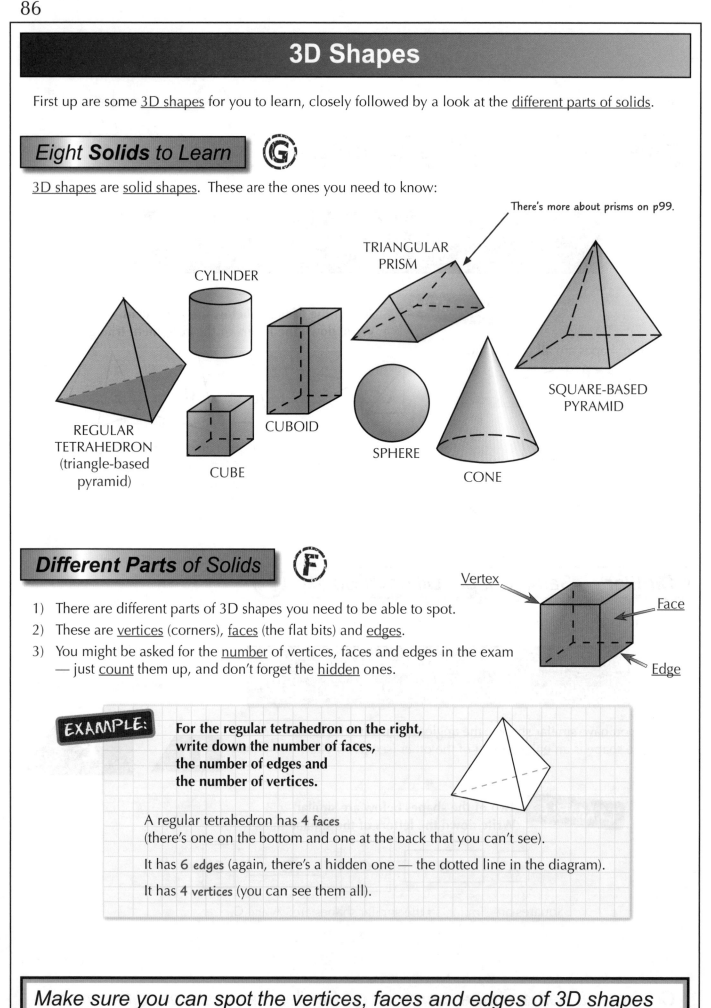

There's more about prisms on p99.

CYLINDER

TRIANGULAR PRISM

SQUARE-BASED PYRAMID

REGULAR TETRAHEDRON
(triangle-based pyramid)

CUBOID

CUBE

SPHERE

CONE

Different Parts of Solids Ⓕ

1) There are different parts of 3D shapes you need to be able to spot.
2) These are <u>vertices</u> (corners), <u>faces</u> (the flat bits) and <u>edges</u>.
3) You might be asked for the <u>number</u> of vertices, faces and edges in the exam — just <u>count</u> them up, and don't forget the <u>hidden</u> ones.

<u>Vertex</u>

<u>Face</u>

<u>Edge</u>

EXAMPLE: **For the regular tetrahedron on the right, write down the number of faces, the number of edges and the number of vertices.**

A regular tetrahedron has **4 faces**
(there's one on the bottom and one at the back that you can't see).

It has **6 edges** (again, there's a hidden one — the dotted line in the diagram).

It has **4 vertices** (you can see them all).

Make sure you can spot the vertices, faces and edges of 3D shapes

Remember — 1 <u>vertex</u>, 2 <u>vertices</u>. They're funny words, designed to confuse you, so don't let them catch you out. You need to know the names of all the solid shapes above too.

Projections

Projections are just different views of a 3D solid shape — looking at it from the front, the side and the top.

The Three Different Projections D

There are three different types of projections — front elevations, side elevations and plans (elevation is just another word for projection).

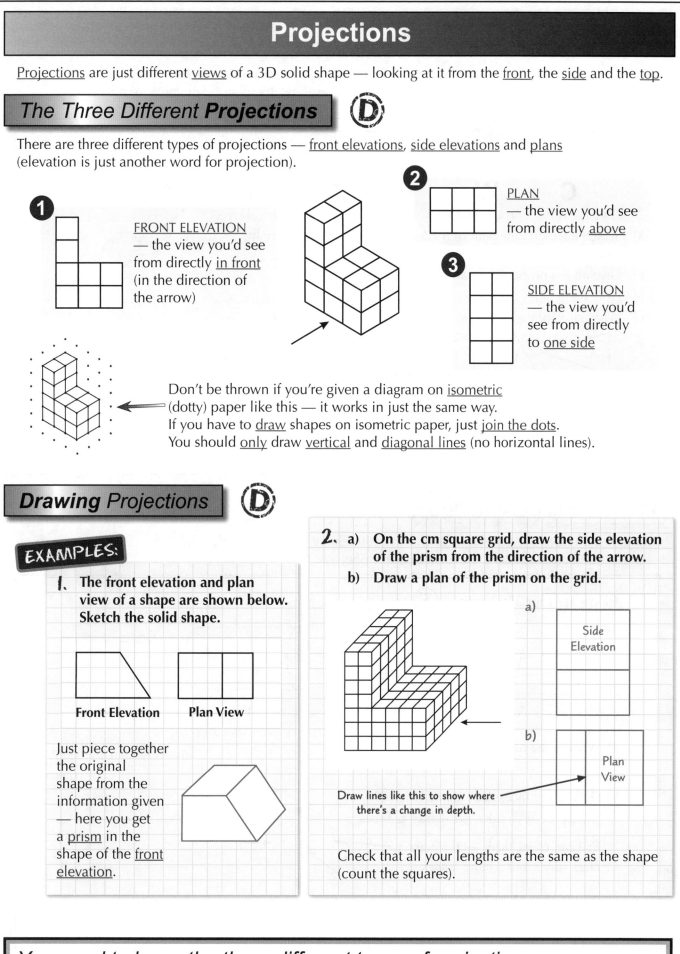

① FRONT ELEVATION — the view you'd see from directly in front (in the direction of the arrow)

② PLAN — the view you'd see from directly above

③ SIDE ELEVATION — the view you'd see from directly to one side

Don't be thrown if you're given a diagram on isometric (dotty) paper like this — it works in just the same way.
If you have to draw shapes on isometric paper, just join the dots.
You should only draw vertical and diagonal lines (no horizontal lines).

Drawing Projections D

EXAMPLES:

1. The front elevation and plan view of a shape are shown below. Sketch the solid shape.

Front Elevation   **Plan View**

Just piece together the original shape from the information given — here you get a prism in the shape of the front elevation.

2. a) On the cm square grid, draw the side elevation of the prism from the direction of the arrow.
b) Draw a plan of the prism on the grid.

a) Side Elevation

b) Plan View

Draw lines like this to show where there's a change in depth.

Check that all your lengths are the same as the shape (count the squares).

You need to know the three different types of projections

Projection questions aren't too bad — just take your time and sketch the diagrams carefully.
Watch out for questions on isometric paper — they might look confusing,
but they can actually be easier than other questions.

Warm-up and Worked Exam Questions

Nothing too tricky so far in this section. Now it's time for some warm-up questions to get your brain ticking — before moving on to the exam-style questions. It's all good practice for the big day.

Warm-up Questions

1) Copy the letters below and mark in all the lines of symmetry.

C   W   I   D   Q

2) Give the order of rotational symmetry for each of the letters above.

3) Sketch a tessellating pattern made up of the shape to the right.
You should draw at least five shapes.

4) Give all the properties of an equilateral triangle.

5) How many lines of symmetry does a kite have?

6) Which two of these shapes are: a) similar? b) congruent?

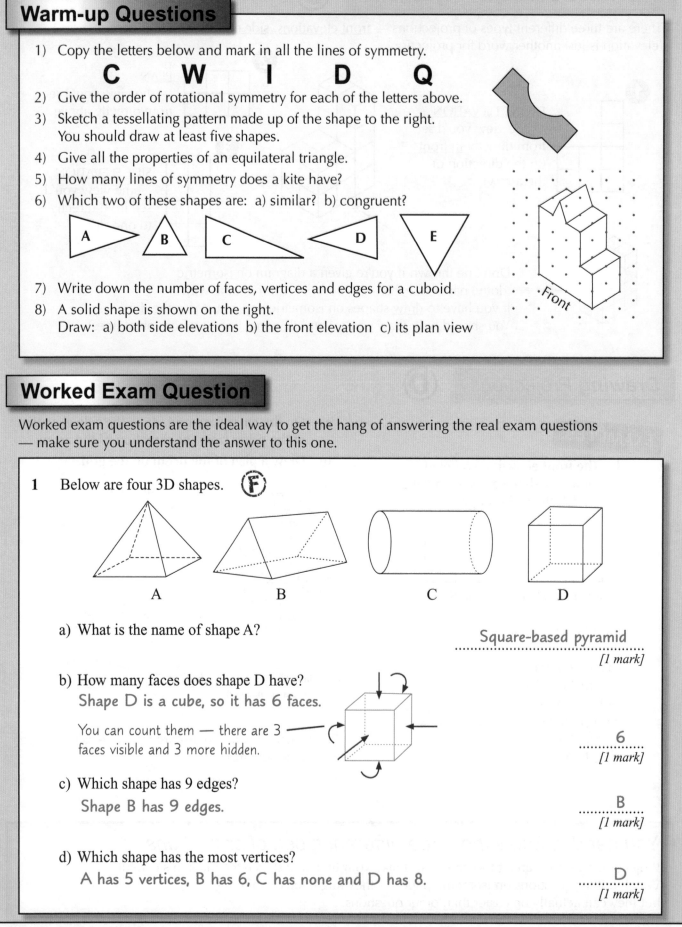

7) Write down the number of faces, vertices and edges for a cuboid.

8) A solid shape is shown on the right.
Draw: a) both side elevations b) the front elevation c) its plan view

Worked Exam Question

Worked exam questions are the ideal way to get the hang of answering the real exam questions — make sure you understand the answer to this one.

1 Below are four 3D shapes. **(F)**

A   B   C   D

a) What is the name of shape A?

Square-based pyramid
[1 mark]

b) How many faces does shape D have?
Shape D is a cube, so it has 6 faces.

You can count them — there are 3 faces visible and 3 more hidden.

6
[1 mark]

c) Which shape has 9 edges?
Shape B has 9 edges.

B
[1 mark]

d) Which shape has the most vertices?
A has 5 vertices, B has 6, C has none and D has 8.

D
[1 mark]

SECTION FOUR — SHAPES AND AREA

Exam Questions

2   Write down the mathematical names of the 3D shapes below.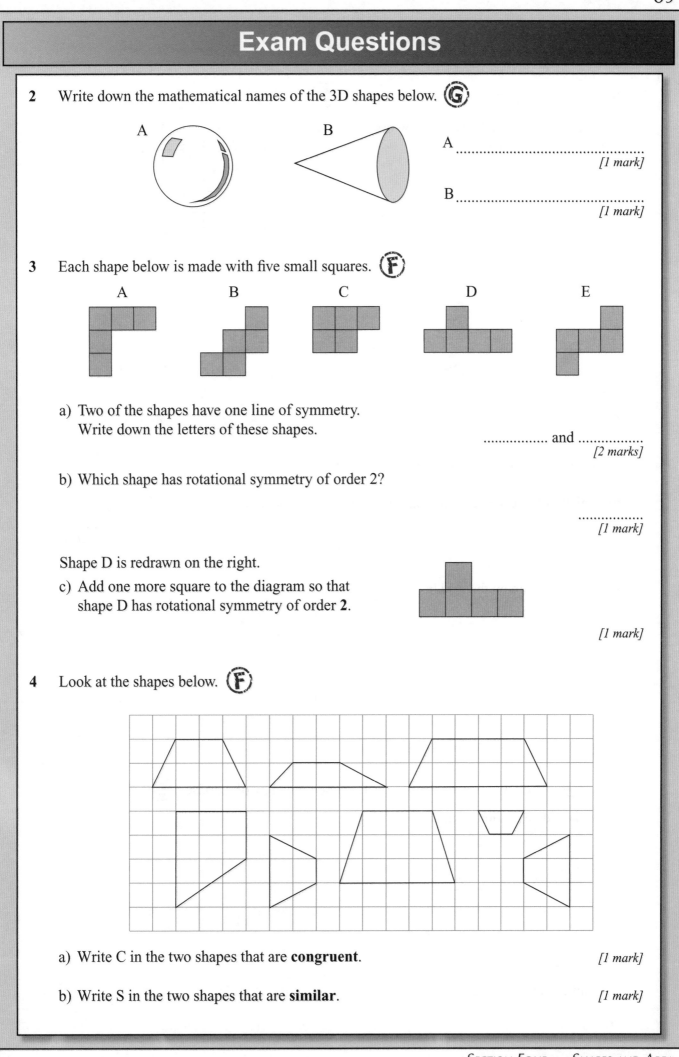

A _____
[1 mark]

B _____
[1 mark]

3   Each shape below is made with five small squares.

A   B   C   D   E

a) Two of the shapes have one line of symmetry.
Write down the letters of these shapes.

.... and ....
[2 marks]

b) Which shape has rotational symmetry of order 2?

....
[1 mark]

Shape D is redrawn on the right.

c) Add one more square to the diagram so that
shape D has rotational symmetry of order **2**.

[1 mark]

4   Look at the shapes below.

a) Write C in the two shapes that are **congruent**.   *[1 mark]*

b) Write S in the two shapes that are **similar**.   *[1 mark]*

Exam Questions

5 Sam buys some garden tiles to make a patio. (F)

Show how the tiles will tessellate by drawing at least four more tiles on the grid.

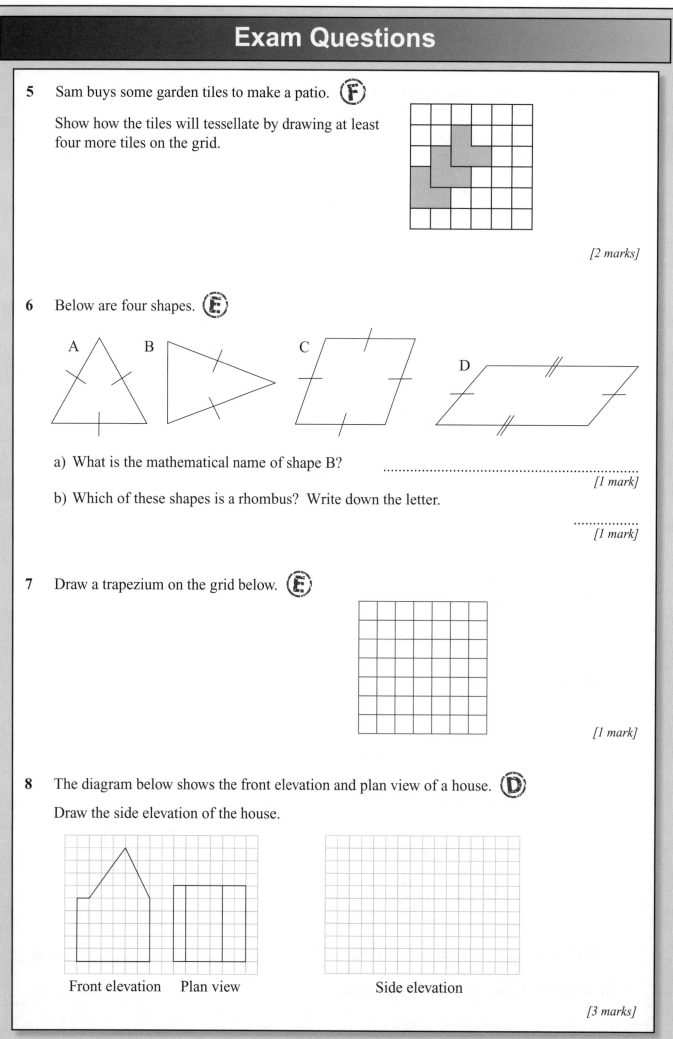

[2 marks]

6 Below are four shapes. (E)

A   B   C   D

a) What is the mathematical name of shape B? ....
[1 mark]

b) Which of these shapes is a rhombus? Write down the letter.
....
[1 mark]

7 Draw a trapezium on the grid below. (E)

[1 mark]

8 The diagram below shows the front elevation and plan view of a house. (D)

Draw the side elevation of the house.

Front elevation   Plan view   Side elevation

[3 marks]

Perimeters

Perimeter is the <u>distance</u> all the way around the <u>outside</u> of a <u>2D shape</u>. It's pretty straightforward if you use the <u>big blob method</u>. So pay attention — this could be easy marks.

Perimeter — *Distance* Around the *Edge* of a Shape   (F)

To find a <u>perimeter</u>, you <u>add up</u> the <u>lengths</u> of all the sides,
but the only <u>reliable</u> way to make sure you get <u>all</u> the sides is this:

> 1) Put a <u>**BIG BLOB**</u> at one corner and then go around the shape.
>
> 2) Write down the <u>**LENGTH**</u> of every side as you go along.
>
> 3) Even sides that seem to have <u>**NO LENGTH GIVEN**</u>
>   — you must <u>work them out</u>.
>
> 4) Keep going until you get back to the <u>**BIG BLOB**</u>.

Yes, I know you think it's <u>yet another fussy method</u>, but believe me, it's so easy to miss a side otherwise.

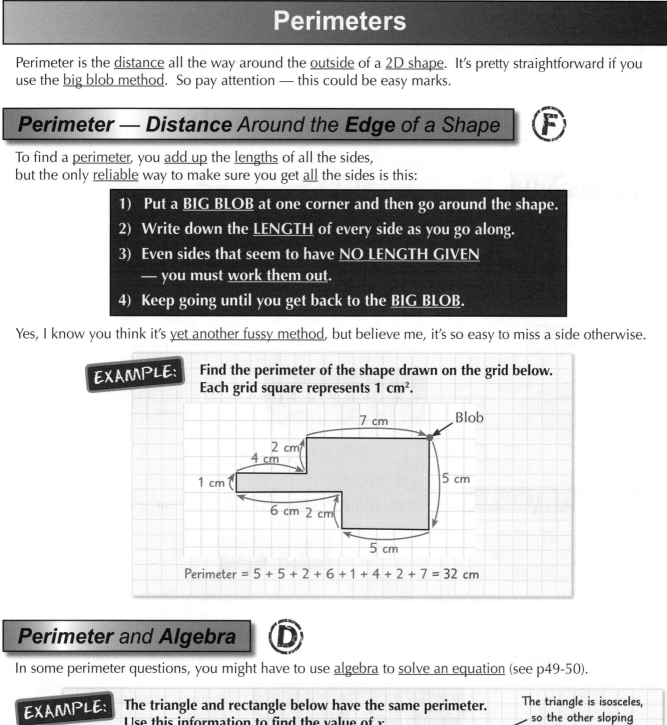

EXAMPLE: **Find the perimeter of the shape drawn on the grid below. Each grid square represents 1 cm².**

Perimeter = 5 + 5 + 2 + 6 + 1 + 4 + 2 + 7 = 32 cm

Perimeter and Algebra   (D)

In some perimeter questions, you might have to use <u>algebra</u> to <u>solve an equation</u> (see p49-50).

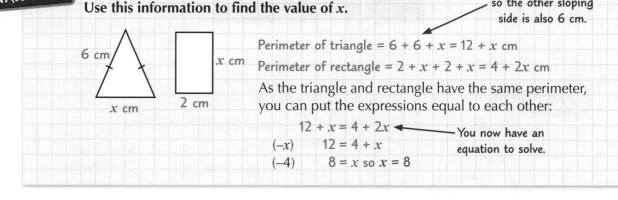

EXAMPLE: **The triangle and rectangle below have the same perimeter. Use this information to find the value of *x*.**

The triangle is isosceles, so the other sloping side is also 6 cm.

Perimeter of triangle = 6 + 6 + x = 12 + x cm
Perimeter of rectangle = 2 + x + 2 + x = 4 + 2x cm

As the triangle and rectangle have the same perimeter, you can put the expressions equal to each other:

$$12 + x = 4 + 2x$$
$$(-x) \quad 12 = 4 + x$$
$$(-4) \quad \quad 8 = x \text{ so } x = 8$$

You now have an equation to solve.

Use the big blob method and you won't go wrong

Perimeter questions can either be really easy (if it's on a cm grid) or pretty tricky (if algebra gets involved) — make sure you can do both types of question.

Areas

On this page are <u>four area formulas</u> you need to <u>learn</u> — rectangles, triangles, parallelograms and trapeziums. Remember that area is measured in <u>square units</u> (e.g. cm², m² or km²).

You Must **Learn** *These Four* **Area Formulas** Ⓓ

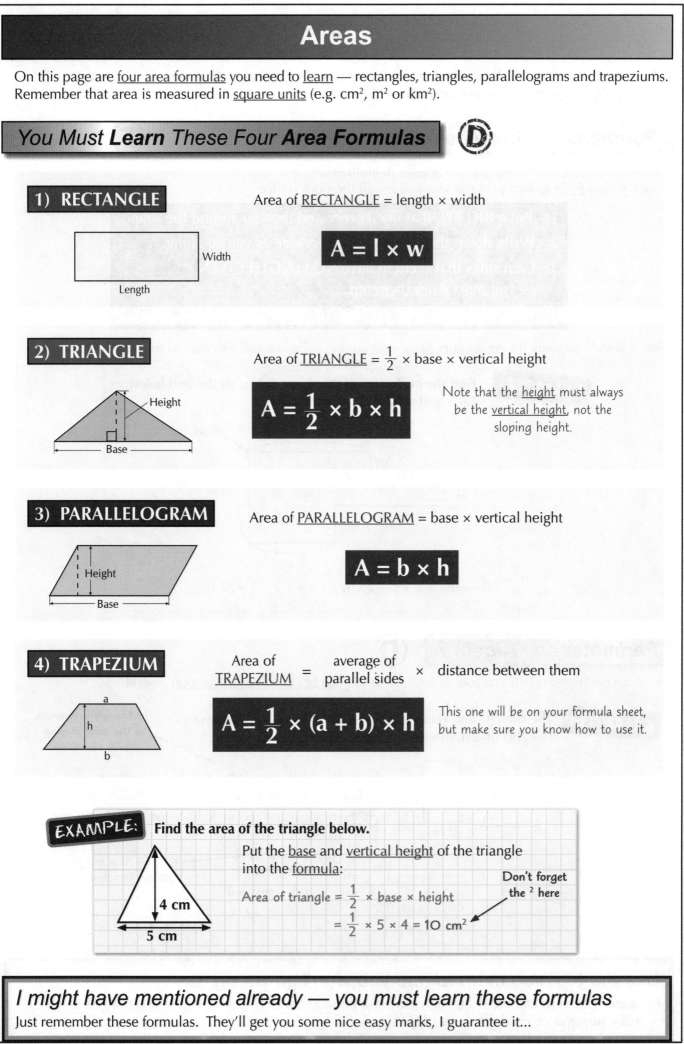

1) RECTANGLE

Area of <u>RECTANGLE</u> = length × width

Width

Length

$$A = l \times w$$

2) TRIANGLE

Area of <u>TRIANGLE</u> = $\frac{1}{2}$ × base × vertical height

Height

Base

$$A = \frac{1}{2} \times b \times h$$

Note that the <u>height</u> must always be the <u>vertical height</u>, not the sloping height.

3) PARALLELOGRAM

Area of <u>PARALLELOGRAM</u> = base × vertical height

Height

Base

$$A = b \times h$$

4) TRAPEZIUM

Area of <u>TRAPEZIUM</u> = average of parallel sides × distance between them

a

h

b

$$A = \frac{1}{2} \times (a + b) \times h$$

This one will be on your formula sheet, but make sure you know how to use it.

EXAMPLE: **Find the area of the triangle below.**

Put the <u>base</u> and <u>vertical height</u> of the triangle into the <u>formula</u>:

4 cm

5 cm

Area of triangle = $\frac{1}{2}$ × base × height

= $\frac{1}{2}$ × 5 × 4 = 10 cm²

Don't forget the ² here

I might have mentioned already — you must learn these formulas

Just remember these formulas. They'll get you some nice easy marks, I guarantee it...

Areas

Make sure you know the <u>formulas</u> for finding the area of <u>rectangles</u> and <u>triangles</u> — you're going to need them again here.

Areas of **More Complicated** Shapes

You often have to find the area of <u>strange-looking</u> shapes in exam questions. What you always find with these questions is that you can break the shape up into <u>simpler ones</u> that you can deal with.

1) <u>SPLIT THEM UP</u> into the two basic shapes: RECTANGLES and TRIANGLES.
2) Work out the area of each bit <u>SEPARATELY</u>.
3) Then <u>ADD THEM ALL TOGETHER</u>.

Basic Rectangle

Basic Triangle

EXAMPLE: **Find the area of the shape below.**

6 cm

3 cm

8 cm

1) Split the shape into a <u>triangle</u> and <u>rectangle</u> as shown and work out the <u>area</u> of each shape:

Area of rectangle = length × width = 8 × 3 = 24 cm²

2) To find the <u>height</u> of the triangle, subtract the height of the rectangle from the total height of the shape (so 6 − 3 = 3).

Area of triangle = $\frac{1}{2}$ × base × height = $\frac{1}{2}$ × 8 × 3 = 12 cm²

Total area of shape = 24 + 12 = **36 cm²**

Area Problems

Once you've worked out the <u>area</u> of a shape, you might have to <u>use</u> the area to <u>answer a question</u> (e.g. find the area of a wall, then work out how many rolls of wallpaper you need to wallpaper it).

EXAMPLE: **Greg is making a stained-glass window in the shape shown below. Coloured glass costs £82 per m². Work out the cost of the glass needed for the window.**

0.6 m

1.2 m

0.8 m

1) First, work out the area of the shape by splitting it into a <u>triangle</u> and <u>rectangle</u> (as shown):

Area of rectangle = length × width = 1.2 × 0.8 = 0.96 m²
Area of triangle = $\frac{1}{2}$ × base × height = $\frac{1}{2}$ × 0.8 × 0.6 = 0.24 m²
Total area of shape = 0.96 + 0.24 = 1.2 m²

2) Then <u>multiply</u> the <u>area</u> by the <u>price</u> to work out the cost:

Cost = area × price per m² = 1.2 × 82 = **£98.40**

Break up complicated shapes into simpler ones

You'd be amazed at the weird and wonderful shapes they can come up with for you to find the area of. But you can always split them up into triangles and rectangles and work out their area in bits.

Circles

There's a surprising number of <u>circle terms</u> you need to know — don't mix them up.

Radius *and* Diameter (F)

The <u>DIAMETER</u> goes <u>right across</u> the circle, passing through the <u>centre</u>.
The <u>RADIUS</u> goes from the <u>centre</u> of the circle to any point on the <u>edge</u>.

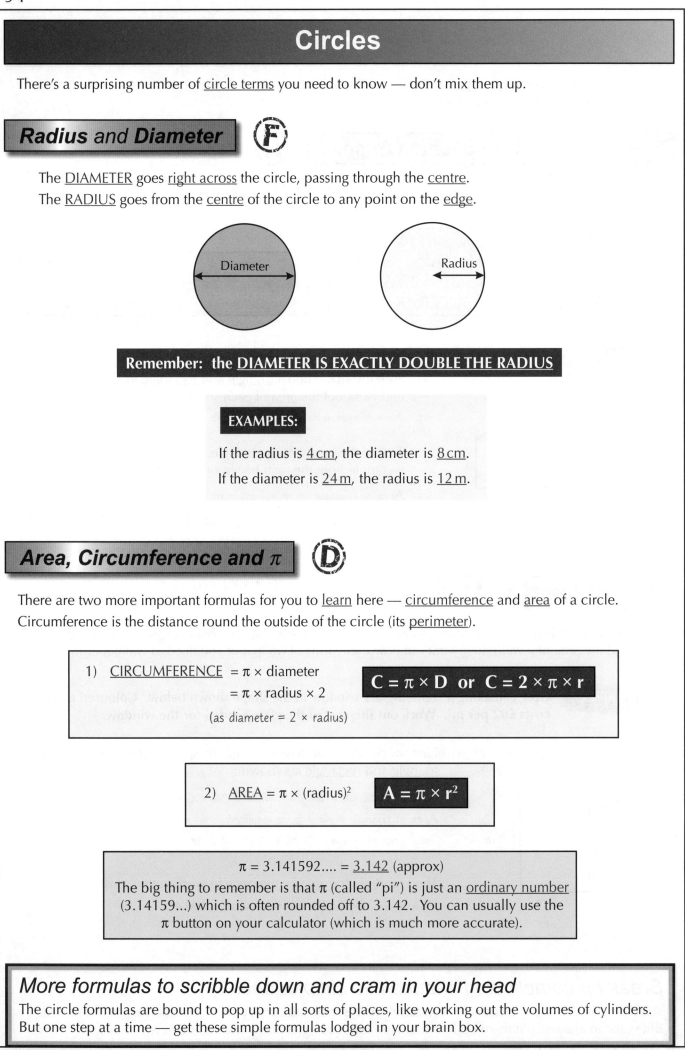

Remember: the <u>DIAMETER IS EXACTLY DOUBLE THE RADIUS</u>

EXAMPLES:

If the radius is <u>4 cm</u>, the diameter is <u>8 cm</u>.
If the diameter is <u>24 m</u>, the radius is <u>12 m</u>.

Area, Circumference *and* π (D)

There are two more important formulas for you to <u>learn</u> here — <u>circumference</u> and <u>area</u> of a circle.
Circumference is the distance round the outside of the circle (its <u>perimeter</u>).

1) <u>CIRCUMFERENCE</u> = π × diameter
 = π × radius × 2
 (as diameter = 2 × radius)

C = π × D or C = 2 × π × r

2) <u>AREA</u> = π × (radius)²

A = π × r²

π = 3.141592.... = <u>3.142</u> (approx)
The big thing to remember is that π (called "pi") is just an <u>ordinary number</u>
(3.14159...) which is often rounded off to 3.142. You can usually use the
π button on your calculator (which is much more accurate).

More formulas to scribble down and cram in your head

The circle formulas are bound to pop up in all sorts of places, like working out the volumes of cylinders.
But one step at a time — get these simple formulas lodged in your brain box.

Circles

Tangents, Chords, Arcs, Sectors and Segments

A TANGENT is a straight line that just touches the outside of a circle.

A CHORD is a line drawn across the inside of a circle.

AN ARC is just part of the circumference of a circle.

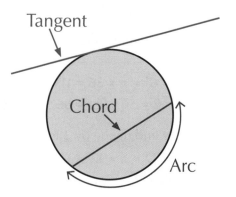

A SECTOR is a WEDGE-SHAPED AREA (like a slice of cake) cut right from the centre.

SEGMENTS are the areas you get when you cut a circle with a chord.

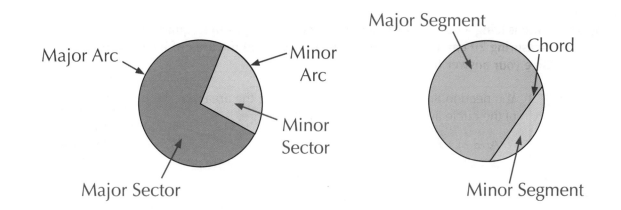

Make sure you know all of these circle terms

There are five on this page and a couple on the previous page — nothing too tricky, so learn them now while you're here. Just don't go mixing up segments and sectors and you'll be fine.

Circle Questions

<u>ALWAYS</u> check that you're using the right value in <u>circle formulas</u> — use the <u>radius</u> to find <u>area</u> and the <u>diameter</u> to find <u>circumference</u>. If you're given the wrong one in the question, <u>multiply</u> or <u>divide</u> by <u>2</u>.

Semicircles and *Quarter Circles*

You might be asked to find the <u>area</u> and <u>perimeter</u> of a <u>semicircle</u> (half circle) or a <u>quarter circle</u>.

> 1) <u>AREA</u>: find the area of the <u>whole circle</u> then <u>divide</u> by <u>2</u> (for a semicircle) or <u>4</u> (for a quarter circle).
> 2) <u>PERIMETER</u>: divide the <u>circumference</u> by <u>2</u> (for a semicircle) or <u>4</u> (for a quarter circle) and <u>add on</u> the <u>straight edges</u> (the <u>diameter</u> for a semicircle or <u>two radiuses</u> for a quarter circle).

EXAMPLE: **Find the area and perimeter of the semicircle shown on the right.**
Give your answers to 2 decimal places.

12 cm

First find the <u>area</u> of the <u>whole circle</u> then <u>divide by 2</u>:

Radius = 12 ÷ 2 = 6 cm

Area of whole circle = $\pi \times r^2 = \pi \times 6^2 = \pi \times 36 = 113.097...$

So area of semicircle = $113.097... \div 2 = 56.548... = 56.55$ cm^2 (2 d.p.)

Find the <u>circumference</u> of the circle and <u>divide by 2</u> to find the curved edge...

Circumference of whole circle = $\pi \times D = \pi \times 12 = 37.699...$

So curved edge = $37.699... \div 2 = 18.849...$

Then <u>add on the diameter</u> to find the total perimeter:

Perimeter = curved edge + diameter = $18.849... + 12 = 30.849... = 30.85$ cm (2 d.p.)

Area Problems with *Circles*

There's a whole range of <u>circle questions</u> you could be asked — but if you <u>learn</u> the <u>circle formulas</u> you should be fine. Make sure you <u>read</u> the questions <u>carefully</u> to find out what you're being asked to do.

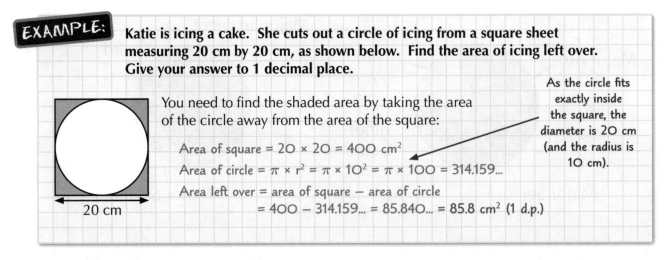

EXAMPLE: **Katie is icing a cake. She cuts out a circle of icing from a square sheet measuring 20 cm by 20 cm, as shown below. Find the area of icing left over. Give your answer to 1 decimal place.**

As the circle fits exactly inside the square, the diameter is 20 cm (and the radius is 10 cm).

You need to find the shaded area by taking the area of the circle away from the area of the square:

Area of square = 20 × 20 = 400 cm^2

Area of circle = $\pi \times r^2 = \pi \times 10^2 = \pi \times 100 = 314.159...$

Area left over = area of square − area of circle
= $400 - 314.159... = 85.840... = 85.8$ cm^2 (1 d.p.)

20 cm

Don't forget the straight edges of semicircles and quarter circles

When you're finding the perimeter of a semicircle or quarter circle, remember to add on the diameter for a semicircle and 2 × the radius for a quarter circle.

Warm-up and Worked Exam Questions

There are lots of formulas in this section. The best way to find out what you know is to practise these questions. If you find you keep forgetting the formulas, you need more practice.

Warm-up Questions

1) Find the perimeter of the shape shown on the right.
2) Give the formulas for:
   a) the area of a rectangle
   b) the circumference of a circle
   c) the area of a parallelogram
3) A triangle has a base of 3 m and a vertical height of 7 m. Calculate its area.
4) A woodworking template has the shape shown: Calculate the area of the template.
5) What is: a) a tangent?   b) a chord?

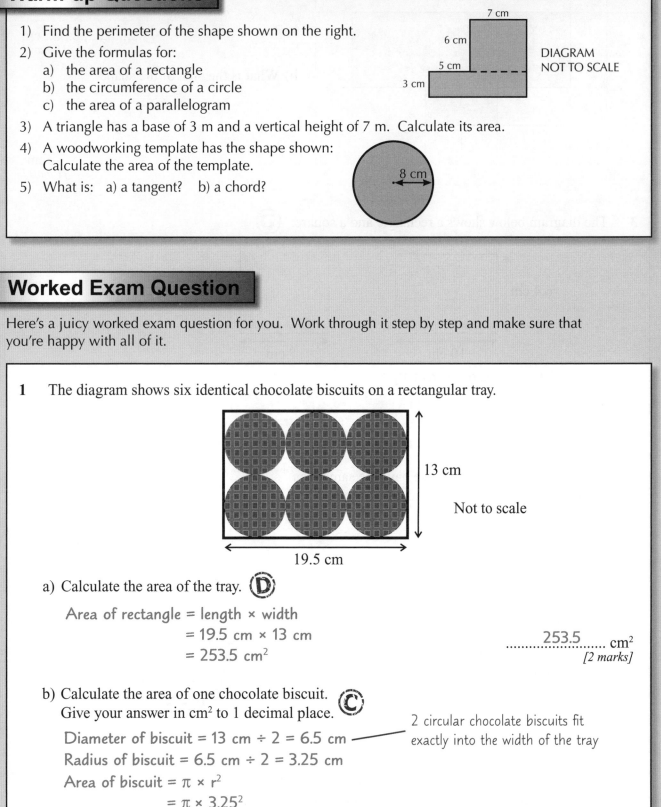

7 cm

6 cm

5 cm

3 cm

DIAGRAM NOT TO SCALE

8 cm

Worked Exam Question

Here's a juicy worked exam question for you. Work through it step by step and make sure that you're happy with all of it.

1 The diagram shows six identical chocolate biscuits on a rectangular tray.

13 cm

Not to scale

19.5 cm

a) Calculate the area of the tray. **(D)**

Area of rectangle = length × width
$$= 19.5 \text{ cm} \times 13 \text{ cm}$$
$$= 253.5 \text{ cm}^2$$

........253.5...... cm²
[2 marks]

b) Calculate the area of one chocolate biscuit. **(C)**
Give your answer in cm² to 1 decimal place.

2 circular chocolate biscuits fit exactly into the width of the tray

Diameter of biscuit = 13 cm ÷ 2 = 6.5 cm
Radius of biscuit = 6.5 cm ÷ 2 = 3.25 cm
Area of biscuit = $\pi \times r^2$
$$= \pi \times 3.25^2$$
$$= 33.183... \text{ cm}^2 = 33.2 \text{ cm}^2 \text{ (to 1 d.p.)}$$

........33.2...... cm²
[3 marks]

Don't forget to round the answer

98

Exam Questions

2 The shape below is drawn on a grid of centimetre squares. Ⓖ

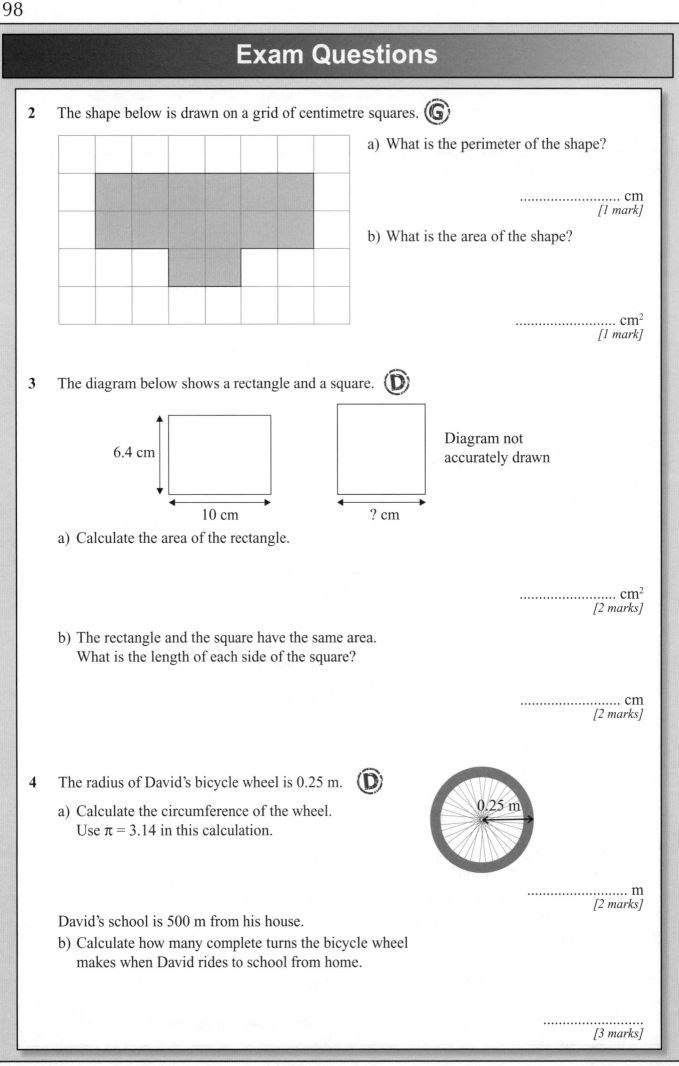

a) What is the perimeter of the shape?

.... cm
[1 mark]

b) What is the area of the shape?

.... cm²
[1 mark]

3 The diagram below shows a rectangle and a square. Ⓓ

6.4 cm

10 cm

? cm

Diagram not accurately drawn

a) Calculate the area of the rectangle.

.... cm²
[2 marks]

b) The rectangle and the square have the same area.
What is the length of each side of the square?

.... cm
[2 marks]

4 The radius of David's bicycle wheel is 0.25 m. Ⓓ

a) Calculate the circumference of the wheel.
Use π = 3.14 in this calculation.

0.25 m

.... m
[2 marks]

David's school is 500 m from his house.

b) Calculate how many complete turns the bicycle wheel
makes when David rides to school from home.

....
[3 marks]

Volume

Remember <u>3D shapes</u>? You came across them on p86. Now it's time to work out their <u>volumes</u>.

LEARN these volume formulas...

Volumes of Cuboids (E)

A <u>cuboid</u> is a <u>rectangular block</u>. Finding its volume is really easy:

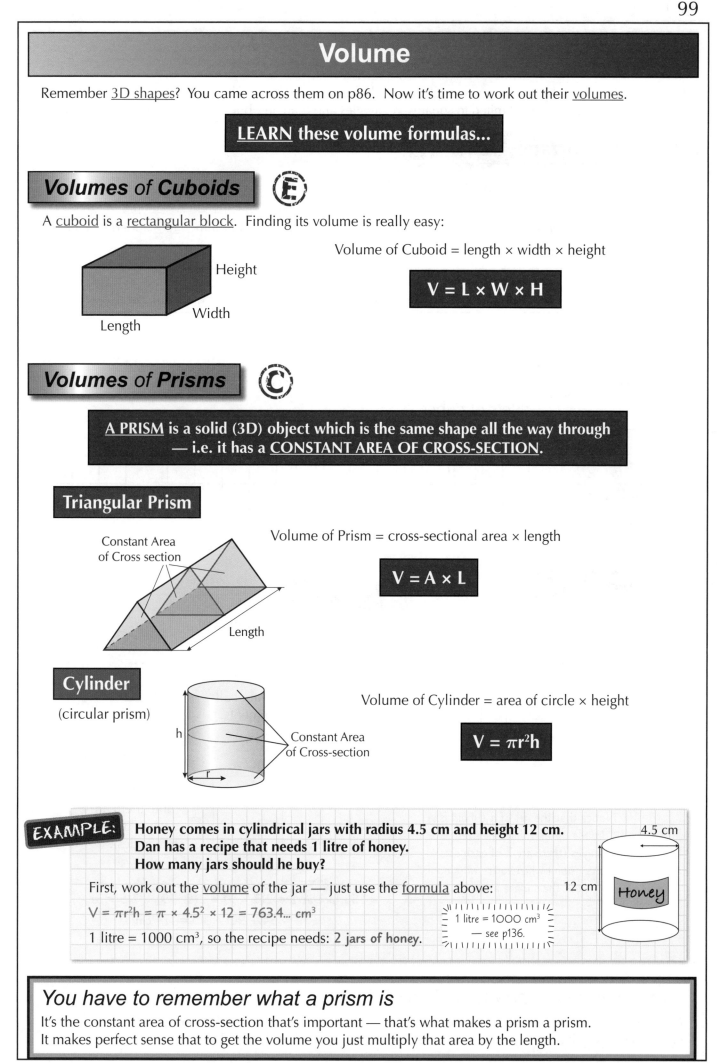

Height

Width

Length

Volume of Cuboid = length × width × height

$$V = L \times W \times H$$

Volumes of Prisms (C)

<u>A PRISM</u> is a solid (3D) object which is the same shape all the way through
— i.e. it has a <u>CONSTANT AREA OF CROSS-SECTION</u>.

Triangular Prism

Constant Area
of Cross section

Length

Volume of Prism = cross-sectional area × length

$$V = A \times L$$

Cylinder

(circular prism)

h

Constant Area
of Cross-section

r

Volume of Cylinder = area of circle × height

$$V = \pi r^2 h$$

EXAMPLE: **Honey comes in cylindrical jars with radius 4.5 cm and height 12 cm.**
Dan has a recipe that needs 1 litre of honey.
How many jars should he buy?

First, work out the <u>volume</u> of the jar — just use the <u>formula</u> above:

$V = \pi r^2 h = \pi \times 4.5^2 \times 12 = 763.4... \text{ cm}^3$

1 litre = 1000 cm³, so the recipe needs: **2 jars of honey.**

4.5 cm

12 cm   Honey

1 litre = 1000 cm³
— see p136.

You have to remember what a prism is

It's the constant area of cross-section that's important — that's what makes a prism a prism.
It makes perfect sense that to get the volume you just multiply that area by the length.

Nets and Surface Area

Pencils and rulers at the ready — you might get to do some drawing over the next two pages. Unfortunately, you're mainly limited to squares, rectangles and triangles, but you might get the odd circle too.

Nets and Surface Area (D)

1) A NET is just a hollow 3D shape folded out flat.

2) There's often more than one net that can be drawn for a 3D shape (see the cube example below).

3) SURFACE AREA only applies to solid 3D objects — it's the total area of all the faces added together.

4) There are two ways to find the surface area:

> 1) Work out the area of each face and add them all together (don't forget the hidden faces).
>
> 2) Sketch the net, then find the area of the net (this is the method we'll use on these pages).

Remember — SURFACE AREA OF SOLID = AREA OF NET.

Cubes (E)

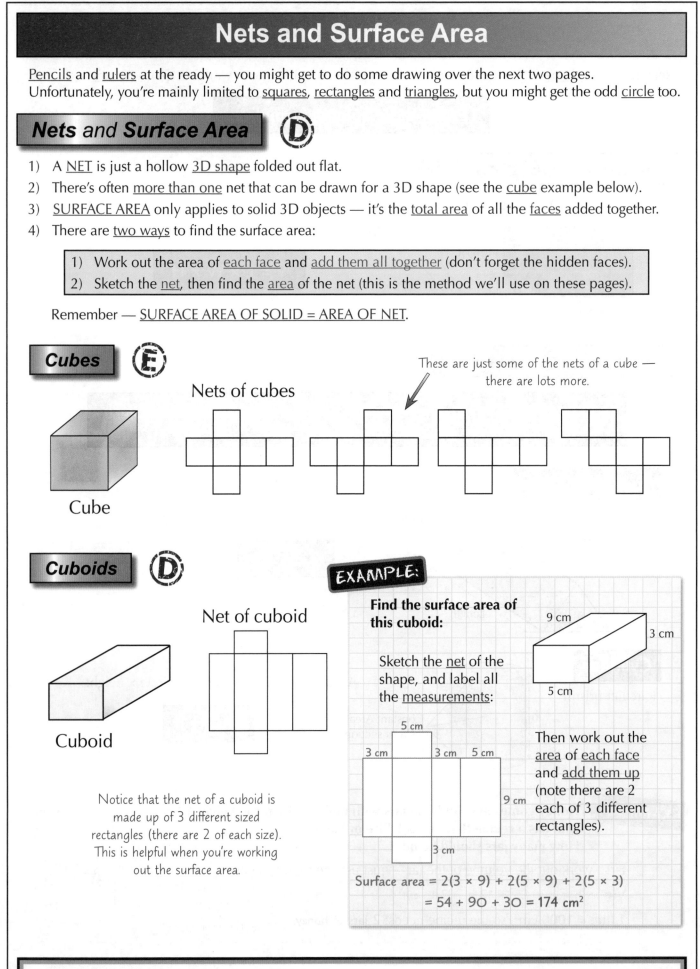

Nets of cubes

These are just some of the nets of a cube — there are lots more.

Cube

Cuboids (D)

Net of cuboid

Cuboid

Notice that the net of a cuboid is made up of 3 different sized rectangles (there are 2 of each size). This is helpful when you're working out the surface area.

EXAMPLE:

Find the surface area of this cuboid:

9 cm

3 cm

5 cm

Sketch the net of the shape, and label all the measurements:

5 cm

3 cm   3 cm   5 cm

9 cm

3 cm

Then work out the area of each face and add them up (note there are 2 each of 3 different rectangles).

Surface area = 2(3 × 9) + 2(5 × 9) + 2(5 × 3)
= 54 + 90 + 30 = 174 cm²

Just imagine the shape unfolded

Nets are useful when working out surface area. A net is just all the sides folded out flat, which makes it easier to see which shapes you need to calculate the areas of. Then just add them up.

Nets and Surface Area

Another page on <u>nets</u> and <u>surface area</u> — and now things get really exciting.
It's time for <u>pyramids</u>, <u>prisms</u> and <u>cylinders</u>.

Triangular Prisms and Pyramids

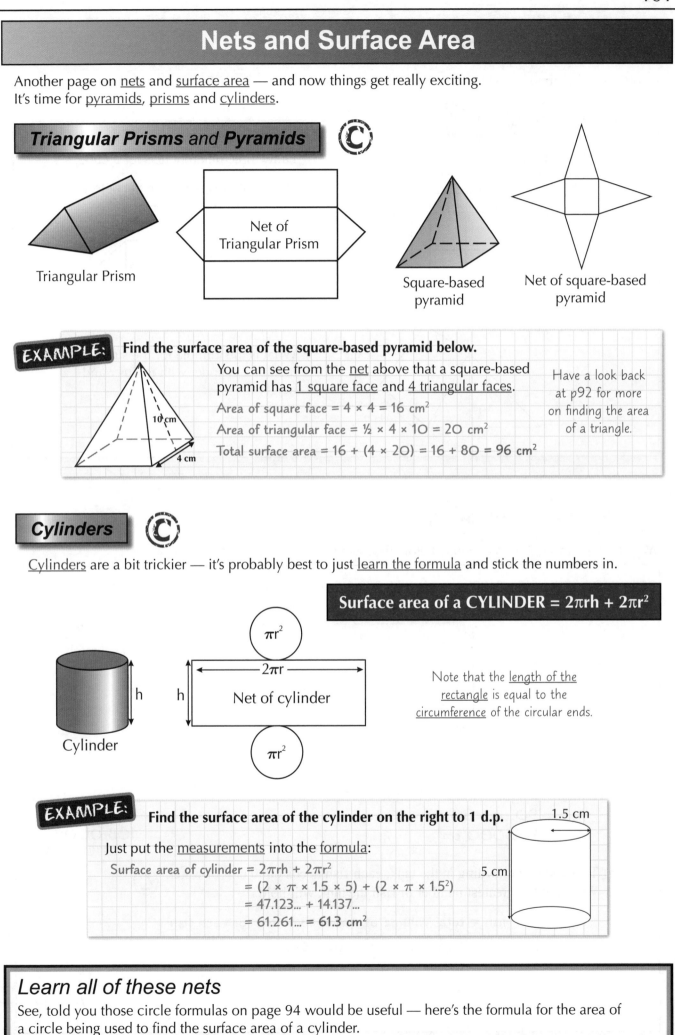

Triangular Prism

Net of Triangular Prism

Square-based pyramid

Net of square-based pyramid

EXAMPLE: **Find the surface area of the square-based pyramid below.**

You can see from the <u>net</u> above that a square-based pyramid has <u>1 square face</u> and <u>4 triangular faces</u>.

Area of square face = 4 × 4 = 16 cm²

Area of triangular face = ½ × 4 × 10 = 20 cm²

Total surface area = 16 + (4 × 20) = 16 + 80 = **96 cm²**

Have a look back at p92 for more on finding the area of a triangle.

10 cm

4 cm

Cylinders

<u>Cylinders</u> are a bit trickier — it's probably best to just <u>learn the formula</u> and stick the numbers in.

Surface area of a CYLINDER = 2πrh + 2πr²

πr²

2πr

Net of cylinder

πr²

h

h

Cylinder

Note that the <u>length of the rectangle</u> is equal to the <u>circumference</u> of the circular ends.

EXAMPLE: **Find the surface area of the cylinder on the right to 1 d.p.**

Just put the <u>measurements</u> into the <u>formula</u>:

Surface area of cylinder = 2πrh + 2πr²
= (2 × π × 1.5 × 5) + (2 × π × 1.5²)
= 47.123... + 14.137...
= 61.261... = **61.3 cm²**

1.5 cm

5 cm

Learn all of these nets

See, told you those circle formulas on page 94 would be useful — here's the formula for the area of a circle being used to find the surface area of a cylinder.

Warm-up and Worked Exam Questions

If you can't remember the volume formulas, have a look back at page 99 before you start these.

Warm-up Questions

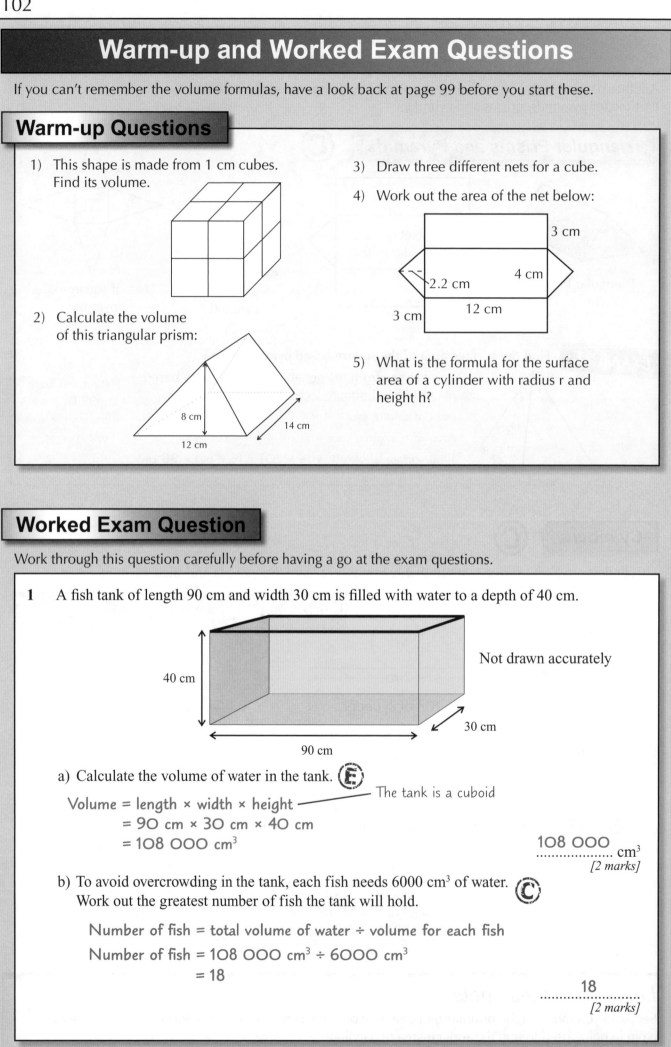

1) This shape is made from 1 cm cubes. Find its volume.

2) Calculate the volume of this triangular prism:

8 cm
14 cm
12 cm

3) Draw three different nets for a cube.

4) Work out the area of the net below:

3 cm
2.2 cm
4 cm
3 cm
12 cm

5) What is the formula for the surface area of a cylinder with radius r and height h?

Worked Exam Question

Work through this question carefully before having a go at the exam questions.

1 A fish tank of length 90 cm and width 30 cm is filled with water to a depth of 40 cm.

40 cm

Not drawn accurately

30 cm

90 cm

a) Calculate the volume of water in the tank. (F)

Volume = length × width × height — The tank is a cuboid
= 90 cm × 30 cm × 40 cm
= 108 000 cm³

108 000 .... cm³
[2 marks]

b) To avoid overcrowding in the tank, each fish needs 6000 cm³ of water. (C) Work out the greatest number of fish the tank will hold.

Number of fish = total volume of water ÷ volume for each fish

Number of fish = 108 000 cm³ ÷ 6000 cm³
= 18

18 ....
[2 marks]

Exam Questions

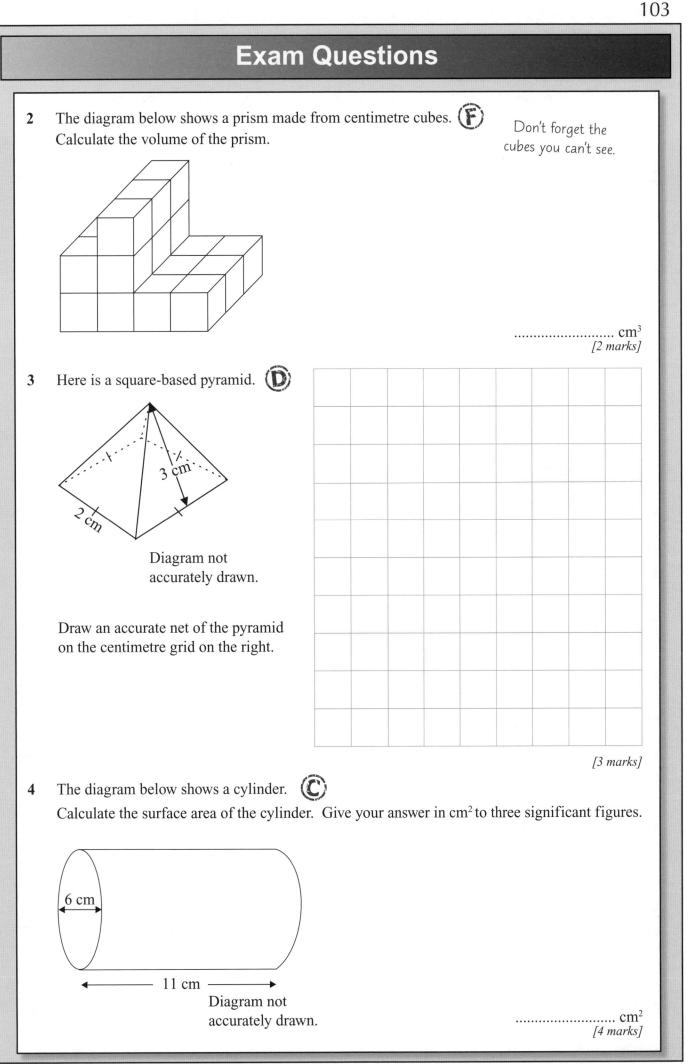

2 The diagram below shows a prism made from centimetre cubes. (F) Calculate the volume of the prism.

Don't forget the cubes you can't see.

.... cm³
[2 marks]

3 Here is a square-based pyramid. (D)

3 cm

2 cm

Diagram not accurately drawn.

Draw an accurate net of the pyramid on the centimetre grid on the right.

[3 marks]

4 The diagram below shows a cylinder. (C)

Calculate the surface area of the cylinder. Give your answer in cm² to three significant figures.

6 cm

11 cm

Diagram not accurately drawn.

.... cm²
[4 marks]

Revision Questions for Section Four

Lots of lovely shapes and formulas to learn in Section 4 — now it's time to see what's sunk in.

- Try these questions and <u>tick off each one</u> when you <u>get it right</u>.
- When you've done <u>all the questions</u> for a topic and are <u>completely happy</u> with it, tick off the topic.

Symmetry and Tessellations (p81-82) ☑

1) For each of the letters below, write down how many lines of symmetry it has and its order of rotational symmetry.

H   Z   T   N   E   ✗   S

2) Sketch a tessellating pattern made up of equilateral triangles. Use at least 6 shapes.

3) Do regular pentagons tessellate?

2D and 3D Shapes (p83-87) ☑

4) Write down the properties of an isosceles triangle.

5) How many lines of symmetry does a rhombus have? What is its order of rotational symmetry?

6) Write down the properties of a parallelogram.

7) What are congruent and similar shapes?

8) Look at the shapes on the right and write down the letters of:
   a) a pair of congruent shapes,
   b) a pair of similar shapes.

9) Write down the number of faces, edges and vertices for the following 3D shapes:
   a) a square-based pyramid   b) a cone   c) a triangular prism.

10) What is a plan view?

11) On squared paper, draw the front elevation (from the direction of the arrow), side elevation and plan view of the shape on the right.

Perimeter and Area (p91-93) ☑

12) Find the perimeter of an equilateral triangle with side length 7 cm.

13) Find the area of a rectangle measuring 4 cm by 8 cm.

14) Write down the formula for finding the area of a trapezium.

15) Find the area of a parallelogram with base 9 cm and vertical height 4 cm.

16) Find the area of the shape on the right.

Circles (p94-96) ☑

17) What is the radius of a circle with diameter 18 mm?

18) Find the area and circumference of a circle with radius 7 cm to 2 decimal places.

19) Draw a circle and label an arc, a sector and a segment.

20) Find the area and perimeter of a quarter circle with radius 3 cm to 2 decimal places.

Volume, Nets and Surface Area (p99-101) ☑

21) Write down the formula for the volume of a cylinder with radius r and height h.

22) A pentagonal prism has a cross-sectional area of 24 cm² and a length of 15 cm. Find its volume.

23) Find the surface area of a cube with side length 5 cm.

24) Find the surface area of a cylinder with height 8 cm and radius 2 cm, to 1 d.p.

Lines and Angles

Some basic, but very important, facts about angles before we start on the rest of the section.

Four Special Angles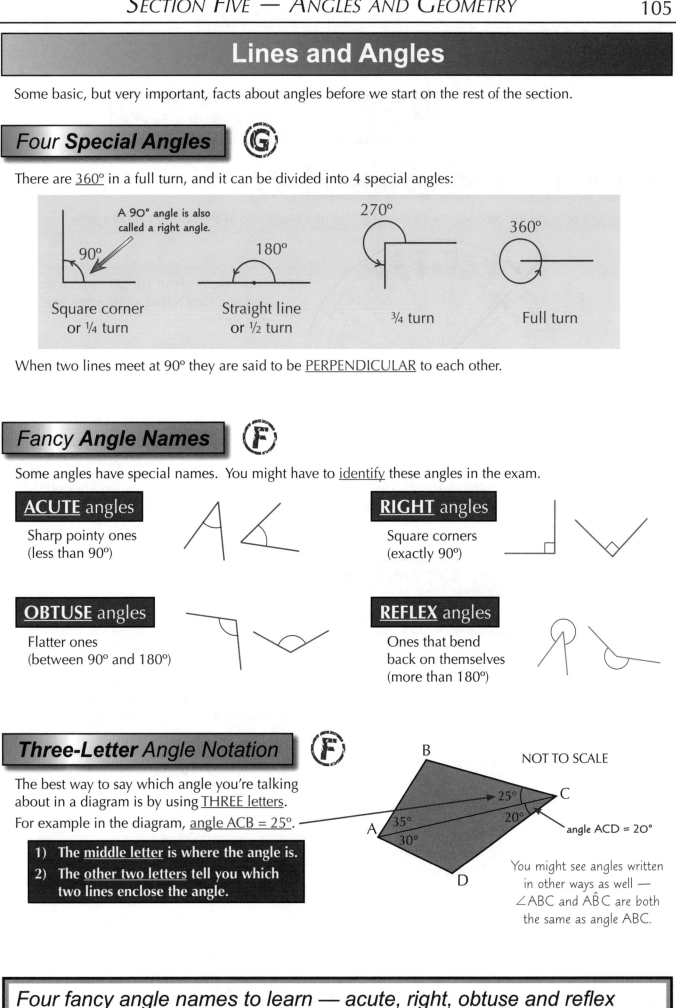

There are <u>360°</u> in a full turn, and it can be divided into 4 special angles:

A 90° angle is also called a right angle.

90°

Square corner or ¼ turn

180°

Straight line or ½ turn

270°

¾ turn

360°

Full turn

When two lines meet at 90° they are said to be <u>PERPENDICULAR</u> to each other.

Fancy Angle Names

Some angles have special names. You might have to <u>identify</u> these angles in the exam.

ACUTE angles

Sharp pointy ones (less than 90°)

RIGHT angles

Square corners (exactly 90°)

OBTUSE angles

Flatter ones (between 90° and 180°)

REFLEX angles

Ones that bend back on themselves (more than 180°)

Three-Letter Angle Notation

The best way to say which angle you're talking about in a diagram is by using <u>THREE letters</u>. For example in the diagram, <u>angle ACB = 25°</u>.

1) The <u>middle letter</u> is where the angle is.
2) The <u>other two letters</u> tell you which two lines enclose the angle.

NOT TO SCALE

25°

20°

angle ACD = 20°

35°

30°

You might see angles written in other ways as well — ∠ABC and AB̂C are both the same as angle ABC.

Four fancy angle names to learn — acute, right, obtuse and reflex

In the exams, it's pretty likely that angles will be referred to using three-letter notation — so make sure you know how to use it. And make sure you've learnt the four special angles too.

Measuring & Drawing Angles

The <u>2 big mistakes</u> that people make with protractors ➡️

1) Not putting the <u>0° line</u> at the <u>start</u> position.
2) Reading from the <u>WRONG SCALE</u>.

Measuring Angles with a *Protractor* (F)

1) <u>ALWAYS</u> position the protractor with the <u>base line</u> of it along one of the lines, as shown here:

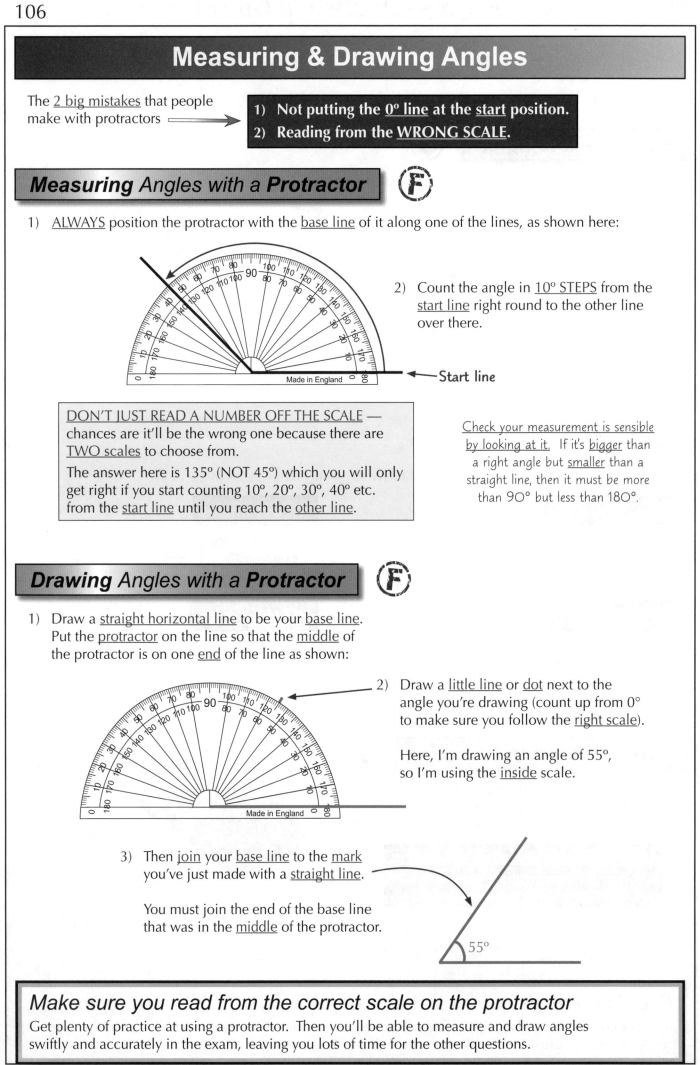

Made in England

2) Count the angle in <u>10° STEPS</u> from the <u>start line</u> right round to the other line over there.

←— **Start line**

<u>DON'T JUST READ A NUMBER OFF THE SCALE</u> — chances are it'll be the wrong one because there are <u>TWO scales</u> to choose from.

The answer here is 135° (NOT 45°) which you will only get right if you start counting 10°, 20°, 30°, 40° etc. from the <u>start line</u> until you reach the <u>other line</u>.

<u>Check your measurement is sensible by looking at it.</u> If it's <u>bigger</u> than a right angle but <u>smaller</u> than a straight line, then it must be more than 90° but less than 180°.

Drawing Angles with a *Protractor* (F)

1) Draw a <u>straight horizontal line</u> to be your <u>base line</u>. Put the <u>protractor</u> on the line so that the <u>middle</u> of the protractor is on one <u>end</u> of the line as shown:

Made in England

2) Draw a <u>little line</u> or <u>dot</u> next to the angle you're drawing (count up from 0° to make sure you follow the <u>right scale</u>).

Here, I'm drawing an angle of 55°, so I'm using the <u>inside</u> scale.

3) Then <u>join</u> your <u>base line</u> to the <u>mark</u> you've just made with a <u>straight line</u>.

You must join the end of the base line that was in the <u>middle</u> of the protractor.

55°

Make sure you read from the correct scale on the protractor

Get plenty of practice at using a protractor. Then you'll be able to measure and draw angles swiftly and accurately in the exam, leaving you lots of time for the other questions.

Five Angle Rules

I'm afraid that when it comes to angles, it's all rules, rules, rules. Still, you've got to learn them.

5 Simple Rules — That's All

1) Angles in a <u>triangle</u> add up to 180°.

$$a + b + c = 180°$$

2) Angles on a <u>straight line</u> add up to 180°.

$$a + b + c = 180°$$

3) Angles in a <u>quadrilateral</u> add up to 360°.

Remember that a quadrilateral is a <u>4-sided</u> shape.

$$a + b + c + d = 360°$$

4) Angles <u>round a point</u> add up to 360°.

$$a + b + c + d = 360°$$

5) <u>Isosceles triangles</u> have <u>2 sides</u> the same and <u>2 angles</u> the same.

These dashes indicate two sides the same length.

These angles are the same.

In an isosceles triangle, you only need to know one angle to be able to find the other two.

There are some examples of using these rules on the next page.

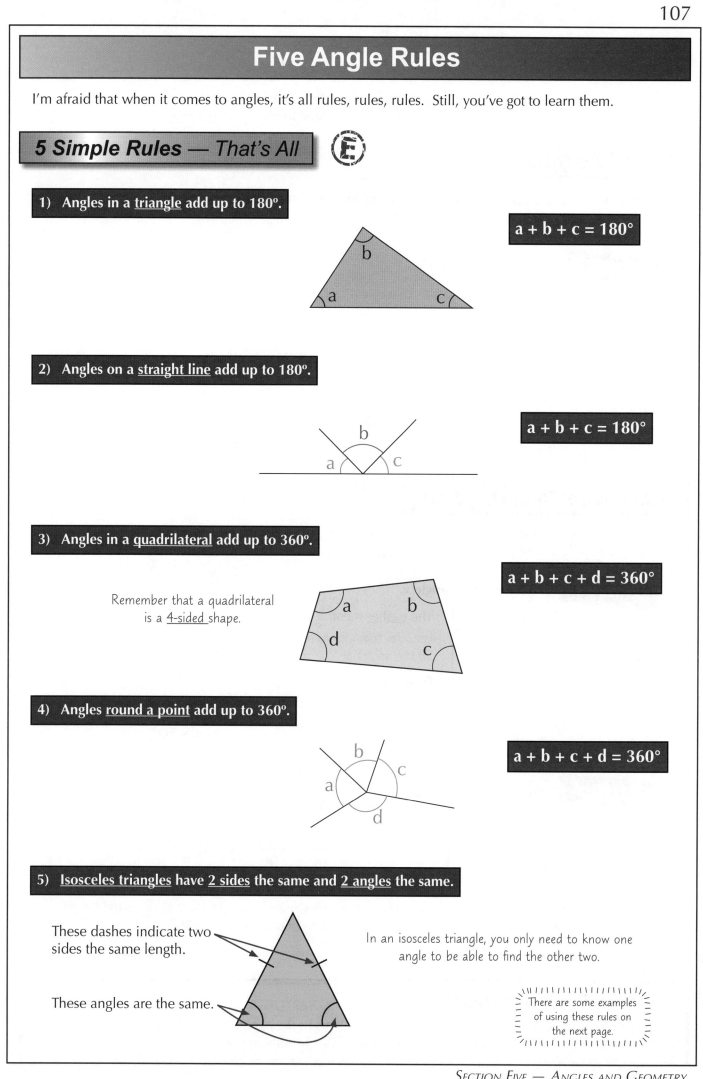

Five Angle Rules

Right, by now you should know the <u>five angle rules</u> (if you're not sure about them, go back over the previous page until you know them really well). Now it's time to see them <u>in action</u>.

Using One Rule E

It's a good idea to <u>write down</u> the <u>rules</u> you're using when finding missing angles
— it helps you <u>keep track</u> of what you're doing.

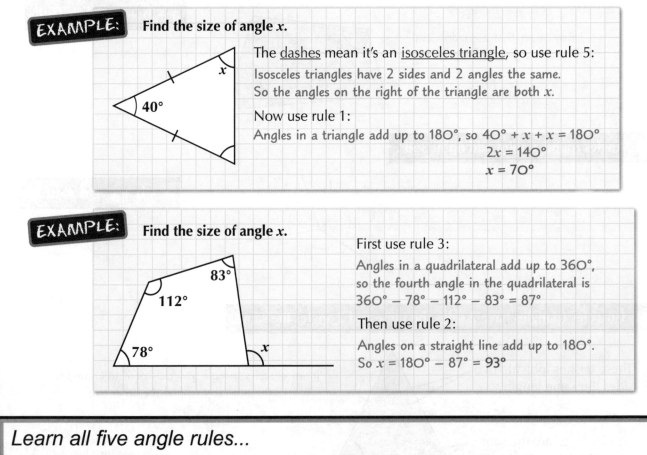

EXAMPLE: **Find the size of angle x.**

x

167°

52°

Use rule 4:

Angles round a point add up to 360°,
so $x + 52° + 90° + 167° = 360°$

$x = 360° - 52° - 90° - 167° = 51°$

Remember — this little
square means that it's a
right angle (90°).

Using More Than One Rule E

It's a bit trickier when you have to use <u>more than one</u> rule — but <u>writing down</u> the rules is a big help again.
The best method is to find <u>whatever angles you can</u> until you can work out the ones you're looking for.

EXAMPLE: **Find the size of angle x.**

40°

x

The <u>dashes</u> mean it's an <u>isosceles triangle</u>, so use rule 5:
Isosceles triangles have 2 sides and 2 angles the same.
So the angles on the right of the triangle are both *x*.

Now use rule 1:
Angles in a triangle add up to 180°, so $40° + x + x = 180°$
$2x = 140°$
$x = 70°$

EXAMPLE: **Find the size of angle x.**

83°

112°

78°

x

First use rule 3:

Angles in a quadrilateral add up to 360°,
so the fourth angle in the quadrilateral is
$360° - 78° - 112° - 83° = 87°$

Then use rule 2:

Angles on a straight line add up to 180°.
So $x = 180° - 87° = 93°$

Learn all five angle rules...

...because they're the only way to solve geometry problems in the exam, I'm afraid. There's nothing too tricky here though — just take some time to make sure it's all firmly in your head.

Parallel Lines

Parallel lines are always the <u>same distance apart</u>. This page is all about them.

*Angles Around **Parallel Lines*** Ⓓ

You also need
to know what
<u>perpendicular lines</u> are
— they meet at <u>90°</u>.

When a line crosses two <u>parallel lines</u>...

1) The two bunches of angles are <u>the same</u>.
2) There are <u>only two different angles</u>: <u>a small one</u> and <u>a big one</u>
3) These <u>ALWAYS ADD UP TO 180°</u>. E.g. 30° and 150° below

The two lines with the <u>arrows</u> on are <u>parallel</u>:

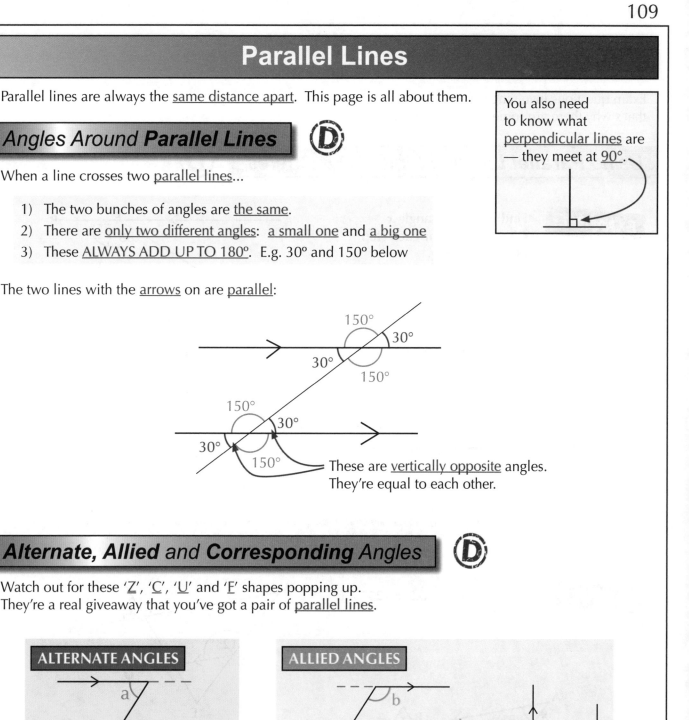

These are <u>vertically opposite</u> angles.
They're equal to each other.

*Alternate, Allied **and Corresponding** Angles* Ⓓ

Watch out for these '<u>Z</u>', '<u>C</u>', '<u>U</u>' and '<u>F</u>' shapes popping up.
They're a real giveaway that you've got a pair of <u>parallel lines</u>.

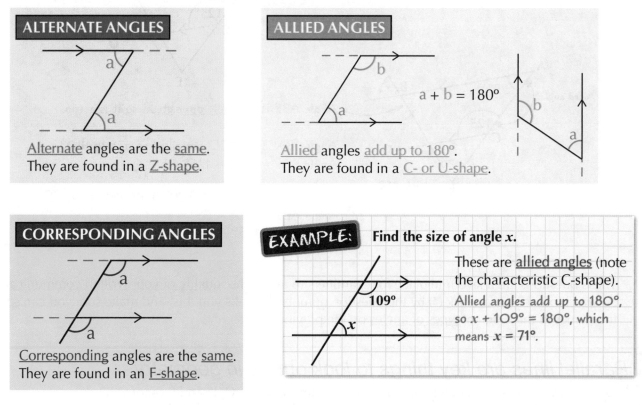

ALTERNATE ANGLES

<u>Alternate</u> angles are the <u>same</u>.
They are found in a <u>Z-shape</u>.

ALLIED ANGLES

a + b = 180°

<u>Allied</u> angles <u>add up to 180°</u>.
They are found in a <u>C- or U-shape</u>.

CORRESPONDING ANGLES

<u>Corresponding</u> angles are the <u>same</u>.
They are found in an <u>F-shape</u>.

EXAMPLE: **Find the size of angle x.**

These are <u>allied angles</u> (note
the characteristic C-shape).

Allied angles add up to 180°,
so *x* + 109° = 180°, which
means *x* = 71°.

Parallel Lines

Exam questions involving parallel lines can be pretty involved —
that's why there are some big examples below.

Using **Parallel Lines** to **Figure Out Angles** (D)

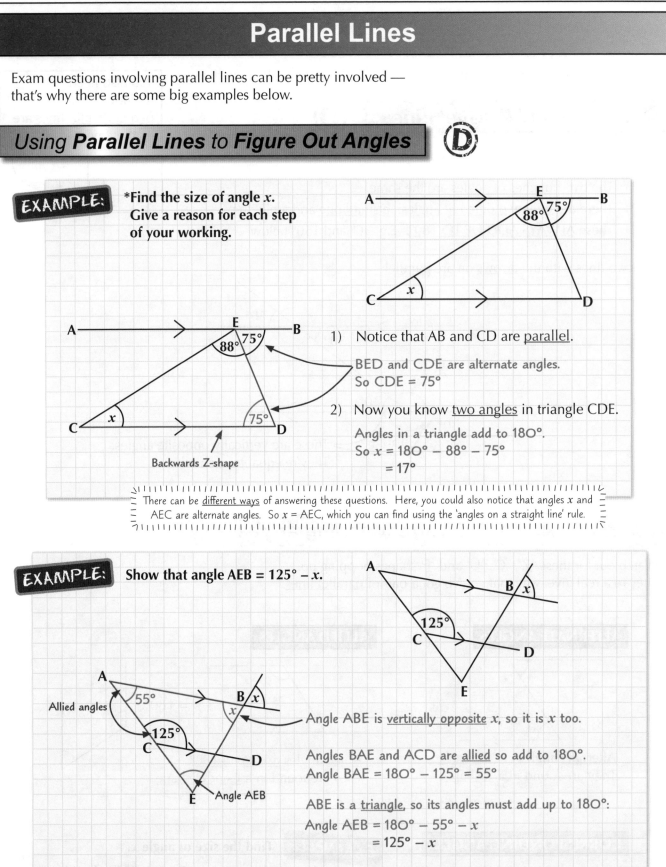

EXAMPLE: *Find the size of angle x.
Give a reason for each step
of your working.

1) Notice that AB and CD are parallel.

    BED and CDE are alternate angles.
    So CDE = 75°

2) Now you know two angles in triangle CDE.

    Angles in a triangle add to 180°.
    So x = 180° − 88° − 75°
          = 17°

Backwards Z-shape

There can be different ways of answering these questions. Here, you could also notice that angles x and AEC are alternate angles. So x = AEC, which you can find using the 'angles on a straight line' rule.

EXAMPLE: Show that angle AEB = 125° − x.

Allied angles

Angle ABE is vertically opposite x, so it is x too.

Angles BAE and ACD are allied so add to 180°.
Angle BAE = 180° − 125° = 55°

ABE is a triangle, so its angles must add up to 180°:
Angle AEB = 180° − 55° − x
          = 125° − x

These questions are a favourite place for examiners to assess the 'quality of your written communication'.
So watch out for the asterisk (*) by the question number that tells you this, and make sure you can spell
the names of the different types of angles on the previous page.

Parallel lines are key things to look out for in geometry

Keep your eyes open for parallel lines and those all-important Z, C, U and F shapes.
Extending the lines can make seeing these things (and therefore your life) a lot easier.

Warm-up and Worked Exam Questions

Oh look at all those lovely diagrams. But don't just look at them — you need to work through
the questions one by one and polish all those geometry skills.

Warm-up Questions

1) Write down one example of:
    a) an acute angle
    b) an obtuse angle
    c) a reflex angle
    d) a right angle.

2) Measure these angles accurately
    with a protractor:

    a)

    b)

    c)

    d)

3) Use a protractor to accurately
    draw these angles:
    a) 35°   b) 150°   c) 80°

4) Work out angles x and y in the diagram below.

    70° y x

5) The diagram shown below has one angle
    given as 60°. Find the other two marked angles.

    60° a b

Worked Exam Question

There'll probably be a question in the exams that asks you to find angles. That means you have to
remember all the different angle rules and practise using them in the right places...

*1   *VXY* is an isosceles triangle. *VX* = *XY*. (E)
     UVY is a right-angled triangle.
     XZ is a straight line.

     Diagram not
     accurately drawn

     Work out the size of angle *UYZ*.
     Give reasons for each stage of your working.

     Angles in a triangle add up to 180°,
     so angle VYX + angle YVX + 96° = 180°

     180° − 96° = 84°, so angle VYX = 84° ÷ 2 = 42°,
     because isosceles triangles have 2 equal angles.

     Angles on a straight line add up to 180°,
     so angle UYZ = 180° − 90° − 42° = 48°.

     If you're asked to <u>work out</u> the size of an angle, don't be tempted to just
     measure it — the diagrams aren't drawn accurately.

     ....48............°
                                    [4 marks]

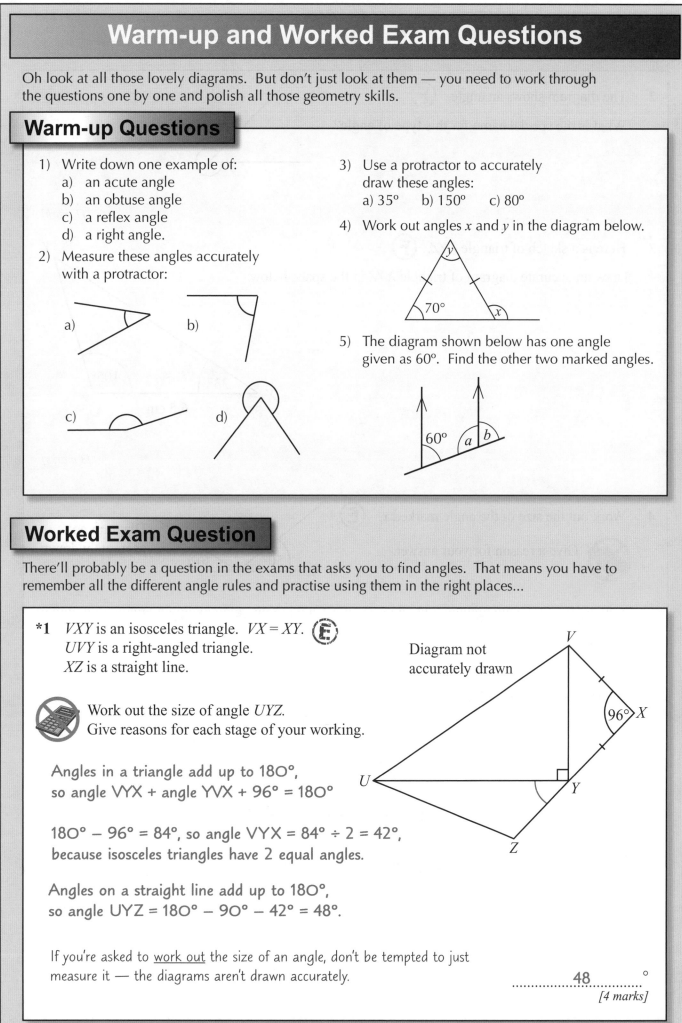

Exam Questions

2 The diagram shows an angle. 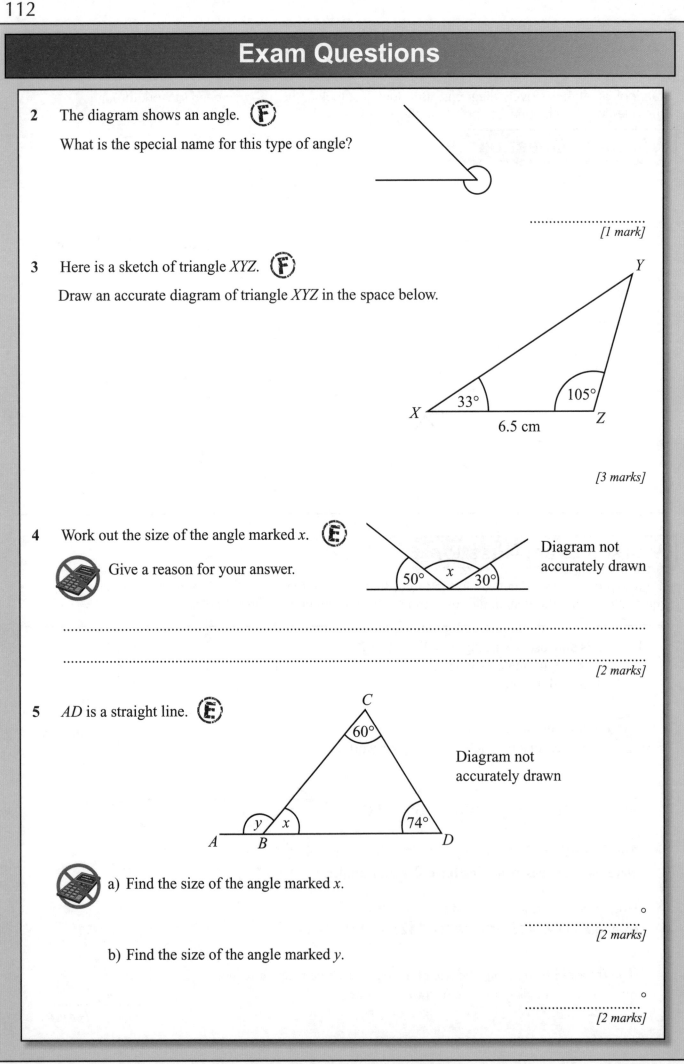 **(F)**

What is the special name for this type of angle?

....

[1 mark]

3 Here is a sketch of triangle *XYZ*. **(F)**

Draw an accurate diagram of triangle *XYZ* in the space below.

[3 marks]

4 Work out the size of the angle marked *x*. **(E)**

Give a reason for your answer.

Diagram not accurately drawn

....

....

[2 marks]

5 *AD* is a straight line. **(E)**

Diagram not accurately drawn

a) Find the size of the angle marked *x*.

.... °

[2 marks]

b) Find the size of the angle marked *y*.

.... °

[2 marks]

Exam Questions

6 *ABC* is an isosceles triangle. **(E)**
AB = *BC*.
AD is a straight line.

Work out the size of angle *BCD*.

Diagram not accurately drawn

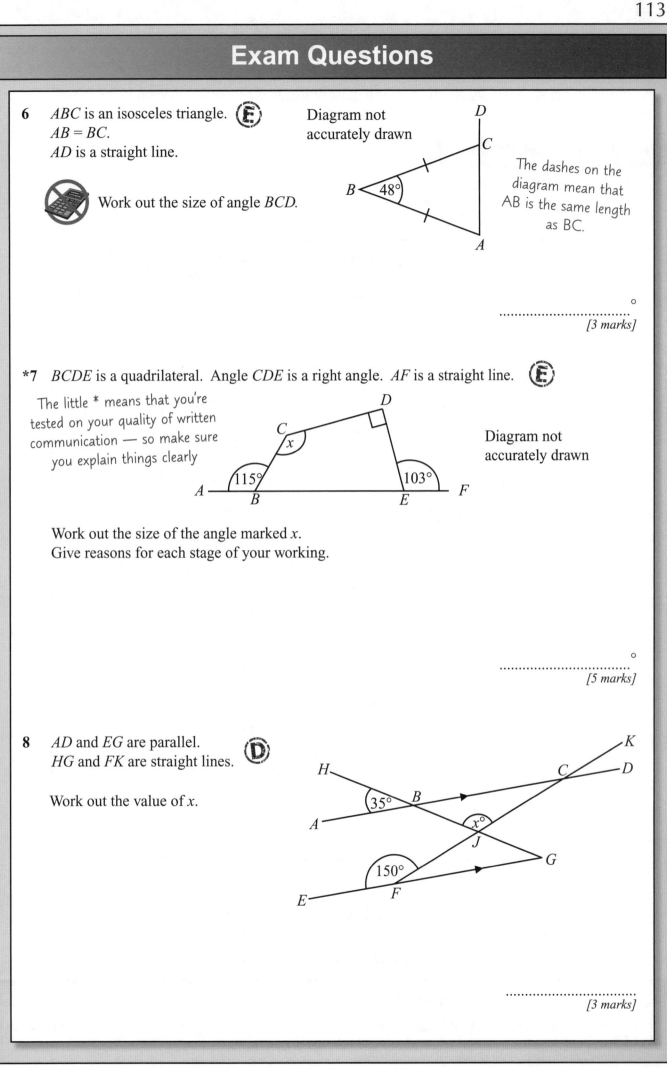

The dashes on the diagram mean that AB is the same length as BC.

.... °
[3 marks]

*7 *BCDE* is a quadrilateral. Angle *CDE* is a right angle. *AF* is a straight line. **(E)**

The little * means that you're tested on your quality of written communication — so make sure you explain things clearly

Diagram not accurately drawn

Work out the size of the angle marked *x*.
Give reasons for each stage of your working.

.... °
[5 marks]

8 *AD* and *EG* are parallel. **(D)**
HG and *FK* are straight lines.

Work out the value of *x*.

....
[3 marks]

Polygons

A polygon is a <u>many-sided shape</u>, and can be <u>regular</u> or <u>irregular</u>. A regular polygon is one where all the sides and angles are the <u>same</u> (in an irregular polygon, the sides and angles are <u>different</u>).

Regular Polygons (E)

Learn the names of these <u>regular polygons</u> and how many <u>sides</u> they have (remember that all the <u>sides</u> and <u>angles</u> in a regular polygon are the <u>same</u>). Make sure you know all the facts about symmetry too.

SHAPE	NAME	NUMBER OF SIDES	NUMBER OF LINES OF SYMMETRY	ORDER OF ROTATIONAL SYMMETRY
	<u>EQUILATERAL TRIANGLE</u>	<u>3 sides</u>	<u>3 lines</u> of symmetry	Rotational symmetry <u>order 3</u>
	<u>SQUARE</u> (regular quadrilateral)	<u>4 sides</u>	<u>4 lines</u> of symmetry	Rotational symmetry <u>order 4</u>
	<u>PENTAGON</u>	<u>5 sides</u>	<u>5 lines</u> of symmetry	Rotational symmetry <u>order 5</u>
	<u>HEXAGON</u>	<u>6 sides</u>	<u>6 lines</u> of symmetry	Rotational symmetry <u>order 6</u>
	<u>HEPTAGON</u>	<u>7 sides</u>	<u>7 lines</u> of symmetry	Rotational symmetry <u>order 7</u>
	<u>OCTAGON</u>	<u>8 sides</u>	<u>8 lines</u> of symmetry	Rotational symmetry <u>order 8</u>
	<u>DECAGON</u>	<u>10 sides</u>	<u>10 lines</u> of symmetry	Rotational symmetry <u>order 10</u>

Seven types of polygon to learn

Remember — the shapes above are <u>regular</u> polygons. <u>Irregular</u> polygons have the same number of sides, but fewer lines of symmetry and a lower order of rotational symmetry.

Polygons and Angles

You're not finished with polygons yet — there are a few things you need to know about their <u>angles</u>, and then there are some <u>formulas</u> to learn too. Nothing too tricky though.

Polygons Have *Interior* and *Exterior* Angles ⒟

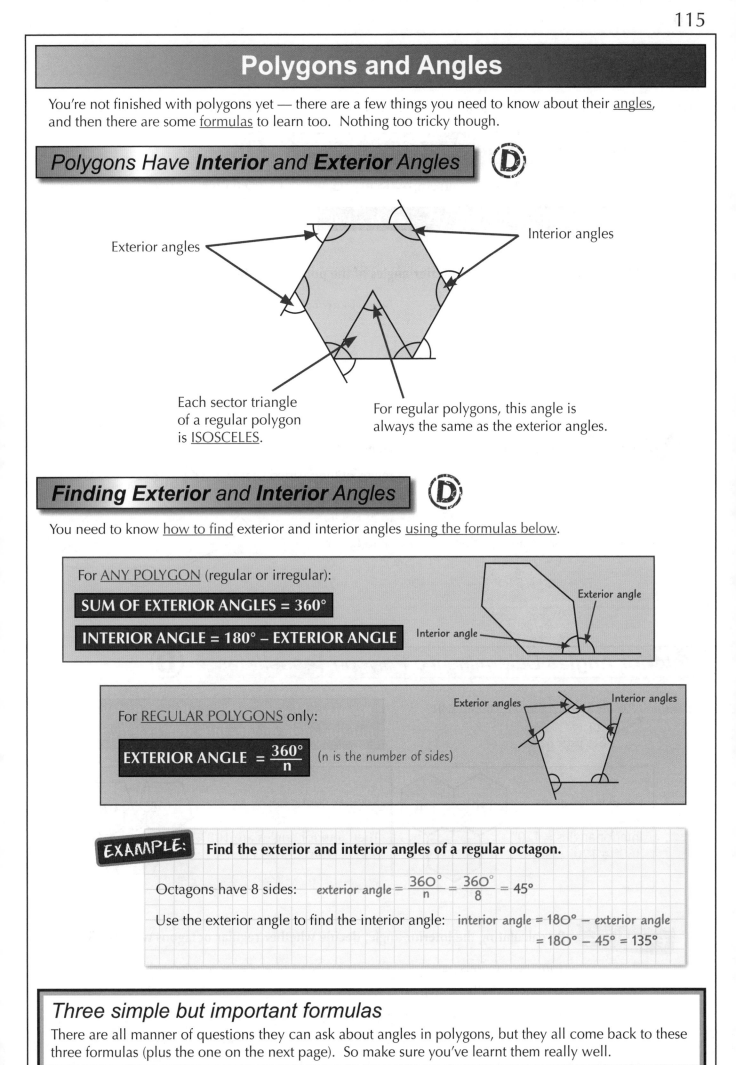

Exterior angles

Interior angles

Each sector triangle of a regular polygon is <u>ISOSCELES</u>.

For regular polygons, this angle is always the same as the exterior angles.

Finding *Exterior* and *Interior* Angles ⒟

You need to know <u>how to find</u> exterior and interior angles <u>using the formulas below</u>.

For <u>ANY POLYGON</u> (regular or irregular):

SUM OF EXTERIOR ANGLES = 360°

INTERIOR ANGLE = 180° – EXTERIOR ANGLE

Exterior angle

Interior angle

For <u>REGULAR POLYGONS</u> only:

EXTERIOR ANGLE $= \dfrac{360°}{n}$ (n is the number of sides)

Exterior angles

Interior angles

EXAMPLE: **Find the exterior and interior angles of a regular octagon.**

Octagons have 8 sides: $\quad$ exterior angle $= \dfrac{360°}{n} = \dfrac{360°}{8} = 45°$

Use the exterior angle to find the interior angle: $\quad$ interior angle = 180° – exterior angle

$$= 180° - 45° = 135°$$

Three simple but important formulas

There are all manner of questions they can ask about angles in polygons, but they all come back to these three formulas (plus the one on the next page). So make sure you've learnt them really well.

Polygons, Angles and Tessellations

Just one more <u>polygon angle formula</u>. Then I'll be unveiling <u>why some polygons tessellate</u> and others don't.

The Tricky One — **Sum** *of* **Interior** *Angles*   (D)

This formula for the <u>sum of the interior angles</u> works for <u>ALL</u> polygons, even irregular ones.

> **SUM OF INTERIOR ANGLES = (n − 2) × 180°**   *(n is the number of sides)*

EXAMPLE: **Find the sum of the interior angles of the polygon on the right.**

The polygon is a hexagon, so n = 6:

Sum of interior angles = (n − 2) × 180°
= (6 − 2) × 180°
= 720°

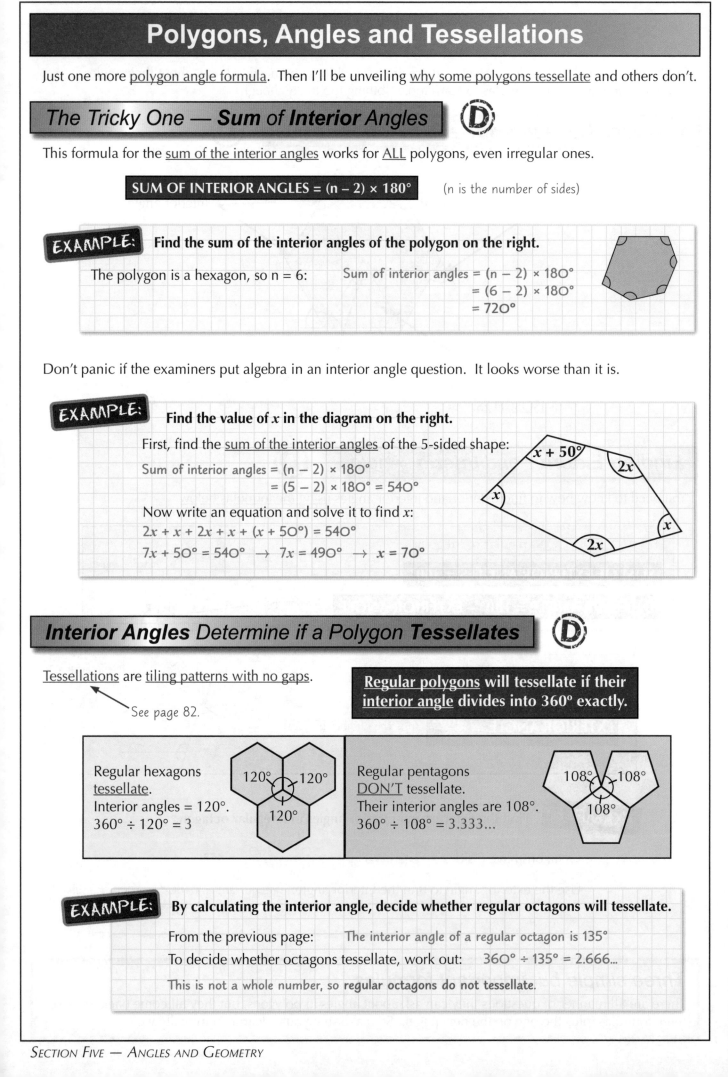

Don't panic if the examiners put algebra in an interior angle question. It looks worse than it is.

EXAMPLE: **Find the value of x in the diagram on the right.**

First, find the <u>sum of the interior angles</u> of the 5-sided shape:

Sum of interior angles = (n − 2) × 180°
= (5 − 2) × 180° = 540°

Now write an equation and solve it to find x:
$2x + x + 2x + x + (x + 50°) = 540°$
$7x + 50° = 540°$ → $7x = 490°$ → $x = 70°$

Interior Angles *Determine if a Polygon* **Tessellates**   (D)

<u>Tessellations</u> are <u>tiling patterns with no gaps</u>.

See page 82.

> **Regular polygons** will tessellate if their
> **interior angle divides into 360° exactly.**

Regular hexagons <u>tessellate</u>.
Interior angles = 120°.
360° ÷ 120° = 3

120° 120°
120°

Regular pentagons <u>DON'T</u> tessellate.
Their interior angles are 108°.
360° ÷ 108° = 3.333...

108° 108°
108°

EXAMPLE: **By calculating the interior angle, decide whether regular octagons will tessellate.**

From the previous page: The interior angle of a regular octagon is 135°

To decide whether octagons tessellate, work out: 360° ÷ 135° = 2.666...

This is not a whole number, so **regular octagons do not tessellate.**

Warm-up and Exam Questions

Now it's time for some warm-up questions to get your brain into gear — before moving on to the exam-style questions. It's great preparation for the real thing.

Warm-up Questions

1) a) What is a regular polygon? b) Name the first six regular polygons.
2) Work out the exterior and interior angles for a regular octagon.
3) Find the sum of the interior angles of a regular heptagon.

Worked Exam Question

Worked exam questions are the ideal way to get the hang of answering the real exam questions.

1 The diagram shows a regular pentagon and an equilateral triangle. Ⓓ

Work out the size of the angle p.

Exterior angle of a regular pentagon = 360° ÷ 5 = 72°

Interior angle of a regular pentagon = 180° − 72° = 108°

Angle in an equilateral triangle = 180° ÷ 3 = 60°

p = 360° − (108° + 60°) ← angles round a point add up to 360°

= 360° − 168° = 192°

Diagram not accurately drawn

....192....°

[4 marks]

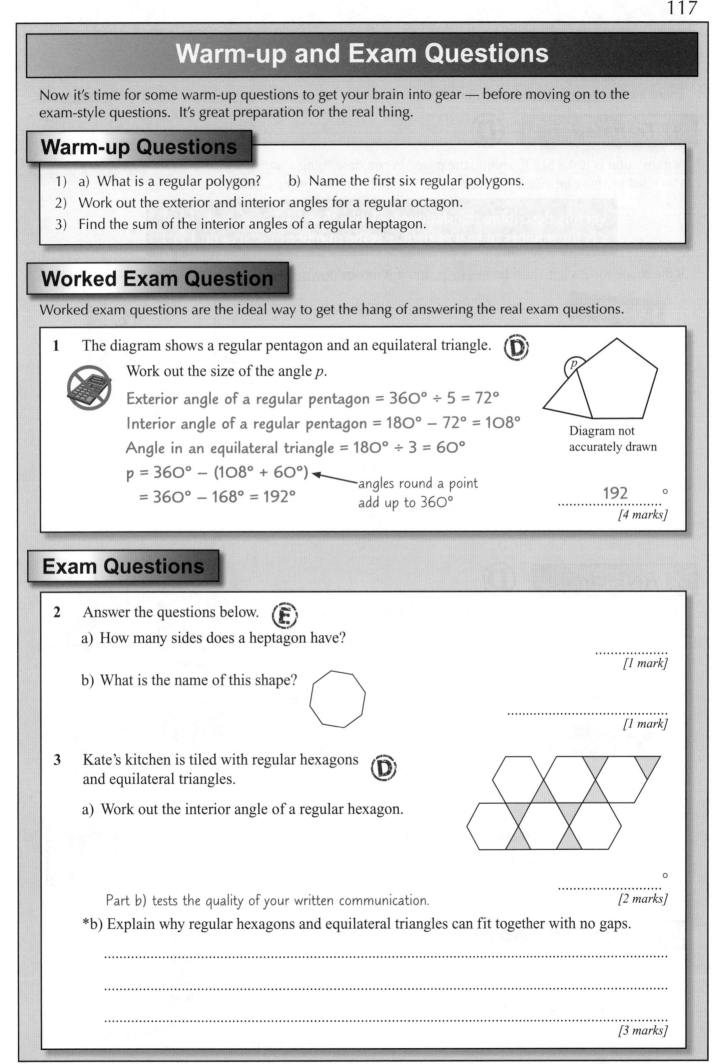

Exam Questions

2 Answer the questions below. Ⓔ

a) How many sides does a heptagon have?

....
[1 mark]

b) What is the name of this shape?

....
[1 mark]

3 Kate's kitchen is tiled with regular hexagons and equilateral triangles. Ⓓ

a) Work out the interior angle of a regular hexagon.

....................°
[2 marks]

Part b) tests the quality of your written communication.

*b) Explain why regular hexagons and equilateral triangles can fit together with no gaps.

....

....

....
[3 marks]

Transformations

There are four <u>transformations</u> you need to know about — <u>translation</u>, <u>reflection</u>, <u>rotation</u> and <u>enlargement</u>.

1) Translations (D)

A translation is just a <u>SLIDE</u> around the page. When describing a translation, you must say <u>how far along</u> and <u>how far up</u> the shape moves using a vector.

> <u>Vectors</u> describing translations look like this. $\longrightarrow$ $\begin{pmatrix} x \\ y \end{pmatrix}$
> x is the number of spaces <u>right</u>, y is the number of spaces <u>up</u>.

If the shape moves <u>left</u> x will be <u>negative</u>, and if it moves <u>down</u> y will be <u>negative</u>.

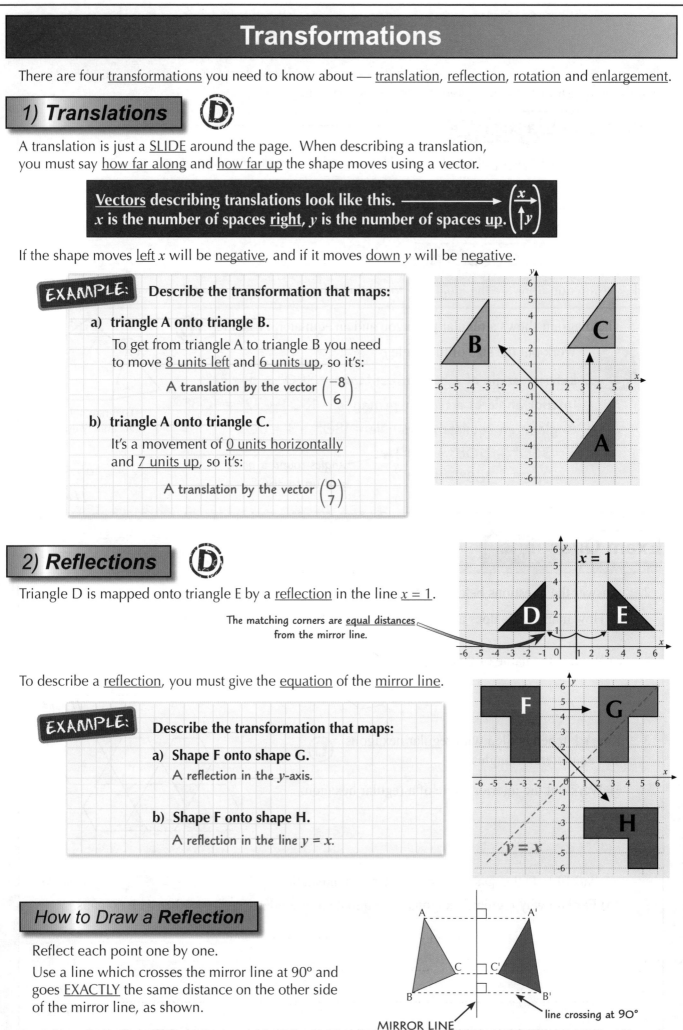

EXAMPLE: **Describe the transformation that maps:**

a) **triangle A onto triangle B.**

To get from triangle A to triangle B you need to move <u>8 units left</u> and <u>6 units up</u>, so it's:

A translation by the vector $\begin{pmatrix} -8 \\ 6 \end{pmatrix}$

b) **triangle A onto triangle C.**

It's a movement of <u>0 units horizontally</u> and <u>7 units up</u>, so it's:

A translation by the vector $\begin{pmatrix} 0 \\ 7 \end{pmatrix}$

2) Reflections (D)

Triangle D is mapped onto triangle E by a <u>reflection</u> in the line $x = 1$.

The matching corners are <u>equal distances</u> from the mirror line.

To describe a <u>reflection</u>, you must give the <u>equation</u> of the <u>mirror line</u>.

EXAMPLE: **Describe the transformation that maps:**

a) **Shape F onto shape G.**

A reflection in the y-axis.

b) **Shape F onto shape H.**

A reflection in the line $y = x$.

How to Draw a *Reflection*

Reflect each point one by one.

Use a line which crosses the mirror line at 90° and goes <u>EXACTLY</u> the same distance on the other side of the mirror line, as shown.

MIRROR LINE

line crossing at 90°

More on Transformations

Transformation number 3 coming up. Rotation.

3) Rotations Ⓓ

To describe a rotation, you need 3 details:

> 1) The angle of rotation (usually 90° or 180°).
> 2) The direction of rotation (clockwise or anticlockwise).
> 3) The centre of rotation

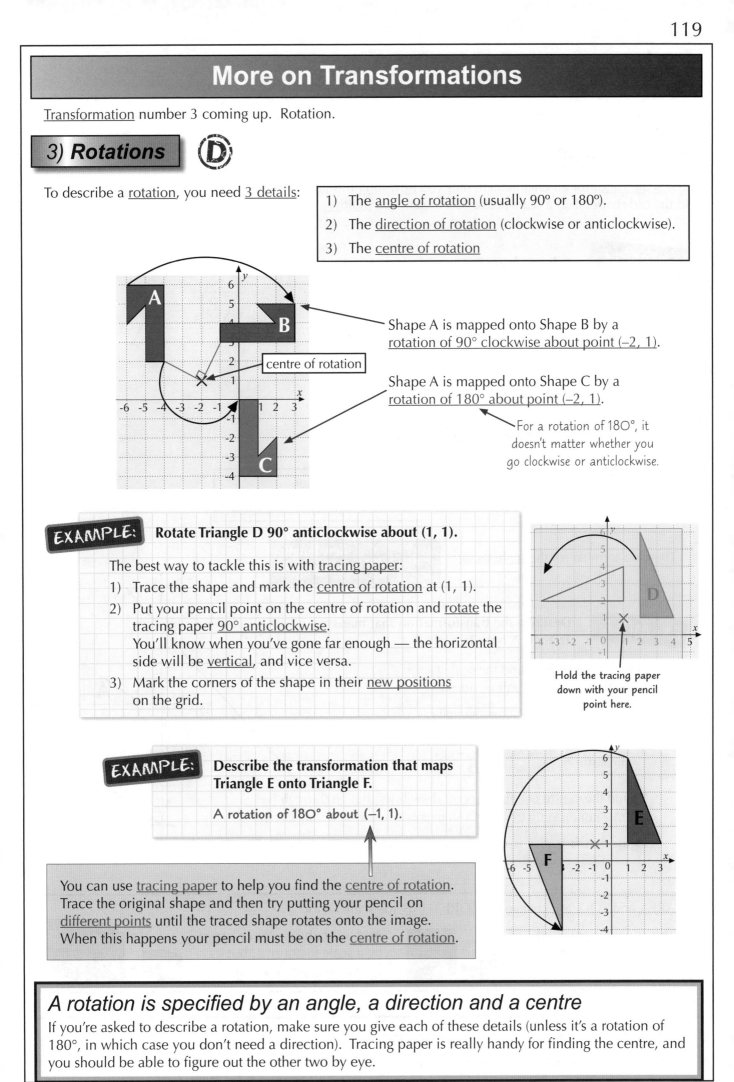

Shape A is mapped onto Shape B by a rotation of 90° clockwise about point (–2, 1).

Shape A is mapped onto Shape C by a rotation of 180° about point (–2, 1).

For a rotation of 180°, it doesn't matter whether you go clockwise or anticlockwise.

EXAMPLE: **Rotate Triangle D 90° anticlockwise about (1, 1).**

The best way to tackle this is with tracing paper:
1) Trace the shape and mark the centre of rotation at (1, 1).
2) Put your pencil point on the centre of rotation and rotate the tracing paper 90° anticlockwise.
   You'll know when you've gone far enough — the horizontal side will be vertical, and vice versa.
3) Mark the corners of the shape in their new positions on the grid.

Hold the tracing paper down with your pencil point here.

EXAMPLE: **Describe the transformation that maps Triangle E onto Triangle F.**

A rotation of 180° about (–1, 1).

You can use tracing paper to help you find the centre of rotation. Trace the original shape and then try putting your pencil on different points until the traced shape rotates onto the image. When this happens your pencil must be on the centre of rotation.

A rotation is specified by an angle, a direction and a centre

If you're asked to describe a rotation, make sure you give each of these details (unless it's a rotation of 180°, in which case you don't need a direction). Tracing paper is really handy for finding the centre, and you should be able to figure out the other two by eye.

Transformations — Enlargements

One more transformation coming up — <u>enlargements</u>. They're the trickiest, but also the most interesting.

4) Enlargements D

The <u>scale factor</u> for an enlargement tells you <u>how long</u> the sides of the new shape are compared to the old shape. E.g. a scale factor of 3 means you <u>multiply</u> each side length by 3.

EXAMPLE: **Enlarge shape P by a scale factor of 3.**

Make each side <u>three times as long</u> as the matching side on shape P. Start with the <u>horizontal</u> and <u>vertical</u> sides.

Take care with the sloping sides — they're much trickier.

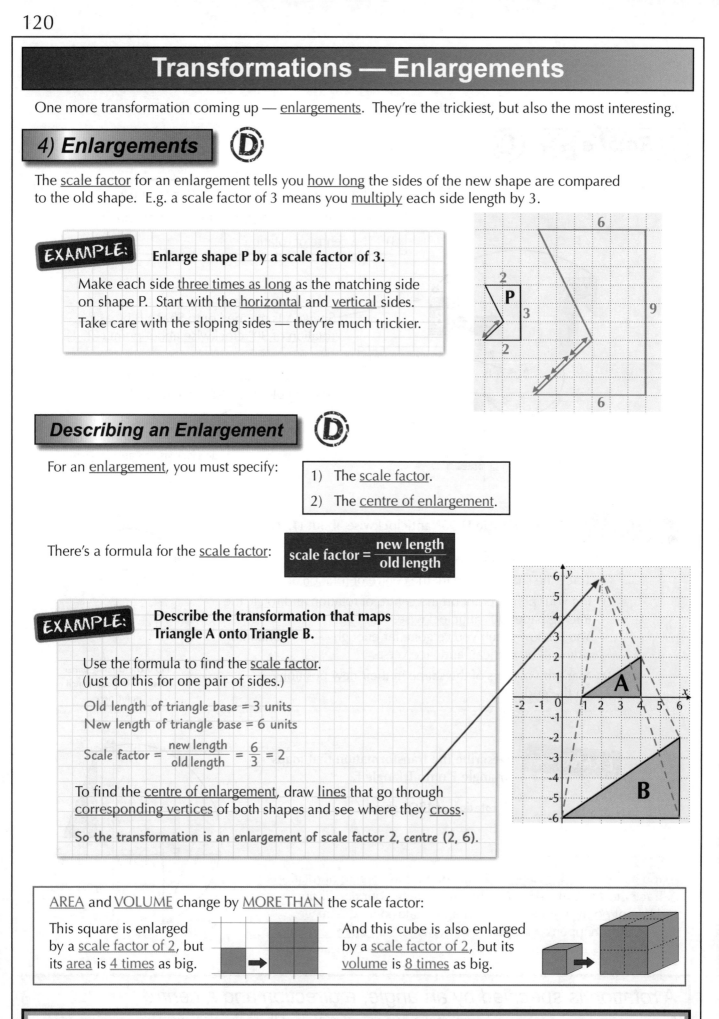

Describing an Enlargement D

For an <u>enlargement</u>, you must specify:

1) The <u>scale factor</u>.
2) The <u>centre of enlargement</u>.

There's a formula for the <u>scale factor</u>:

$$\text{scale factor} = \frac{\text{new length}}{\text{old length}}$$

EXAMPLE: **Describe the transformation that maps Triangle A onto Triangle B.**

Use the formula to find the <u>scale factor</u>. (Just do this for one pair of sides.)

Old length of triangle base = 3 units
New length of triangle base = 6 units

$$\text{Scale factor} = \frac{\text{new length}}{\text{old length}} = \frac{6}{3} = 2$$

To find the <u>centre of enlargement</u>, draw <u>lines</u> that go through <u>corresponding vertices</u> of both shapes and see where they <u>cross</u>.

So the transformation is an enlargement of scale factor 2, centre (2, 6).

<u>AREA</u> and <u>VOLUME</u> change by <u>MORE THAN</u> the scale factor:

This square is enlarged by a <u>scale factor of 2</u>, but its <u>area</u> is <u>4 times</u> as big.

And this cube is also enlarged by a <u>scale factor of 2</u>, but its <u>volume</u> is <u>8 times</u> as big.

Shapes are similar under enlargement

The position and the size change, but the angles and ratios of the sides don't (see p85).

Harder Transformations

Just one more page on transformations, and then you're done. With transformations anyway, not with Maths.

Enlarging a Shape with a Given *Centre of Enlargement*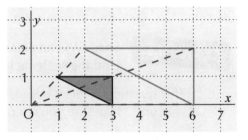

If you're given the <u>centre of enlargement</u>, then it's vitally important <u>where</u> your new shape is on the grid.

> **The <u>scale factor</u> tells you the <u>RELATIVE DISTANCE</u> of the old points and new points from the <u>centre of enlargement</u>.**

So, a <u>scale factor of 2</u> means the corners of the enlarged shape are <u>twice as far from the centre of enlargement</u> as the corners of the original shape. And a <u>scale factor of 3</u> means the corners of the enlarged shape are <u>three times as far from the centre of enlargement</u> as the corners of the old shape.

EXAMPLE: **Enlarge the shaded shape by a scale factor of 2, about centre O.**

The scale factor is 2, so make each corner of the new shape <u>twice as far</u> from O as it is in the original shape.

Combinations of *Transformations*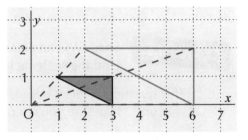

If they're feeling really mean, the examiners might make you do <u>two transformations</u> to the <u>same shape</u>, then ask you to <u>describe</u> the <u>single transformation</u> that would get you to the <u>final shape</u>. It's not as bad as it sounds.

> **Remember to specify <u>ALL</u> the details for the transformation.**

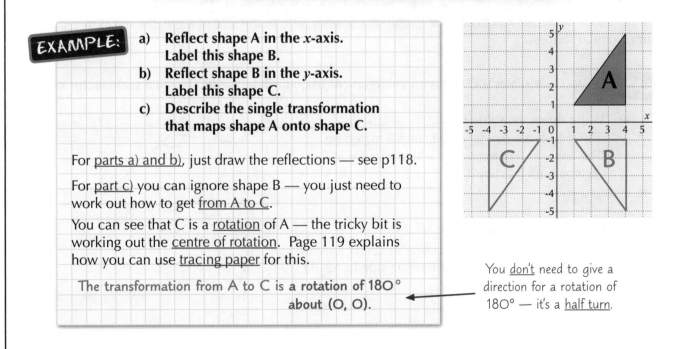

EXAMPLE:
a) **Reflect shape A in the *x*-axis. Label this shape B.**
b) **Reflect shape B in the *y*-axis. Label this shape C.**
c) **Describe the single transformation that maps shape A onto shape C.**

For <u>parts a) and b)</u>, just draw the reflections — see p118.

For <u>part c)</u> you can ignore shape B — you just need to work out how to get <u>from A to C</u>.

You can see that C is a <u>rotation</u> of A — the tricky bit is working out the <u>centre of rotation</u>. Page 119 explains how you can use <u>tracing paper</u> for this.

The transformation from A to C is **a rotation of 180° about (O, O).**

You <u>don't</u> need to give a direction for a rotation of 180° — it's a <u>half turn</u>.

Combinations of transformations aren't as scary as they look

Don't try to cut corners by working out the combined transformation in your head. Draw all the shapes — if you slip up on writing down the transformation these shapes will get you a few marks.

Similar Shape Problems

Similar shapes are <u>exactly the same shape</u>, but are <u>different sizes</u> (they can also be <u>rotated</u> or <u>reflected</u>).

Similar Shapes Have the *Same Angles* Ⓔ

Two shapes are <u>similar</u> if:

> 1) All the <u>angles</u> match up.
> 2) The <u>sides</u> are all enlarged by the <u>same scale factor</u>.

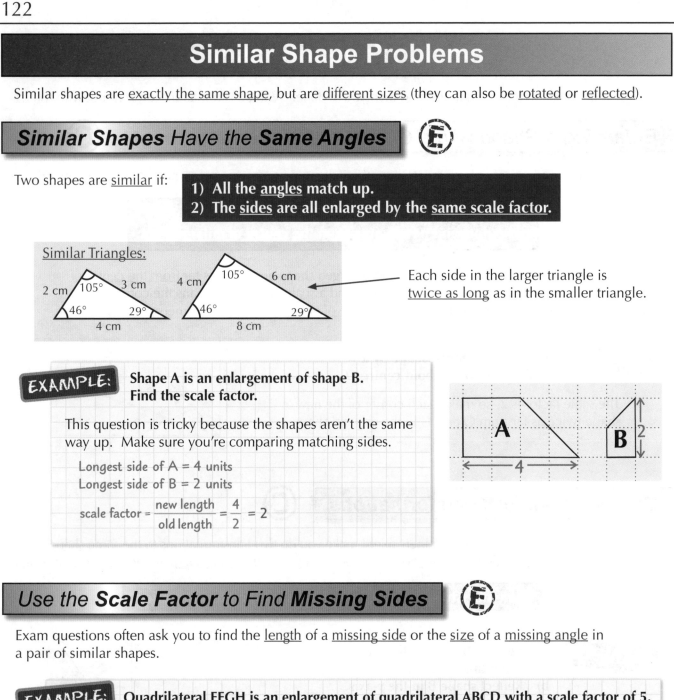

Similar Triangles:

Each side in the larger triangle is <u>twice as long</u> as in the smaller triangle.

EXAMPLE: **Shape A is an enlargement of shape B. Find the scale factor.**

This question is tricky because the shapes aren't the same way up. Make sure you're comparing matching sides.

Longest side of A = 4 units
Longest side of B = 2 units

scale factor = $\dfrac{\text{new length}}{\text{old length}} = \dfrac{4}{2} = 2$

Use the *Scale Factor* to Find *Missing Sides* Ⓔ

Exam questions often ask you to find the <u>length</u> of a <u>missing side</u> or the <u>size</u> of a <u>missing angle</u> in a pair of similar shapes.

EXAMPLE: **Quadrilateral EFGH is an enlargement of quadrilateral ABCD with a scale factor of 5.**

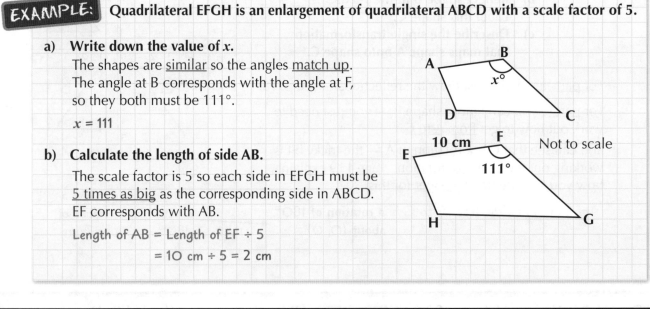

a) **Write down the value of x.**
The shapes are <u>similar</u> so the angles <u>match up</u>.
The angle at B corresponds with the angle at F, so they both must be 111°.

$x = 111$

b) **Calculate the length of side AB.**
The scale factor is 5 so each side in EFGH must be <u>5 times as big</u> as the corresponding side in ABCD.
EF corresponds with AB.

Length of AB = Length of EF ÷ 5
= 10 cm ÷ 5 = 2 cm

Don't get 'similar' and 'congruent' muddled

Congruent shapes are the same size whereas similar shapes have been scaled up or down (see page 85).

Warm-up and Worked Exam Questions

Once you've learnt the facts about transformations, have a go at these questions.
Don't skip the warm-up questions though — they're just as important to get your brain going.

Warm-up Questions

1) Describe fully these 4 transformations:
A→B
B→C
C→A
A→D

2) Enlarge shape **J** by scale factor 4.

3) What scale factor should you apply to line PQ to get line RS?
P ——— Q   R ————————————— S

4) What single transformation will convert shape **E** into shape **F**?

Worked Exam Question

Unfortunately, the helpful blue writing won't be there in the exam, so make the most of it now.

1 The grid shows shape **A** and shape **B**. **(D)**

a) Describe fully the single transformation that maps shape **A** onto shape **B**.

A translation by the vector $\begin{pmatrix} 1 \\ -7 \end{pmatrix}$

[2 marks]

b) Reflect shape **A** in the line $y = x$.
Label the image **C**.

[2 marks]

c) Describe fully the single transformation that maps shape **C** onto shape **B**.

A rotation of 180° around the point (3, −1)

[3 marks]

Because it's a rotation of 180° you don't need to specify the direction of rotation.

Exam Questions

2   Triangle **A** has been drawn on the grid below.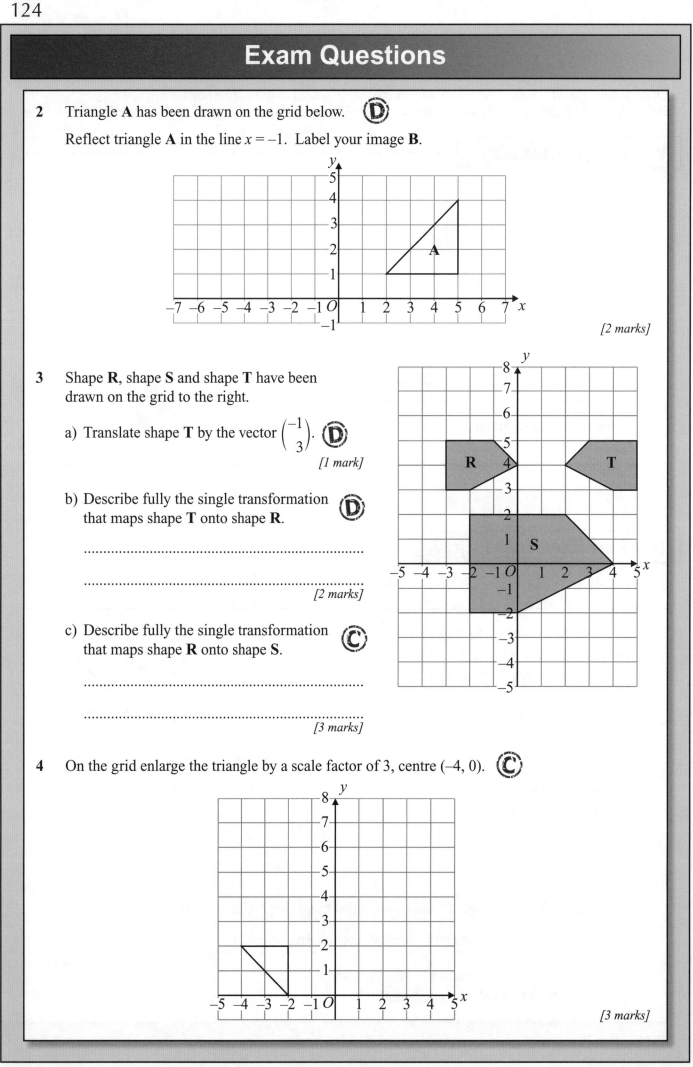

Reflect triangle **A** in the line $x = -1$. Label your image **B**.

[2 marks]

3   Shape **R**, shape **S** and shape **T** have been drawn on the grid to the right.

a) Translate shape **T** by the vector $\begin{pmatrix} -1 \\ 3 \end{pmatrix}$.

[1 mark]

b) Describe fully the single transformation that maps shape **T** onto shape **R**.

....

....

[2 marks]

c) Describe fully the single transformation that maps shape **R** onto shape S.

....

....

[3 marks]

4   On the grid enlarge the triangle by a scale factor of 3, centre $(-4, 0)$.

[3 marks]

Triangle Construction

How you construct a triangle depends on what <u>information you're given</u> about the triangle...

Three sides — Use a *Ruler and Compasses* (E)

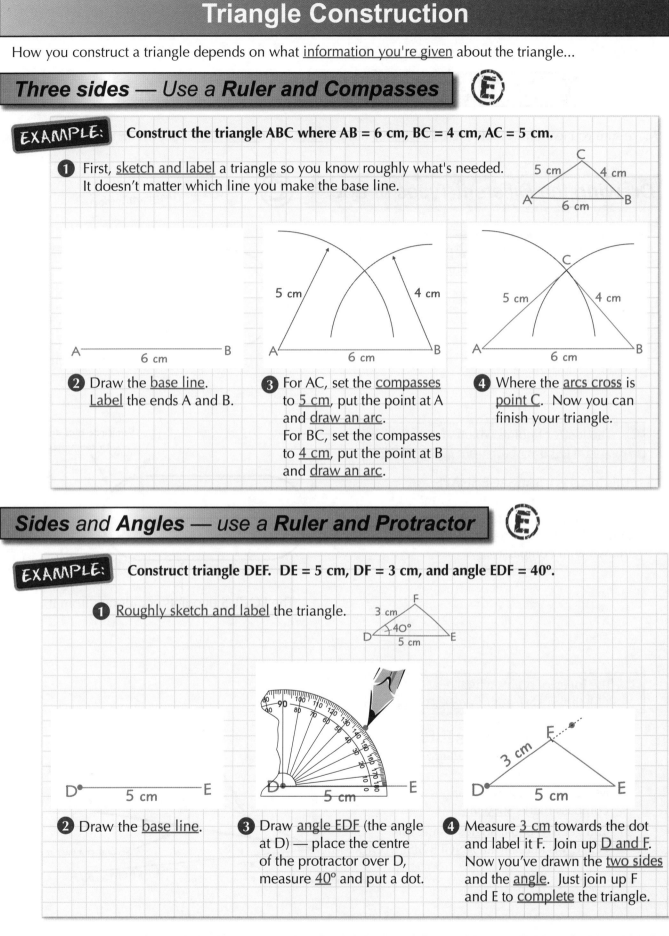

EXAMPLE: Construct the triangle ABC where AB = 6 cm, BC = 4 cm, AC = 5 cm.

❶ First, <u>sketch and label</u> a triangle so you know roughly what's needed. It doesn't matter which line you make the base line.

❷ Draw the <u>base line</u>. <u>Label</u> the ends A and B.

❸ For AC, set the <u>compasses</u> to <u>5 cm</u>, put the point at A and <u>draw an arc</u>. For BC, set the compasses to <u>4 cm</u>, put the point at B and <u>draw an arc</u>.

❹ Where the <u>arcs cross</u> is <u>point C</u>. Now you can finish your triangle.

Sides and Angles — use a *Ruler and Protractor* (E)

EXAMPLE: Construct triangle DEF. DE = 5 cm, DF = 3 cm, and angle EDF = 40°.

❶ <u>Roughly sketch and label</u> the triangle.

❷ Draw the <u>base line</u>.

❸ Draw <u>angle EDF</u> (the angle at D) — place the centre of the protractor over D, measure <u>40°</u> and put a dot.

❹ Measure <u>3 cm</u> towards the dot and label it F. Join up <u>D and F</u>. Now you've drawn the <u>two sides</u> and the <u>angle</u>. Just join up F and E to <u>complete</u> the triangle.

Don't forget your compasses for the exam

Constructing a triangle isn't difficult — learn the methods above, and remember to take your ruler, protractor and compasses with you into the exam. You won't get far without those.

Loci and Constructions

A <u>LOCUS</u> (more maths jargon) is simply:

> **A LINE or REGION that shows <u>all the points which fit in with a given rule</u>.**

Make sure you learn how to do these <u>PROPERLY</u> using a <u>ruler</u> and <u>compasses</u>, as shown on the next few pages.

Drawing the Different Types of Loci

Loci is just the plural of locus.

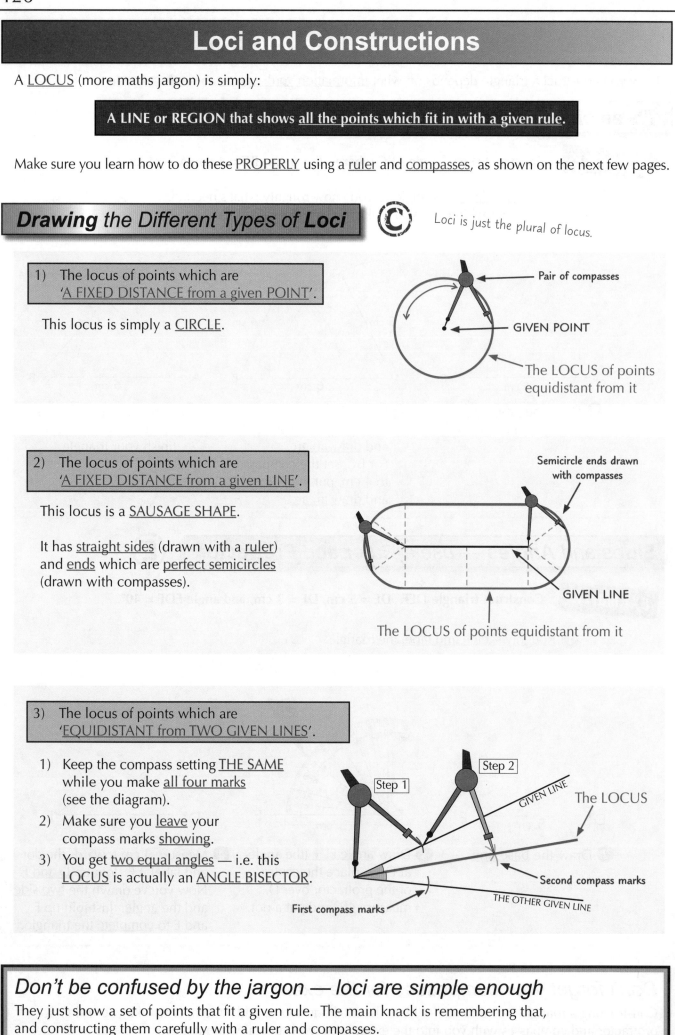

1)   The locus of points which are
     '<u>A FIXED DISTANCE from a given POINT</u>'.

This locus is simply a <u>CIRCLE</u>.

Pair of compasses

GIVEN POINT

The LOCUS of points
equidistant from it

2)   The locus of points which are
     '<u>A FIXED DISTANCE from a given LINE</u>'.

This locus is a <u>SAUSAGE SHAPE</u>.

It has <u>straight sides</u> (drawn with a <u>ruler</u>)
and <u>ends</u> which are <u>perfect semicircles</u>
(drawn with compasses).

Semicircle ends drawn
with compasses

GIVEN LINE

The LOCUS of points equidistant from it

3)   The locus of points which are
     '<u>EQUIDISTANT from TWO GIVEN LINES</u>'.

1)   Keep the compass setting <u>THE SAME</u>
     while you make <u>all four marks</u>
     (see the diagram).
2)   Make sure you <u>leave</u> your
     compass marks <u>showing</u>.
3)   You get <u>two equal angles</u> — i.e. this
     <u>LOCUS</u> is actually an <u>ANGLE BISECTOR</u>.

Step 1

Step 2

GIVEN LINE

The LOCUS

Second compass marks

THE OTHER GIVEN LINE

First compass marks

Don't be confused by the jargon — loci are simple enough

They just show a set of points that fit a given rule. The main knack is remembering that,
and constructing them carefully with a ruler and compasses.

Loci and Constructions

One more locus to learn, then it's on to constructions.

4) The locus of points which are
'EQUIDISTANT from TWO GIVEN POINTS'.

This LOCUS is all points which are the same distance from A as they are from B.

This time the locus is actually the PERPENDICULAR BISECTOR of the line joining the two points.

The perpendicular bisector of line segment AB is a line at right angles to AB, passing through the midpoint of AB. This is the method to use if you're asked to draw it.

(In the diagram below, A and B are the two given points.)

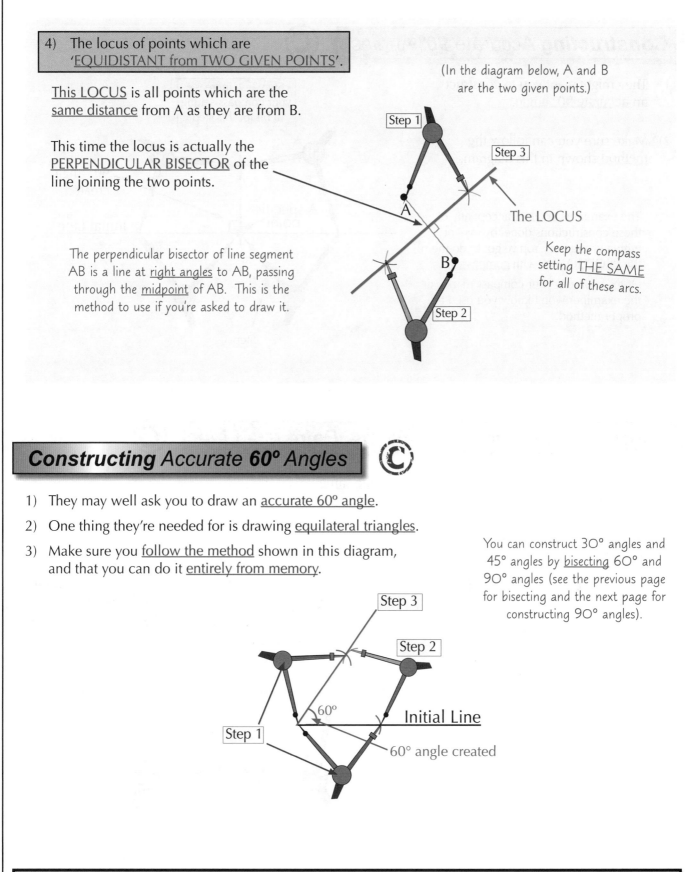

Step 1

Step 3

The LOCUS

Keep the compass setting THE SAME for all of these arcs.

Step 2

Constructing Accurate 60° Angles ©

1) They may well ask you to draw an accurate 60° angle.

2) One thing they're needed for is drawing equilateral triangles.

3) Make sure you follow the method shown in this diagram, and that you can do it entirely from memory.

You can construct 30° angles and 45° angles by bisecting 60° and 90° angles (see the previous page for bisecting and the next page for constructing 90° angles).

Step 3

Step 2

60°

Initial Line

Step 1

60° angle created

Constructions are basically set methods for maths drawings

Nothing too 'mathsy' about this page. That doesn't mean you can skim over it though. Make sure you've got these straight before you go on to the last two constructions on the next page.

Loci and Constructions

There are two more constructions you need to know...

Constructing Accurate 90° Angles

1) They might want you to construct an <u>accurate 90° angle</u>.

2) Make sure you can <u>follow the method</u> shown in this diagram.

> The examiners <u>WON'T</u> accept any of these constructions done 'by eye' or with a protractor. You've got to do them the <u>PROPER WAY</u>, with <u>compasses</u>.
>
> <u>DON'T</u> rub out your compass marks, or the examiner won't know you used the proper method.

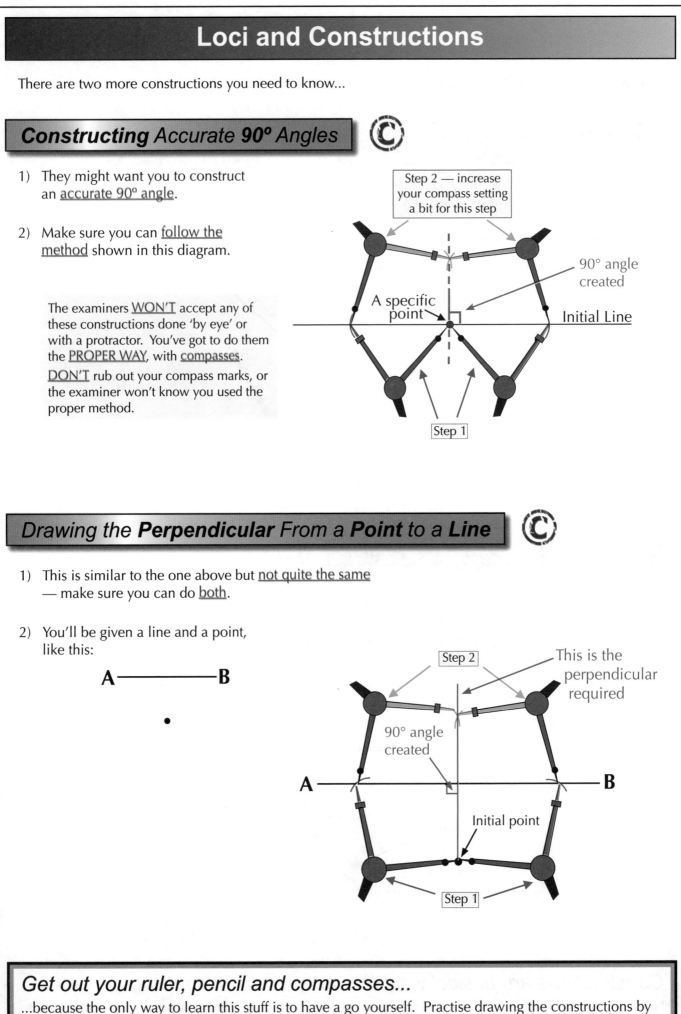

Step 2 — increase your compass setting a bit for this step

90° angle created

A specific point

Initial Line

Step 1

Drawing the Perpendicular From a Point to a Line

1) This is similar to the one above but <u>not quite the same</u> — make sure you can do <u>both</u>.

2) You'll be given a line and a point, like this:

A————B

•

Step 2

This is the perpendicular required

90° angle created

A ———————— B

Initial point

Step 1

Get out your ruler, pencil and compasses...

...because the only way to learn this stuff is to have a go yourself. Practise drawing the constructions by following the instructions first, and then try to do them from memory.

Loci and Constructions — Worked Example

Now you know what <u>loci</u> are, and how to do all the <u>constructions</u> you need, it's time to put them all together.

Finding a **Locus** That Satisfies **More Than One Rule** ©

In the exam, you might be given <u>two conditions</u>, and asked to find the region that satisfies <u>both</u> of them.

There's more on scale drawings on pages 148-149.

EXAMPLE:

Mary is deciding where to plant a tree in her garden.
She makes a plan of her garden with a scale of <u>1 cm = 2 m</u>.
1) Her house runs along <u>side AD</u>. The tree cannot be planted <u>within 4 m</u> of the house.
2) Mary wants the tree to be <u>closer to corner D than to corner B</u>.
Complete the plan to show where Mary can plant the tree.

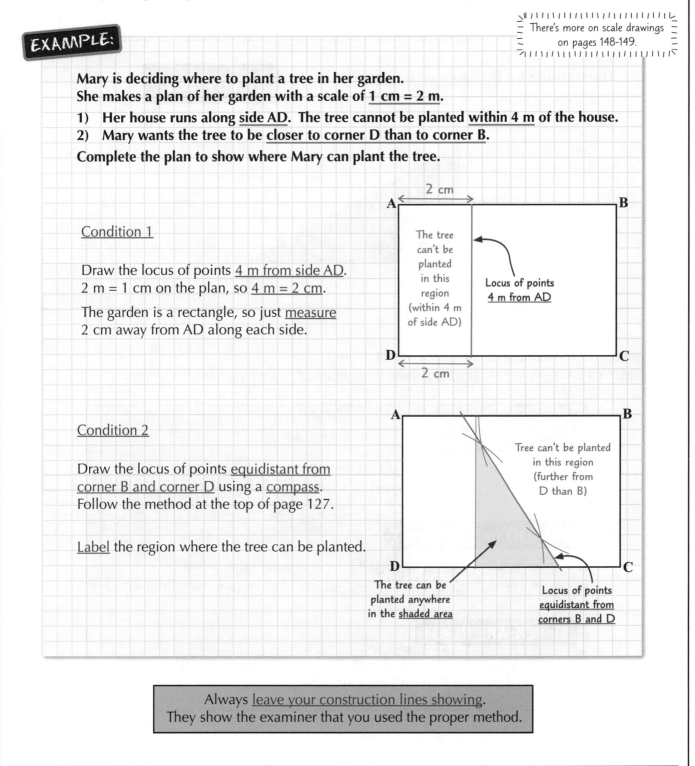

Condition 1

Draw the locus of points <u>4 m from side AD</u>.
2 m = 1 cm on the plan, so <u>4 m = 2 cm</u>.

The garden is a rectangle, so just <u>measure</u> 2 cm away from AD along each side.

2 cm

A — 2 cm — B

The tree can't be planted in this region (within 4 m of side AD)

Locus of points 4 m from AD

D — 2 cm — C

Condition 2

Draw the locus of points <u>equidistant from corner B and corner D</u> using a <u>compass</u>.
Follow the method at the top of page 127.

<u>Label</u> the region where the tree can be planted.

A B

Tree can't be planted in this region (further from D than B)

The tree can be planted anywhere in the <u>shaded area</u>

Locus of points <u>equidistant from corners B and D</u>

D C

Always <u>leave your construction lines showing</u>.
They show the examiner that you used the proper method.

If there are several rules, draw each locus then find the bit you want

Don't panic if you get a wordy loci question — just deal with one condition at a time and work out which bit you need to shade at the end. Make sure you draw your loci accurately — it's really important. Use a ruler and a pair of compasses, otherwise you won't pick up all the marks.

Pythagoras' Theorem

If you've a length to find on a right-angled triangle, then Pythagoras' theorem is what you need.

Pythagoras' Theorem *is Used on* Right-Angled Triangles

Pythagoras' theorem only works for <u>RIGHT-ANGLED TRIANGLES</u>.
It uses <u>two sides</u> to find the <u>third side</u>.

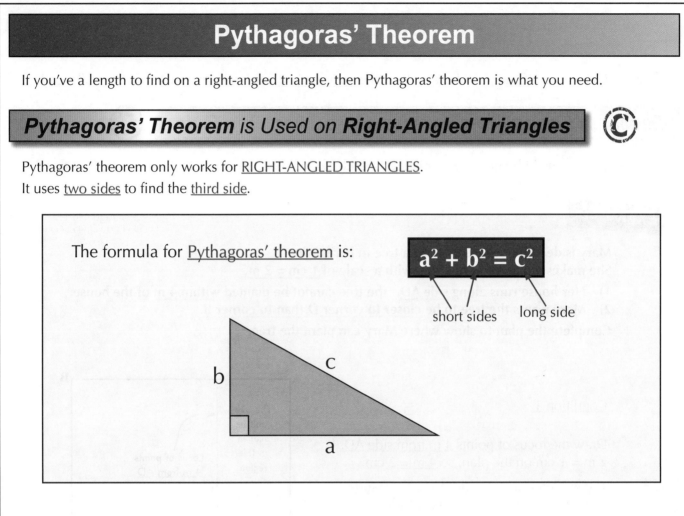

The formula for <u>Pythagoras' theorem</u> is:

$$a^2 + b^2 = c^2$$

short sides   long side

The trouble is, the formula can be quite difficult to use.
<u>Instead</u>, it's a lot better to just <u>remember</u> these <u>three simple steps</u>, which work every time:

1) SQUARE THEM

<u>SQUARE THE TWO NUMBERS</u> that you are given,
(use the x^2 button if you've got your calculator).

2) ADD or SUBTRACT

To find the <u>longest side</u>, <u>ADD</u> the two squared numbers.
To find <u>a shorter side</u>, <u>SUBTRACT</u> the smaller one from the larger.

3) SQUARE ROOT

Once you've got your answer, take the <u>SQUARE ROOT</u>
(use the $\sqrt{\ }$ button on your calculator).

There are some examples using this method on the next page.

Use Pythagoras to find lengths in right-angled triangles

This is probably one of the most famous of all maths formulas. It <u>will</u> be in your exam at some point.
So you really need to learn it and make sure you've practised plenty of questions.

Pythagoras' Theorem

Now that you know the method, it's time for some examples of using Pythagoras' theorem.

EXAMPLE:

Find the length of side PQ in this triangle.

1) <u>Square</u> them: $5^2 = 25$, $12^2 = 144$
2) You want to find the <u>longest side</u>, so <u>ADD</u>: $25 + 144 = 169$
3) <u>Square root</u>: $\sqrt{169} = 13$ cm

Always check the answer's <u>sensible</u> — 13 cm is <u>longer</u> than the other two sides, but not too much longer, so it seems OK.

EXAMPLE:

Find the length of SU to 1 decimal place.

1) <u>Square</u> them: $3^2 = 9$, $6^2 = 36$
2) You want to find a <u>shorter side</u>, so <u>SUBTRACT</u>: $36 - 9 = 27$
3) <u>Square root</u>: $\sqrt{27} = 5.196...$
   $= 5.2$ m (to 1 d.p.)

Check the answer is <u>sensible</u> — yes, it's a bit <u>shorter</u> than the longest side.

EXAMPLE:

Find the length of the line segment shown.

Work out <u>how far across and up</u> it is from <u>A to B</u> and then treat this exactly like a <u>normal triangle</u>.

1) <u>Square</u> them: $3^2 = 9$, $4^2 = 16$
2) You want to find the <u>longest side</u>, so <u>ADD</u>: $9 + 16 = 25$
3) <u>Square root</u>: $\sqrt{25} = 5$

So the length of the line segment = 5 units

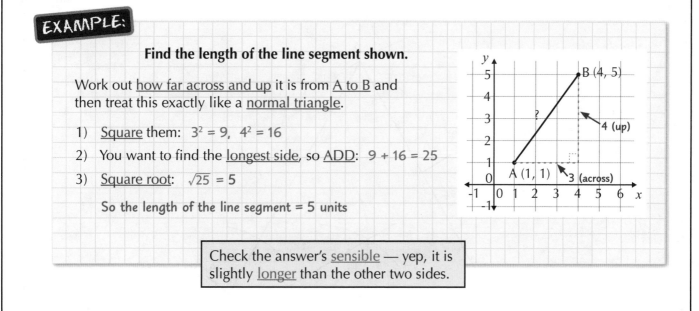

Check the answer's <u>sensible</u> — yep, it is slightly <u>longer</u> than the other two sides.

Don't forget to square the sides

These three examples might look quite tricky at first — but they all use the same method.
It's just a case of adding or subtracting the squares of the sides. Nothing more than that.

Warm-up and Worked Exam Questions

You need to work through these one by one and make sure you really know what you're doing with your ruler and compasses.

Warm-up Questions

1) Using a ruler and compasses construct an equilateral triangle with length of side 4 cm.

2) a) Construct a triangle ABC with side AB = 3 cm, side BC = 4 cm and angle ABC = 90°.
   b) Measure the length of side AC.

3) Draw the locus of the point P that moves around a 3 cm vertical line at a constant distance of 1 cm.

4) Use a ruler and compasses to construct a square with sides 5.5 cm long.

5) Draw a line and a point and construct the perpendicular from the point to the line.

6) In a right-angled triangle, the two shorter sides are 10 cm and 8.4 cm. Find the length of the longest side, correct to 3 significant figures.

Worked Exam Question

A lovely worked exam question here before you get into doing the questions yourself. Make the most of it.

1 Side *BC* of the equilateral triangle *ABC* has been accurately drawn below.

   a) Use a ruler and compasses to complete the accurate drawing of triangle *ABC*. **(E)**

   [1 mark]

   b) Construct the bisector of angle *ACB* of the triangle. **(C)**
      You must show all your construction lines.

   [2 marks]

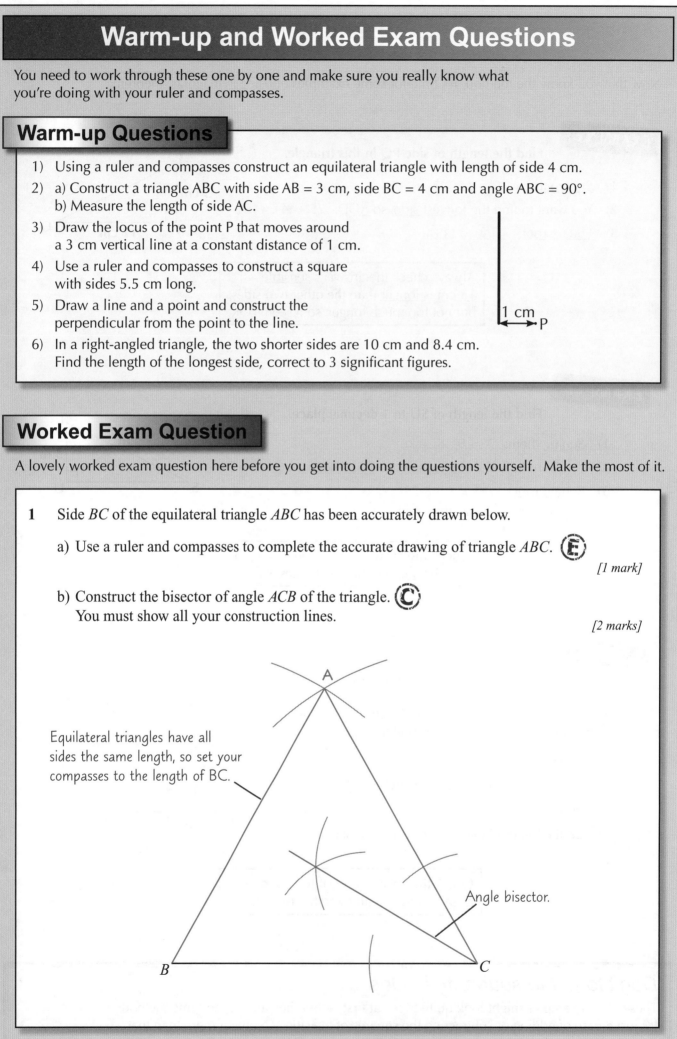

Equilateral triangles have all sides the same length, so set your compasses to the length of BC.

Angle bisector.

Exam Questions

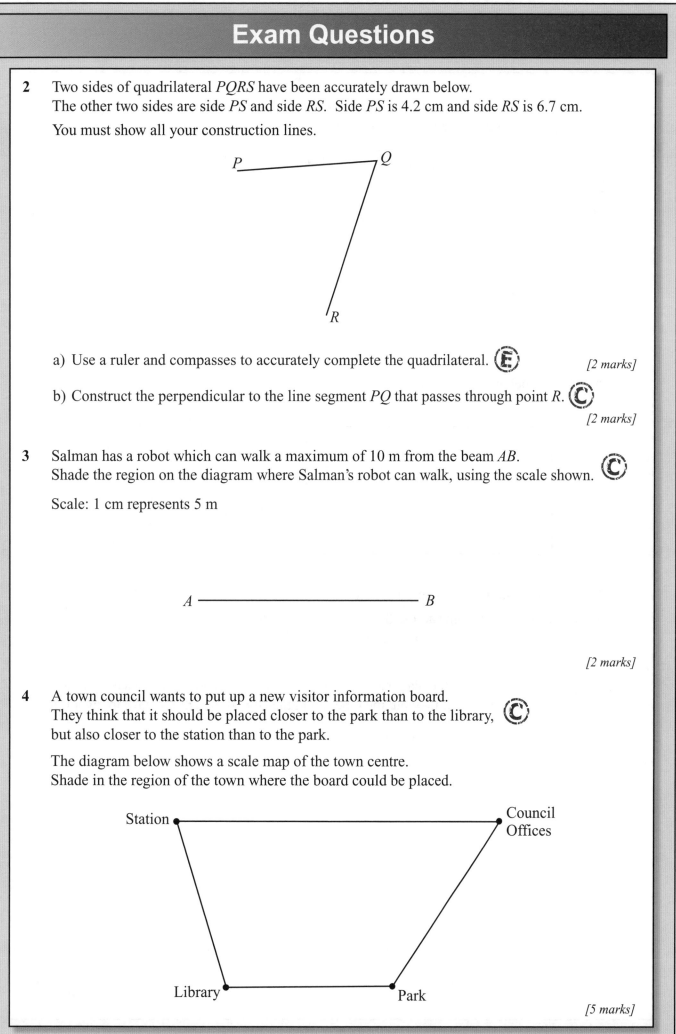

2 Two sides of quadrilateral *PQRS* have been accurately drawn below.
The other two sides are side *PS* and side *RS*. Side *PS* is 4.2 cm and side *RS* is 6.7 cm.

You must show all your construction lines.

a) Use a ruler and compasses to accurately complete the quadrilateral. **E** *[2 marks]*

b) Construct the perpendicular to the line segment *PQ* that passes through point *R*. **C**
[2 marks]

3 Salman has a robot which can walk a maximum of 10 m from the beam *AB*.
Shade the region on the diagram where Salman's robot can walk, using the scale shown. **C**

Scale: 1 cm represents 5 m

A —————————————— *B*

[2 marks]

4 A town council wants to put up a new visitor information board.
They think that it should be placed closer to the park than to the library, **C**
but also closer to the station than to the park.

The diagram below shows a scale map of the town centre.
Shade in the region of the town where the board could be placed.

[5 marks]

Exam Questions

5 A 3.5 m long ladder is resting against a vertical wall and on the horizontal ground. For safety the bottom of the ladder must be at least 2.1 m away from the wall. Ⓒ

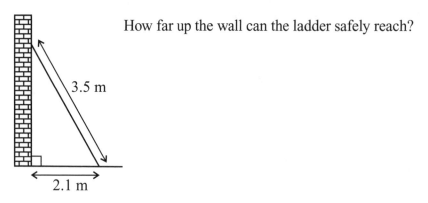

How far up the wall can the ladder safely reach?

.... m
[3 marks]

*6 Alison and Rob want to put a water pipe across their rectangular field. Ⓒ The diagram below shows their field.

The * next to the question number means that you're tested on your quality of written communication.

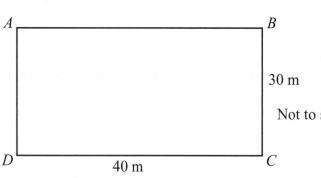

30 m

Not to scale

TIP: Draw a quick sketch using the information that you're given in the question — then you can easily see where to put the numbers in the Pythagoras formula.

The pipe must run from a tap at point *A* to the shed at point *C*.

Alison wants to put the pipe across the diagonal of the field. Rob wants to put the pipe round the edge of the field.

The pipe costs £8.35 per metre.

If they put the pipe across the diagonal they will have to dig a trench and replace the grass, which will cost £202.50. If they put the pipe round the edge they will not have to dig a trench.

Which option is cheaper? Explain your answer.

....

....
[6 marks]

Revision Questions for Section Five

There are lots of opportunities to show off your artistic skills here (as long as you use them to answer the questions).
- Try these questions and <u>tick off each one</u> when you <u>get it right</u>.
- When you've done <u>all the questions</u> for a topic and are <u>completely happy</u> with it, tick off the topic.

Lines and Angles (p105-110) ☑

1) What is the name for an angle larger than 90° but smaller than 180°?

2) What do angles in a quadrilateral add up to?

3) Find the missing angles in the diagrams below.

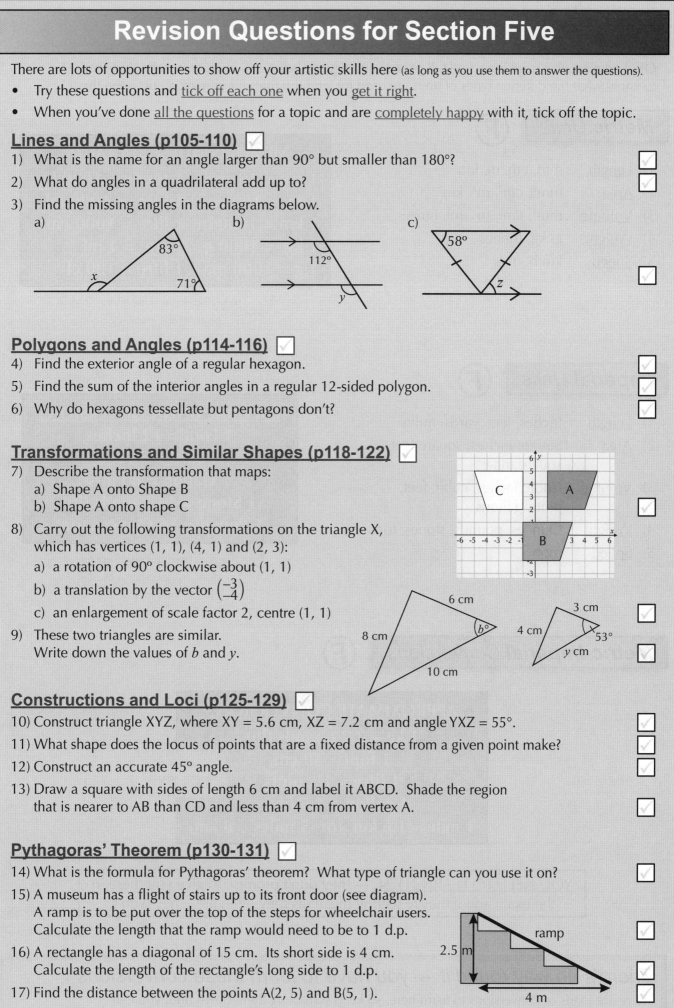

Polygons and Angles (p114-116) ☑

4) Find the exterior angle of a regular hexagon.

5) Find the sum of the interior angles in a regular 12-sided polygon.

6) Why do hexagons tessellate but pentagons don't?

Transformations and Similar Shapes (p118-122) ☑

7) Describe the transformation that maps:
a) Shape A onto Shape B
b) Shape A onto shape C

8) Carry out the following transformations on the triangle X, which has vertices (1, 1), (4, 1) and (2, 3):
a) a rotation of 90° clockwise about (1, 1)
b) a translation by the vector $\binom{-3}{-4}$
c) an enlargement of scale factor 2, centre (1, 1)

9) These two triangles are similar. Write down the values of b and y.

Constructions and Loci (p125-129) ☑

10) Construct triangle XYZ, where XY = 5.6 cm, XZ = 7.2 cm and angle YXZ = 55°.

11) What shape does the locus of points that are a fixed distance from a given point make?

12) Construct an accurate 45° angle.

13) Draw a square with sides of length 6 cm and label it ABCD. Shade the region that is nearer to AB than CD and less than 4 cm from vertex A.

Pythagoras' Theorem (p130-131) ☑

14) What is the formula for Pythagoras' theorem? What type of triangle can you use it on?

15) A museum has a flight of stairs up to its front door (see diagram). A ramp is to be put over the top of the steps for wheelchair users. Calculate the length that the ramp would need to be to 1 d.p.

16) A rectangle has a diagonal of 15 cm. Its short side is 4 cm. Calculate the length of the rectangle's long side to 1 d.p.

17) Find the distance between the points A(2, 5) and B(5, 1).

Metric and Imperial Units

OK, I admit it, this page is packed full of facts and figures, but it's all important stuff. You need to know what the different types of units are used for, and the rules for converting between them.

Metric Units

1) <u>Length</u>   mm, cm, m, km
2) <u>Area</u>   mm², cm², m², km²
3) <u>Volume</u>   mm³, cm³, m³, ml, litres
4) <u>Weight</u>   g, kg, tonnes
5) <u>Speed</u>   km/h, m/s

MEMORISE THESE KEY FACTS:

1 cm = 10 mm	1 tonne = 1000 kg
1 m = 100 cm	1 litre = 1000 ml
1 km = 1000 m	1 litre = 1000 cm³
1 kg = 1000 g	1 cm³ = 1 ml
1 g = 1000 mg	1 litre = 100 cl

You need to know these off by heart for the exam.

Imperial Units

1) <u>Length</u>   Inches, feet, yards, miles
2) <u>Area</u>   Square inches, square feet, square miles
3) <u>Volume</u>   Cubic inches, cubic feet, pints, gallons
4) <u>Weight</u>   Ounces, pounds, stones, tons
5) <u>Speed</u>   mph

IMPERIAL UNIT CONVERSIONS

1 Foot = 12 Inches
1 Yard = 3 Feet
1 Gallon = 8 Pints
1 Stone = 14 Pounds (lb)
1 Pound = 16 Ounces (oz)

You don't need to know these for the exam, but you should be able to <u>use</u> the conversions.

Metric-Imperial Conversions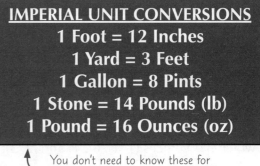

APPROXIMATE CONVERSIONS

1 kg ≈ 2.2 pounds (lb)
1 foot ≈ 30 cm
1 litre ≈ 1¾ pints
1 gallon ≈ 4.5 litres
1 mile ≈ 1.6 km (or 5 miles ≈ 8 km)

≈ means roughly equal to

<u>YOU NEED TO LEARN THESE</u> — they don't promise to give you these in the exam and if they're feeling mean (as they often are), they won't.

There's no way round it — you have to learn these conversions

There are loads of conversions to learn here, so keep scribbling them down until you can remember every single one. You won't get far with a conversion question without them.

Converting Units

Conversion factors are a really good way of dealing with all sorts of questions — and the method is really easy. Learn the 3 steps and you'll then be ready to tackle any conversion question.

3 Step Method

1) Find the **CONVERSION FACTOR** (always easy).

2) **Multiply AND divide by it.**

3) Choose the **COMMON-SENSE ANSWER**.

EXAMPLES:

1. A boat is docked in the harbour at Grange-over-Sands. It is 18.6 m in length. How long is this in cm?

1) Find the conversion factor

$1 \text{ m} = 100 \text{ cm}$

Conversion factor = 100

2) Multiply and divide by it

$18.6 \times 100 = 1860$ — makes sense
$18.6 \div 100 = 0.186$ — ridiculous

3) Choose the sensible answer

$18.6 \text{ m} = 1860 \text{ cm}$

2. The villages of Birdingbury and Fenny Compton are 12 miles from each other. How far is this in km?

1) Find the conversion factor

$1 \text{ mile} \approx 1.6 \text{ km}$
⟶ Conversion factor = 1.6

2) Multiply and divide by it

$12 \times 1.6 = 19.2$
$12 \div 1.6 = 7.5$

3) Choose the sensible answer —
1 mile is about 1.6 km so there should be more km than miles

$12 \text{ miles} \approx 19.2 \text{ km}$

Just three steps to simple conversions

Read through these examples and make sure you fully understand how to use conversion factors. There are another two examples on the next page just so you can be extra sure.

Converting Units

Finding the conversion factor is the key to these questions.
You can even use this method to convert speeds.

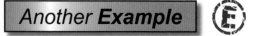
Another Example Ⓔ

EXAMPLE:

Lisa buys a carton of her favourite gooseberry and lime juice.
The carton has a volume of 1500 cm³. What is its volume in pints?

1) First convert cm³ to litres — the numbers are simple here.

$$1000 \text{ cm}^3 = 1 \text{ litre} \longrightarrow 1500 \text{ cm}^3 = 1.5 \text{ litres}$$

2) Then convert litres to pints — use the conversion factor.

$$1 \text{ litre} \approx 1.75 \text{ pints} \longrightarrow \text{Conversion factor} = 1.75$$

3) Multiply and divide by it.

$$1.5 \times 1.75 = 2.625$$
$$1.5 \div 1.75 = 0.8571...$$

4) Choose the sensible answer — 1 litre is about 1.75 pints so there should be more pints than litres.

$$1.5 \text{ litres} \approx 2.625 \text{ pints}$$

Converting Speeds Ⓓ

Don't panic if you're asked to convert a speed from, say, miles per hour (mph) to km per hour (km/h)...
...if the time part of the units stays the same then it's really just a distance conversion in disguise.

EXAMPLE:

Sophie is driving at 60 km/h. The speed limit is 40 mph.
Is she breaking the speed limit?

1) First convert 60 km into miles:

$$1 \text{ mile} \approx 1.6 \text{ km} \longrightarrow \text{Conversion factor} = 1.6$$
$$60 \times 1.6 = 96 \text{ — too big}$$
$$60 \div 1.6 = 37.5 \text{ — makes sense,}$$
$$\text{so } 60 \text{ km} = 37.5 \text{ miles}$$

2) Add in the 'per hour' bit to get the speed:

$$60 \text{ km/h} = 37.5 \text{ mph}$$
So, Sophie isn't breaking the 40 mph speed limit.

Just remember to use your common sense

These questions are just straightforward multiplying and dividing. The only tricky bit is knowing which gives you the right answer. But use your common sense and you'll be fine.

More Conversions

This page is a bit <u>trickier</u>, but everything comes down to the same old <u>conversion factors</u>...

Converting *Areas* and *Volumes* — *Tricky* (D)

Be really <u>careful</u> — 1 m = 100 cm <u>DOES NOT</u> mean 1 m² = 100 cm² or 1 m³ = 100 cm³.
You won't slip up if you <u>LEARN THESE RULES</u>:

> **AREA**
> Units come with a <u>2</u>, e.g. mm², cm², m²
> — <u>use the conversion factor 2 times</u>.

> **VOLUME**
> Units come with a <u>3</u>, e.g. mm³, cm³, m³
> — <u>use the conversion factor 3 times</u>.

> <u>Multiply AND divide</u> the correct number of times, then pick the <u>sensible</u> answer — use the <u>rule</u> linking the units to decide whether your answer should be <u>bigger or smaller</u> than what you started with.

EXAMPLES:

1. **The area of the top of a table is 0.6 m². Find its area in cm².**

1) Find the <u>conversion factor</u>: 1 m = 100 cm ⟶ Conversion factor = 100

2) It's an area — multiply and divide <u>twice</u> by conversion factor:
   0.6 × 100 × 100 = 6000 — **makes sense**
   0.6 ÷ 100 ÷ 100 = 0.00006 — **too small**

3) Choose the <u>sensible</u> answer: 0.6 m² = **6000 cm²**
   1 m = 100 cm so expect more cm than m.

2. **A glass has a volume of 72 000 mm³. What is its volume in cm³?**

1) Find the <u>conversion factor</u>: 1 cm = 10 mm ⟶ Conversion factor = 10

2) It's a volume — multiply and divide <u>3 times</u> by conversion factor:
   72 000 × 10 × 10 × 10 = 72 000 000 — **too big**
   72 000 ÷ 10 ÷ 10 ÷ 10 = 72 — **makes sense**

3) Choose the <u>sensible</u> answer: 72 000 mm³ = **72 cm³**
   1 cm = 10 mm, so expect fewer cm than mm.

Remember — area comes with a 2 and volume comes with a 3

Learn the rules for converting between units for areas and volumes. Always check your answer to see if it is sensible or not. Common sense will get you a long way.

Warm-up and Worked Exam Questions

So, here's the first lot of questions for Section Six. By now you should know the drill — get your brain ticking with the warm-up questions, have a good read over the worked question, and then test yourself with the exam questions. It's perfect, foolproof practice for the exam.

Warm-up Questions

1) a) How many cm is 2 metres?
   b) How many mm is 6.5 cm?

2) a) How many kg is 250 g?
   b) How many litres is 1500 cm³?

3) A machine was found to weigh 0.16 tonnes. What is this in kg?

4) A rod is 46 inches long. What is this in feet and inches?

5) a) Roughly how many km is 200 miles?
   b) Roughly how many feet is 120 cm?

6) Convert these measurements:
   a) 23 m² to cm²
   b) 34 500 cm² to m².

Worked Exam Question

Understanding the solution below will really help you with the exam questions that follow.

1 1 litre = $1\frac{3}{4}$ pints. **(D)**

Margaret needs 14 pints of water to fill up her fish tank.
She uses three different sized containers to fill up the tank.
One container holds 520 ml, one holds 540 ml and the other holds 720 ml.

To fill up the tank, Margaret uses the 720 ml container three times,
the 520 ml container five times and then the 540 ml container to finish filling it up.

How many times must she fill up the 540 ml container to finish filling the fish tank?

1 litre = 1.75 pints   Start by finding the conversion
⟶ Conversion factor = 1.75   factor for litres and pints

14 × 1.75 = 24.5 litres
14 ÷ 1.75 = 8 litres
So 14 pints = 8 litres = 8000 ml of water needed

There should be fewer
litres than pints   (3 × 720 ml) + (5 × 520 ml) = 4760 ml already in tank

8000 ml − 4760 ml = 3240 ml left to fill
3240 ÷ 540 = 6 times

6
....
[5 marks]

Exam Questions

2 Emma is having a party with some friends. (E)

She has 2.5 litres of orange juice.
How many 250 ml cups can be filled from it?

.... cups
[3 marks]

3 Nicole wants to post some books to a friend in another country.
Each book weighs 1.5 lb and each package can hold a maximum weight of 2500 g. (E)

How many books can she send in one package?   Hint — you need to do a conversion between kilograms and pounds

....
[4 marks]

4 The formula to convert kilometres (k) to miles (m) is $m = \dfrac{5k}{8}$.

a) How many miles is 40 km? (E)

.... miles
[2 marks]

b) A giraffe can run at 60 km/h.
Show that a giraffe could be outrun by a camel running at 40 mph. (D)

[2 marks]

5 The playing surface of a snooker table has an area of 39 200 cm². (D)

Convert the area of the snooker table into m².

.... m²
[2 marks]

Reading Scales

You can pick up some <u>easy marks</u> if you get a question asking you to read a scale.
The <u>same rules</u> apply to scales measuring <u>lengths</u>, <u>weights</u>, <u>volumes</u>, <u>speeds</u> and <u>temperatures</u>,
so learn them and those marks are yours.

How to Read a Scale Ⓖ

All scales consist of a <u>line divided into intervals</u> like this:

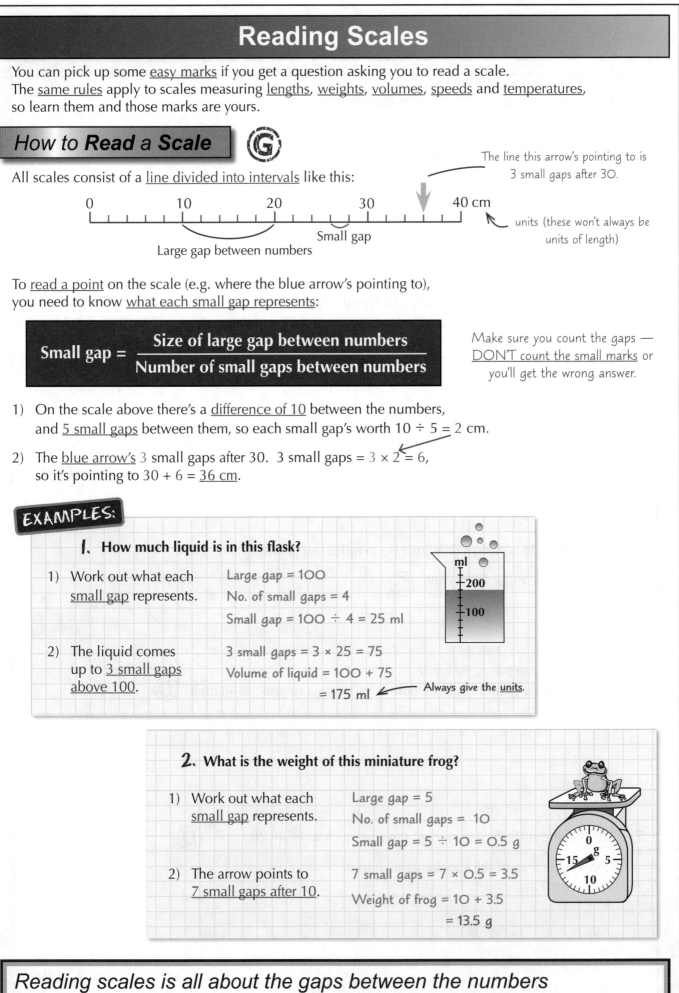

The line this arrow's pointing to is
3 small gaps after 30.

0   10   20   30   40 cm

Large gap between numbers

Small gap

units (these won't always be
units of length)

To <u>read a point</u> on the scale (e.g. where the blue arrow's pointing to),
you need to know <u>what each small gap represents</u>:

$$\text{Small gap} = \frac{\text{Size of large gap between numbers}}{\text{Number of small gaps between numbers}}$$

Make sure you count the gaps —
<u>DON'T count the small marks</u> or
you'll get the wrong answer.

1) On the scale above there's a <u>difference of 10</u> between the numbers,
    and <u>5 small gaps</u> between them, so each small gap's worth 10 ÷ 5 = 2 cm.

2) The <u>blue arrow's</u> 3 small gaps after 30. 3 small gaps = 3 × 2 = 6,
    so it's pointing to 30 + 6 = <u>36 cm</u>.

EXAMPLES:

1. How much liquid is in this flask?

1) Work out what each
    <u>small gap</u> represents.

    Large gap = 100
    No. of small gaps = 4
    Small gap = 100 ÷ 4 = 25 ml

2) The liquid comes
    up to <u>3 small gaps
    above 100</u>.

    3 small gaps = 3 × 25 = 75
    Volume of liquid = 100 + 75
    = 175 ml ← Always give the <u>units</u>.

ml
200
100

2. What is the weight of this miniature frog?

1) Work out what each
    <u>small gap</u> represents.

    Large gap = 5
    No. of small gaps = 10
    Small gap = 5 ÷ 10 = 0.5 g

2) The arrow points to
    <u>7 small gaps after 10</u>.

    7 small gaps = 7 × 0.5 = 3.5
    Weight of frog = 10 + 3.5
    = 13.5 g

Reading scales is all about the gaps between the numbers

Once you've figured out what each gap on the scale means, it makes it much easier to read off the
measurement. Remember to divide the size of the large gap by the number of smaller gaps.

Rounding and Estimating Measurements

You need to be able to <u>measure things accurately</u>, and <u>make estimates</u> of lengths and heights.
There's nothing too tricky here, so have a good read through and you'll be sorted.

Measuring a Line with a Ruler Ⓖ

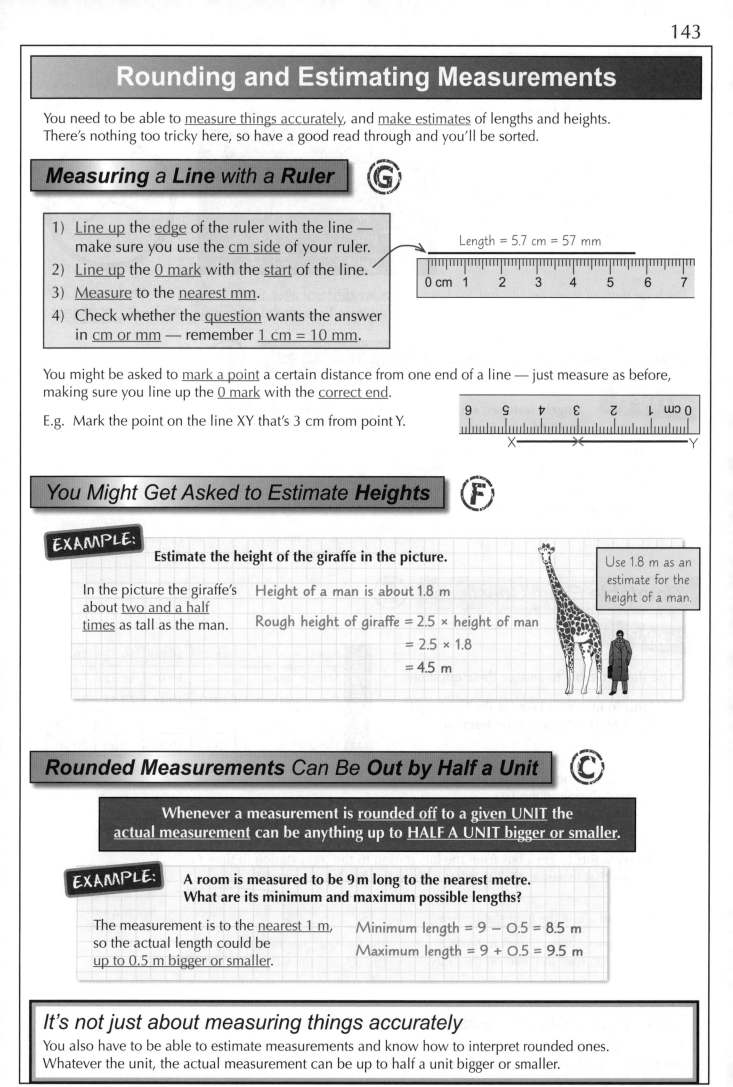

1) <u>Line up</u> the <u>edge</u> of the ruler with the line —
make sure you use the <u>cm side</u> of your ruler.
2) <u>Line up</u> the <u>0 mark</u> with the <u>start</u> of the line.
3) <u>Measure</u> to the <u>nearest mm</u>.
4) Check whether the <u>question</u> wants the answer
in <u>cm or mm</u> — remember <u>1 cm = 10 mm</u>.

Length = 5.7 cm = 57 mm

0 cm 1 2 3 4 5 6 7

You might be asked to <u>mark a point</u> a certain distance from one end of a line — just measure as before,
making sure you line up the <u>0 mark</u> with the <u>correct end</u>.

E.g. Mark the point on the line XY that's 3 cm from point Y.

0 cm 1 2 3 4 5 6

X———✕———Y

You Might Get Asked to Estimate Heights Ⓕ

EXAMPLE:

Estimate the height of the giraffe in the picture.

In the picture the giraffe's about <u>two and a half times</u> as tall as the man.

Height of a man is about 1.8 m

Rough height of giraffe = 2.5 × height of man
= 2.5 × 1.8
= 4.5 m

Use 1.8 m as an estimate for the height of a man.

Rounded Measurements Can Be Out by Half a Unit Ⓒ

**Whenever a measurement is <u>rounded off</u> to a <u>given UNIT</u> the
<u>actual measurement</u> can be anything up to <u>HALF A UNIT bigger or smaller</u>.**

EXAMPLE: **A room is measured to be 9 m long to the nearest metre.
What are its minimum and maximum possible lengths?**

The measurement is to the <u>nearest 1 m</u>,
so the actual length could be
<u>up to 0.5 m bigger or smaller</u>.

Minimum length = 9 − 0.5 = 8.5 m
Maximum length = 9 + 0.5 = 9.5 m

It's not just about measuring things accurately

You also have to be able to estimate measurements and know how to interpret rounded ones.
Whatever the unit, the actual measurement can be up to half a unit bigger or smaller.

Reading Timetables

I'm sure you're a dab hand at reading <u>clocks</u>, but here's a quick reminder...

<u>am</u> means <u>morning</u>.
<u>pm</u> means <u>afternoon or evening</u>.

<u>12 am</u> (<u>00:00</u>) means <u>midnight</u>.
<u>12 pm</u> (<u>12:00</u>) means <u>noon</u>.

12-hour clock	24-hour clock
12.00 am	00:00
1.12 am	01:12
12.15 pm	12:15
1.47 pm	13:47
11.32 pm	23:32

The hour parts of times on 12- and 24- hour clocks are <u>different after 1 pm</u>:
<u>add 12 hours</u> to go from <u>12-hour to 24-hour</u>, and subtract 12 to go the other way.

$$3.24 \text{ pm} \overset{+\,12\,h}{\underset{-\,12\,h}{\rightleftarrows}} 15{:}24$$

Break *Time* Calculations into *Simple Stages* (F)

EXAMPLE:

Angela watched a film that started at 7.20 pm and finished at 10.05 pm. How long was the film in minutes?

1) Split the time between 7.20 pm and 10.05 pm into <u>simple stages</u>.

7.20 pm $\xrightarrow{+\,2\text{ hours}}$ 9.20 pm $\xrightarrow{+\,40\text{ minutes}}$ 10.00 pm $\xrightarrow{+\,5\text{ minutes}}$ 10.05 pm

2) <u>Convert</u> the hours to minutes.

2 hours = 2 × 60 = 120 minutes

3) <u>Add</u> to get the total minutes.

120 + 40 + 5 = 165 minutes

Avoid calculators — the decimal answers they give are confusing, e.g. 2.5 hours = 2 hours 30 mins, NOT 2 hours 50 mins.

Timetable Exam Questions (E)

EXAMPLE:

Use the timetable to answer these questions:

a) How long does it take for the bus to get from <u>Market Street</u> to the hospital?

Bus Timetable				
Bus Station	18 45	19 00	19 15	19 30
Market Street	18 52	19 07	19 22	19 37
Long Lane Shops	19 01	19 16	19 31	19 46
Train Station	19 11	19 26	19 41	19 56
Hospital	19 23	19 38	19 53	20 08

Read times from the <u>same column</u> (I've used the 1st) — break the <u>time</u> into <u>stages</u>.

Market Street   Hospital
18:52 $\xrightarrow{+\,8\text{ mins}}$ 19:00 $\xrightarrow{+\,23\text{ mins}}$ 19:23

8 + 23 = 31 minutes

b) Harry wants to get a bus from the <u>bus station</u> to the <u>train station</u> in time for a train that leaves at <u>19:30</u>. What is the latest bus he can catch?

1) Read along the <u>train station</u> row.

2) Move up this column to the <u>bus station</u> row and read off the entry.

19 11   (19 26)   19 41   19 56

This is the latest time he could arrive before 19:30.

The bus that gets to the train station at 19:26 leaves the bus station at 19:00.

This page might look easy, but make sure you learn it all

It's easy to go wrong when you're using your calculator for time questions, so be careful.
Always try to split questions down into easier stages — that way you'll make fewer mistakes.

Warm-up and Worked Exam Questions

Time for some more warm-up questions to run over the basics of reading scales, measuring, estimating and reading timetables — all of which are likely to come up in the exams.

Warm-up Questions

1) Look at the diagrams below:

   a) What speed does the speedometer show?   b) What temperature does the thermometer show?

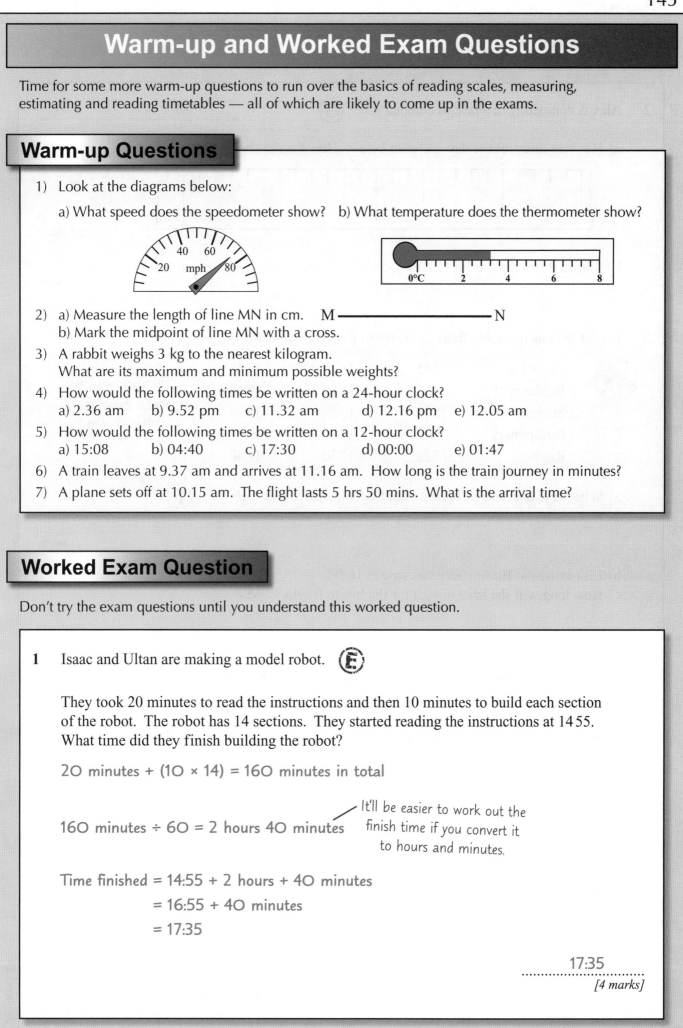

2) a) Measure the length of line MN in cm.   M————————————N
   b) Mark the midpoint of line MN with a cross.

3) A rabbit weighs 3 kg to the nearest kilogram.
   What are its maximum and minimum possible weights?

4) How would the following times be written on a 24-hour clock?
   a) 2.36 am   b) 9.52 pm   c) 11.32 am   d) 12.16 pm   e) 12.05 am

5) How would the following times be written on a 12-hour clock?
   a) 15:08   b) 04:40   c) 17:30   d) 00:00   e) 01:47

6) A train leaves at 9.37 am and arrives at 11.16 am. How long is the train journey in minutes?

7) A plane sets off at 10.15 am. The flight lasts 5 hrs 50 mins. What is the arrival time?

Worked Exam Question

Don't try the exam questions until you understand this worked question.

1 Isaac and Ultan are making a model robot. (E)

They took 20 minutes to read the instructions and then 10 minutes to build each section of the robot. The robot has 14 sections. They started reading the instructions at 1455. What time did they finish building the robot?

20 minutes + (10 × 14) = 160 minutes in total

160 minutes ÷ 60 = 2 hours 40 minutes   It'll be easier to work out the finish time if you convert it to hours and minutes.

Time finished = 14:55 + 2 hours + 40 minutes
= 16:55 + 40 minutes
= 17:35

17:35
....
[4 marks]

Exam Questions

2 Alex is competing in a javelin competition. Ⓖ

a) Her best throw of the day is shown below. How far did she throw?

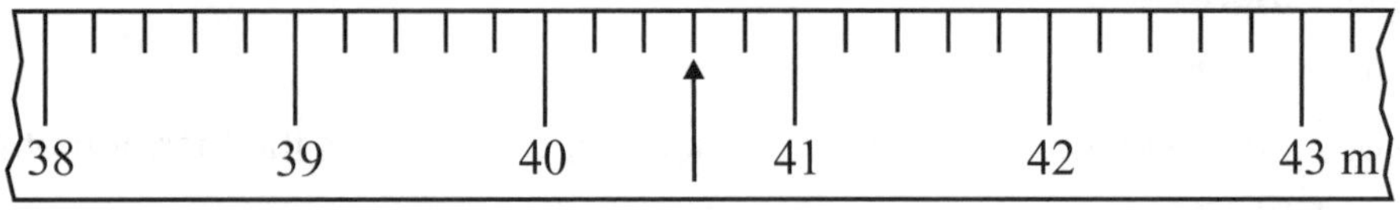

.... m
[1 mark]

b) Martin threw the javelin 42.4 m. Mark his throw on the tape measure above.

[1 mark]

3 Part of the bus timetable from Coventry to Rugby is shown below.

Coventry	1445	1615	1745
Bubbenhall	–	1640	1810
Stretton	1514	1654	1824
Birdingbury	–	1704	–
Rugby	1535	1730	1840

The dashes on the timetable mean the bus doesn't stop.

a) What time does the 1615 bus from Coventry leave Birdingbury? Ⓕ

....
[1 mark]

b) Lisa arrives at Birdingbury bus stop at 1658.
How long will she have to wait for the bus to Rugby? Ⓕ

.... minutes
[1 mark]

The 1615 bus from Coventry continues to Lutterworth after Rugby.
It arrives in Lutterworth at 1815.

c) Anne lives in Bubbenhall. If she catches this bus from her home, Ⓔ
how long will it take her to get to Lutterworth?

.... h .... mins
[2 marks]

4 Joseph is weighing himself. His scales give his weight to the nearest kilogram. Ⓒ

According to his scales, Joseph is 57 kg.
What are the minimum and maximum weights that he could be?

Minimum weight: .... kg

Maximum weight: .... kg
[2 marks]

Compass Directions and Bearings

Compass points and bearings both describe the <u>direction</u> of something.
You'll have seen a <u>compass</u> before — make sure you <u>know</u> all <u>8 directions</u>.
<u>Bearings</u> are trickier but really useful — they can
describe <u>any direction</u>, not just the 8 compass points.

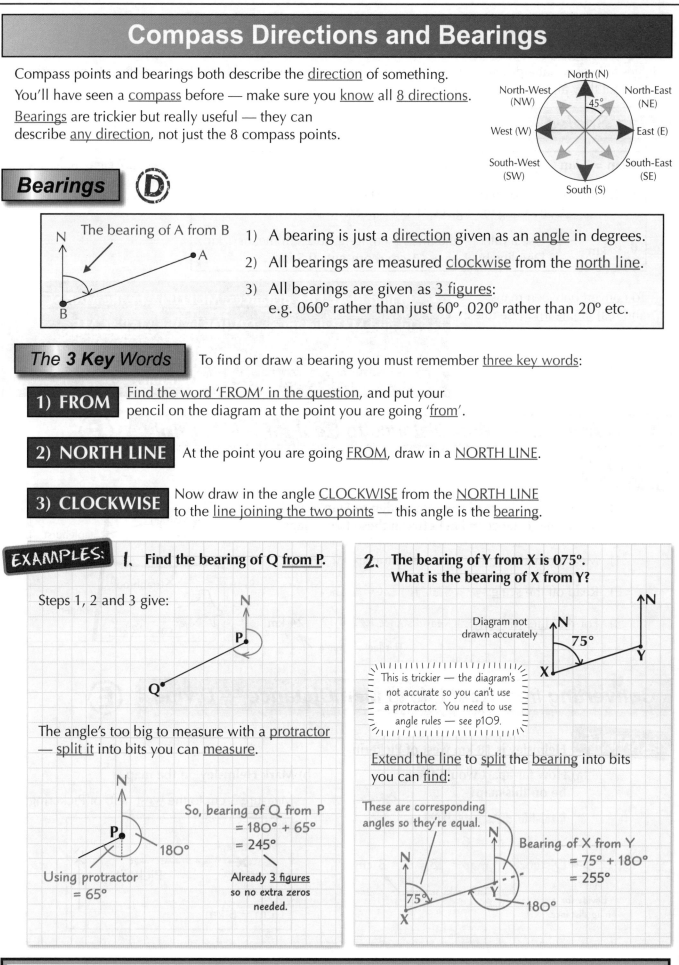

Bearings (D)

1) A bearing is just a <u>direction</u> given as an <u>angle</u> in degrees.

2) All bearings are measured <u>clockwise</u> from the <u>north line</u>.

3) All bearings are given as <u>3 figures</u>:
e.g. 060° rather than just 60°, 020° rather than 20° etc.

The 3 Key Words

To find or draw a bearing you must remember <u>three key words</u>:

1) FROM <u>Find the word 'FROM' in the question</u>, and put your
pencil on the diagram at the point you are going '<u>from</u>'.

2) NORTH LINE At the point you are going <u>FROM</u>, draw in a <u>NORTH LINE</u>.

3) CLOCKWISE Now draw in the angle <u>CLOCKWISE</u> from the <u>NORTH LINE</u>
to the <u>line joining the two points</u> — this angle is the <u>bearing</u>.

EXAMPLES:

1. Find the bearing of Q from P.

Steps 1, 2 and 3 give:

The angle's too big to measure with a <u>protractor</u>
— <u>split it</u> into bits you can <u>measure</u>.

Using protractor = 65°

So, bearing of Q from P
= 180° + 65°
= 245°

Already <u>3 figures</u>
so no extra zeros
needed.

**2. The bearing of Y from X is 075°.
What is the bearing of X from Y?**

Diagram not drawn accurately

This is trickier — the diagram's
not accurate so you can't use
a protractor. You need to use
angle rules — see p109.

<u>Extend the line</u> to <u>split</u> the <u>bearing</u> into bits
you can <u>find</u>:

These are corresponding
angles so they're equal.

Bearing of X from Y
= 75° + 180°
= 255°

From... North line... Clockwise — that's all you need to remember

This is a straightforward exam question. Make sure you get the bearing from the right place,
draw a north line and measure clockwise. From, North line, Clockwise — am I going on...

Maps and Scale Drawings

Scales tell you what a <u>distance</u> on a <u>map</u> or <u>drawing</u> represents in <u>real life</u>. They can be written in various ways, but they all boil down to something like "<u>1 cm represents 5 km</u>".

Map Scales

1 cm = 3 km — "**1 cm represents 3 km**"

1 : 2000 — 1 cm on the map means 2000 cm in real life.
Converting to m gives "**1 cm represents 20 m**".

Use a ruler — the line's 2 cm long, so 2 cm means 1 km.
Dividing by 2 gives "**1 cm represents 0.5 km**".

|————————|
0   km   1

See p137 for a reminder about conversions.

To <u>convert</u> between <u>maps</u> and <u>real life</u>, <u>learn</u> these rules:

- To find **REAL-LIFE** distances, **MULTIPLY** by the **MAP SCALE**.
- To find **MAP** distances, **DIVIDE** by the **MAP SCALE**.
- Make sure your map scale is of the form "**1 cm = ...**"
- Always check your answer looks **sensible**.

Converting *from* **Map** *Distance to* **Real Life** — **Multiply** Ⓔ

EXAMPLE:

This map shows the road from Benenden to Staplehurst.
Work out the distance in km between these two villages.

1) Measure with a <u>ruler</u>:   Distance on map = 2 cm

2) Read off the <u>scale</u>:   Scale is 1 cm = 12 km

3) For <u>real life</u>, <u>multiply</u>:   Real distance is: 2 × 12 = 24 km

This looks sensible.

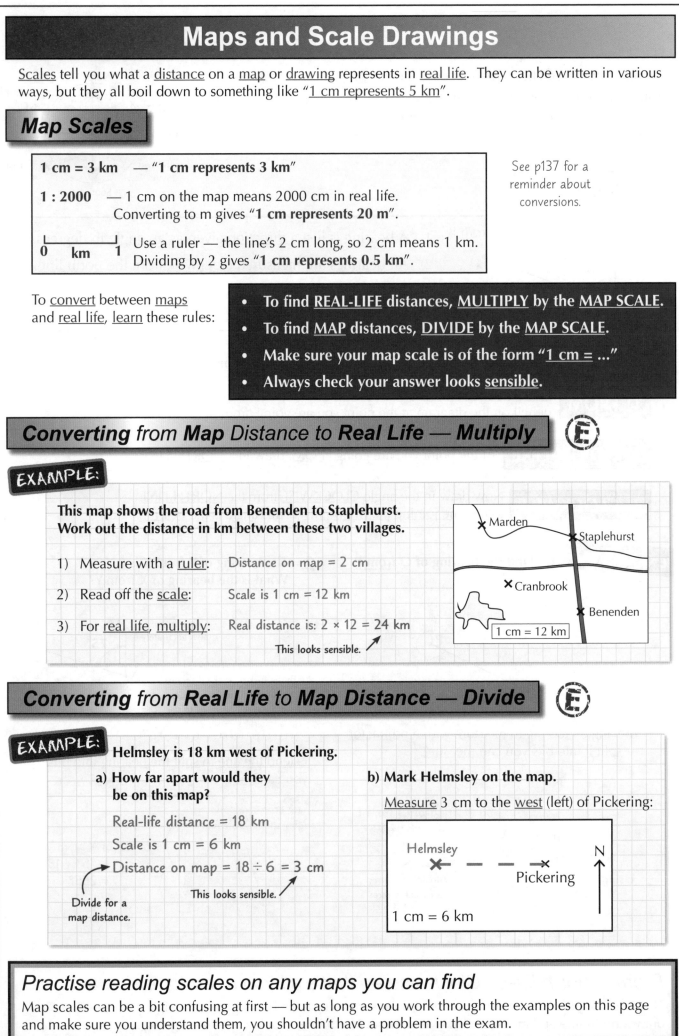

Marden ✕
✕ Staplehurst
✕ Cranbrook
✕ Benenden
1 cm = 12 km

Converting *from* **Real Life** *to* **Map Distance** — **Divide** Ⓔ

EXAMPLE:

Helmsley is 18 km west of Pickering.

a) How far apart would they be on this map?

Real-life distance = 18 km

Scale is 1 cm = 6 km

Distance on map = 18 ÷ 6 = 3 cm

This looks sensible.

Divide for a map distance.

b) Mark Helmsley on the map.

<u>Measure</u> 3 cm to the <u>west</u> (left) of Pickering:

Helmsley ✕ – – – – ✕ Pickering   N ↑

1 cm = 6 km

Practise reading scales on any maps you can find

Map scales can be a bit confusing at first — but as long as you work through the examples on this page and make sure you understand them, you shouldn't have a problem in the exam.

Maps and Scale Drawings

Scale Drawings (E)

Scale drawings work just like maps. To convert between real life and scale drawings, just replace the word 'map' with 'drawing' in the rules on the previous page.

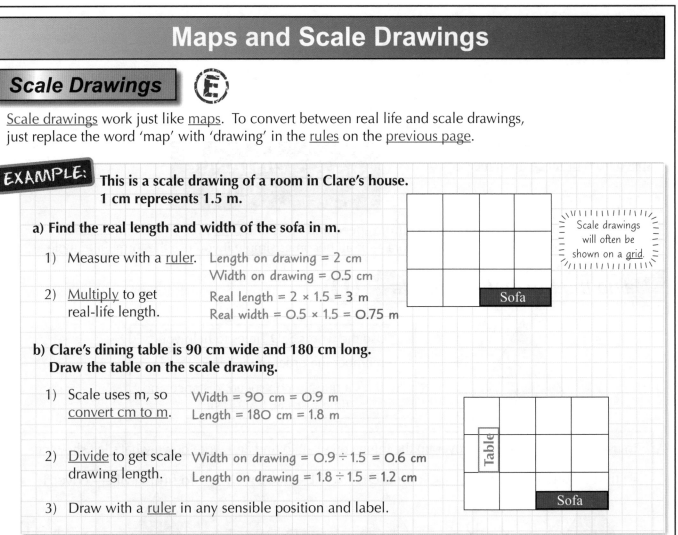

EXAMPLE:
This is a scale drawing of a room in Clare's house.
1 cm represents 1.5 m.

a) Find the real length and width of the sofa in m.

1) Measure with a ruler.　Length on drawing = 2 cm
　　　　　　　　　　　　　Width on drawing = 0.5 cm

2) Multiply to get　　　　Real length = 2 × 1.5 = 3 m
　real-life length.　　　Real width = 0.5 × 1.5 = 0.75 m

> Scale drawings will often be shown on a grid.

Sofa

b) Clare's dining table is 90 cm wide and 180 cm long.
Draw the table on the scale drawing.

1) Scale uses m, so　　　Width = 90 cm = 0.9 m
　convert cm to m.　　　Length = 180 cm = 1.8 m

2) Divide to get scale　　Width on drawing = 0.9 ÷ 1.5 = 0.6 cm
　drawing length.　　　Length on drawing = 1.8 ÷ 1.5 = 1.2 cm

3) Draw with a ruler in any sensible position and label.

Table | Sofa

Map Questions Using *Bearings* (D)

EXAMPLE:
Liam walks 1.2 km from the car park on a bearing of 120°.

a) Mark his position on the map.

1) Work out how many　　1 cm = 20 000 cm
　km 1 cm represents.　　= 200 m = 0.2 km.
　　　　　　　　　　　　So 1 cm = 0.2 km

2) Divide to get the
　distance on the map.　Distance walked on map = 1.2 ÷ 0.2
　　　　　　　　　　　　　　　= 6 cm

3) Mark a point 6 cm away, 120° clockwise from the north line.

> See p106 for how to draw an angle.

N　　　　　　　Scale = 1 : 20 000
　　　120°　　　　●Farm
Car
park　　　　　　　　6 cm

b) How far is he from the farm in km?

1) Measure the distance between Liam and the farm.　Distance between Liam and farm = 4 cm

2) For real life, multiply:　　　　　　　　　　　Real distance = 4 × 0.2 = 0.8 km

Bearings can crop up in map questions too

You could come across questions which ask you to use maps and bearings. Just remember to measure your angle from the north line. Take a look back at page 147 if you get a bit stuck.

Speed

Learn the formula triangle on this page and you'll be ready to tackle any speed question that comes along...

Speed = *Distance* ÷ *Time*   ⓓ

This is the basic formula for calculating speed from distance and time:

$$SPEED = \frac{DISTANCE}{TIME}$$

You also need to be able to find distance from speed and time, and time from speed and distance. But fear not, there's no need for any algebra — it all becomes simple if you use the formula triangle...

Using the Formula Triangle   ⓓ

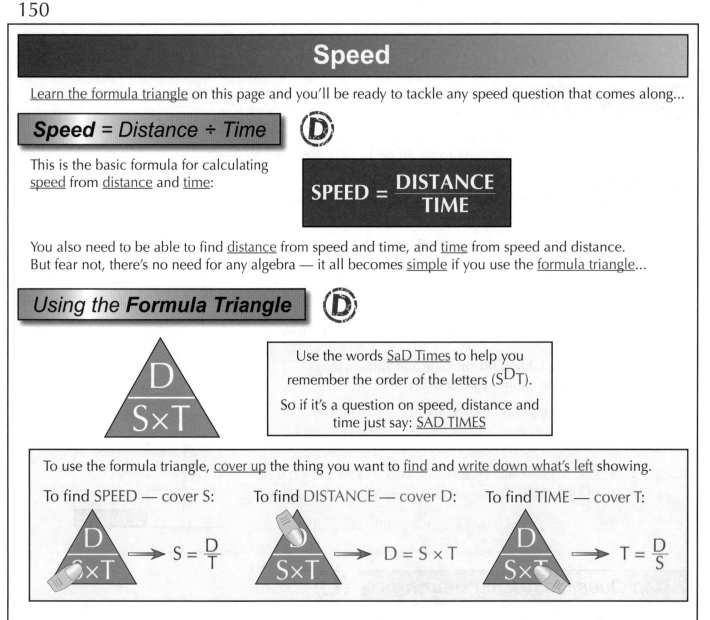

Use the words SaD Times to help you remember the order of the letters (S^DT).

So if it's a question on speed, distance and time just say: SAD TIMES

To use the formula triangle, cover up the thing you want to find and write down what's left showing.

To find SPEED — cover S: $S = \frac{D}{T}$ To find DISTANCE — cover D: $D = S \times T$ To find TIME — cover T: $T = \frac{D}{S}$

Always give units when you do speed calculations — the units you get out depend on the units you put in, e.g. distance in km and time in hours gives speed in km per hour (km/h).

EXAMPLES:

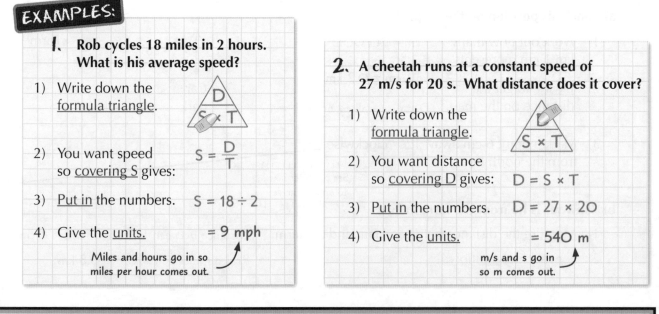

1. Rob cycles 18 miles in 2 hours. What is his average speed?
1) Write down the formula triangle.
2) You want speed so covering S gives: $S = \frac{D}{T}$
3) Put in the numbers. $S = 18 \div 2$
4) Give the units. $= 9$ mph
Miles and hours go in so miles per hour comes out.

2. A cheetah runs at a constant speed of 27 m/s for 20 s. What distance does it cover?
1) Write down the formula triangle.
2) You want distance so covering D gives: $D = S \times T$
3) Put in the numbers. $D = 27 \times 20$
4) Give the units. $= 540$ m
m/s and s go in so m comes out.

Formula triangles are incredibly useful

With this method you don't need to worry about changing the subjects of formulas. So make sure you understand formula triangles and use them whenever you can.

Warm-up and Worked Exam Questions

Last lot of questions for Section Six. Make sure you're concentrating now...

Warm-up Questions

1) Draw a dot on a piece of paper to represent home, and then draw 2 lines, one going out in a south-westerly direction and the other on a bearing of 080°.

2) a) Work out the length in m of the runway shown here:
   b) How many cm on the map would a 500 m runway be?

Island   Runway

Scale: 1 cm to 200 m

3) A car travels at an average speed of 45 km/h for 1.5 hours. How far has it travelled?

Worked Exam Question

This worked exam question will help you see how to put all the facts you've learnt to good use.

1   Beatrix is on a cycling holiday in Cumbria.
    The table below shows the distances between each of the places she visits. ©

Keswick		
26 miles	Windermere	
33 miles	5 miles	Hawkshead

On Friday it took her 45 minutes to cycle from Hawkshead to Windermere.

On Saturday, she then cycled from Windermere, arriving in Keswick $3\frac{1}{4}$ hours later.

On Sunday, she cycled back from Keswick to Hawkshead in 3 hours 45 minutes.

*a) On which day did Beatrix ride her bike at the fastest average speed?

The little * means that you're tested on your quality of written communication — so make sure you explain things clearly

Friday: speed = 5 ÷ 0.75 = 6.66... mph
Saturday: speed = 26 ÷ 3.25 = 8 mph
Sunday: speed = 33 ÷ 3.75 = 8.8 mph

Write all the times in hours to get speed in mph

So Beatrix rode fastest on Sunday

..............Sunday..........
[4 marks]

b) Work out Beatrix's average cycling speed over all three days.
   Give your answer to 1 decimal place.

Overall speed = $\frac{distance}{time} = \frac{5+26+33}{0.75+3.25+3.75} = \frac{64}{7.75} = 8.3$ mph (to 1 d.p.)

.............8.3............. mph
[3 marks]

152

Exam Questions

2 This map is drawn to a scale of 1 cm to 2 km. (E)

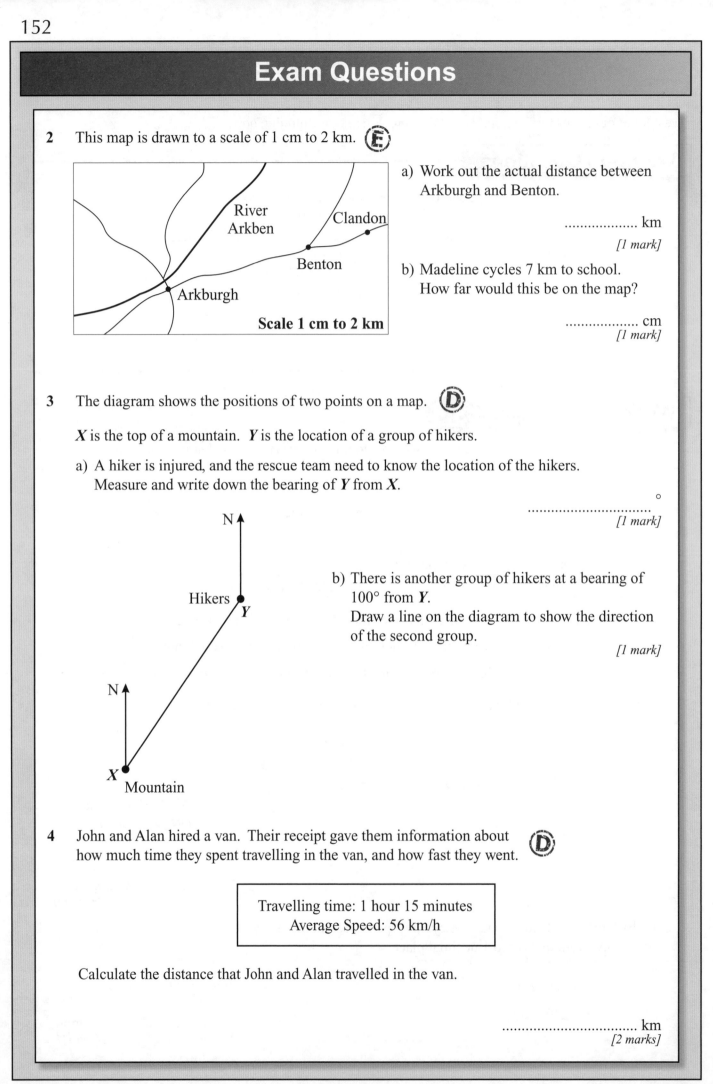

River
Arkben

Clandon

Benton

Arkburgh

Scale 1 cm to 2 km

a) Work out the actual distance between Arkburgh and Benton.

.... km
[1 mark]

b) Madeline cycles 7 km to school. How far would this be on the map?

.... cm
[1 mark]

3 The diagram shows the positions of two points on a map. (D)

X is the top of a mountain. *Y* is the location of a group of hikers.

a) A hiker is injured, and the rescue team need to know the location of the hikers. Measure and write down the bearing of *Y* from *X*.

°
....
[1 mark]

N

Hikers
Y

b) There is another group of hikers at a bearing of 100° from *Y*.
Draw a line on the diagram to show the direction of the second group.

[1 mark]

N

X
Mountain

4 John and Alan hired a van. Their receipt gave them information about how much time they spent travelling in the van, and how fast they went. (D)

Travelling time: 1 hour 15 minutes
Average Speed: 56 km/h

Calculate the distance that John and Alan travelled in the van.

.... km
[2 marks]

Revision Questions for Section Six

When you think you've got the measure of this section, use this page to make sure it's all sorted.

- Try these questions and <u>tick off each one</u> when you <u>get it right</u>.
- When you've done <u>all the questions</u> for a topic and are <u>completely happy</u> with it, tick off the topic.

Converting Units (p136-139) ☑

1) Fill in these gaps.
   - 1 m = ... cm
   - 1 litre = ... cm³
   - 1000 m = ... km
   - 1 litre ≈ .... pints
   - 1 mile ≈ .... km
   - 1 foot ≈ .... cm
2) What is the 3-step method for converting units?
3) Kevin is filling in a form to join the gym and needs to give his weight in kg. He knows he weighs 143 pounds. How much is this in kg?
4) How many times do you multiply or divide by the conversion factor when converting the units of an area? How many times for a volume?
5) A bath holds 120 litres of water. What is its volume in m³?

Reading Scales and Measuring (p142-143) ☑

6) How do you work out what a small gap on a scale stands for?
7) Using the ruler shown, what is the length of this key?

8) Measure the length of line AB in cm. Mark the point that's 1.3 cm from point B with a cross.

A———————————————B

9) By comparing with the man's height, estimate the height of this giant snail.

Reading Timetables (p144) ☑

10) Write a) 4.20 pm as a 24-hour clock time, b) 07:52 as a 12-hour clock time.
11) Using the timetable, how many minutes does the journey from Edinburgh to York last for?
12) Jane lives in Berwick and needs to be in Durham by 1.30 pm. a) What is the latest train she can catch? She lives 20 minutes' walk from the train station. b) What is the latest time she should leave the house?

Train Timetable			
Edinburgh	11 14	11 37	12 04
Berwick	11 55	12 18	12 45
Newcastle	12 43	13 06	13 33
Durham	12 57	13 20	13 47
Darlington	13 15	13 38	14 05
York	13 45	14 08	14 35

Bearings, Maps and Scale Drawings (p147-149) ☑

13) What are the three key words to remember when you're working with bearings?
14) How do you use a map scale to go from a real-life distance to a map distance, and vice versa?
15) Bobby is planning the layout of a new car park for his local supermarket, shown on the right. Draw a plan of the car park using a scale of 1 cm = 5 m.
16) What is the bearing of Port Q from Port P?
17) The map on the right has a scale of 1 cm = 10 miles. A ship is 15 miles from Port P on a bearing of 230° from Port P. Mark the ship on the map.

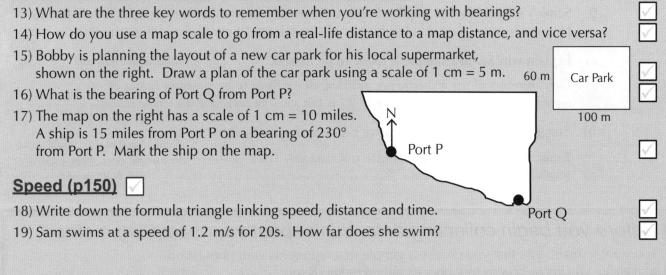

60 m Car Park
100 m

Speed (p150) ☑

18) Write down the formula triangle linking speed, distance and time.
19) Sam swims at a speed of 1.2 m/s for 20s. How far does she swim?

Collecting Data

Data is just information. Before you collect any data, you need to think carefully about where to get it from.

Choose Your *Sample Carefully*

1) For any statistical project, you need to find out about a group of people or things. E.g. all the pupils in a school, or all the trees in a forest. This whole group is called the POPULATION.

2) Information can be collected by doing a survey — you can record observations yourself, or ask people to fill in a questionnaire.

3) Often you can't survey the whole population, e.g. because it's too big. So you select a smaller group from the population, called a SAMPLE, instead.

4) It's really important that your sample fairly represents the WHOLE population. This allows you to apply any conclusions from your survey to the whole population.

You Need to Spot Problems *with* Sampling Methods

A BIASED sample (or survey) is one that doesn't properly represent the whole population.

To SPOT BIAS, you need to think about:
1) WHEN, WHERE and HOW the sample is taken.
2) HOW MANY members are in it.

- If any groups are left out of the sample, there can be BIAS in things like age, gender, or different interests.

- If the sample is too small, it's also likely to be biased.

Bigger populations need bigger samples to represent them.

If possible, the best way to AVOID BIAS is to select a large sample at random from the whole population.

EXAMPLES:

1. Tina wants to find out how often people travel by train.
She decides to ask the people waiting for trains at her local train station one morning.
Give one reason why this might not be a suitable sample to choose.

The sample is biased because there won't be anyone who never uses the train and there will probably be a lot of people who use the train regularly.

Think about when, where and how Tina selects her sample.

Or you could say that the sample is only taken at one particular place and time, so won't represent everyone.

2. Samir's school has 800 pupils. Samir is interested in whether these pupils would like to have more music lessons. For his sample he selects 10 members of the school orchestra.

a) Explain why Samir's sample is likely to be biased.

Think about how and how many.

Only members of the orchestra are included, so it's likely to be biased in favour of more music lessons. And a sample of 10 is too small to represent the whole school.

b) Suggest how Samir could improve his sampling method.

Samir should choose a larger sample and randomly select pupils from the whole school.

Think about how he could make the sample more representative of all 800 pupils.

Before you begin collecting data, think about your sampling method

You want to make sure that you choose a sample that represents your population.
Make sure you know how to spot poor sampling methods too.

Collecting Data

You can <u>collect</u> your data in several different ways — you can record it in a <u>table</u> or design a <u>questionnaire</u>. Then you're ready to <u>process</u> it and <u>interpret</u> it.

*You Can **Record** Your **Data** in a **Table*** Ⓓ

1) <u>Data-collection tables</u> (or <u>sheets</u>) should look like this table:

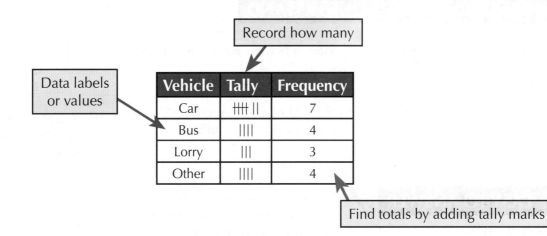

2) The <u>first column</u> can contain <u>words</u> (like the table above) or <u>numbers</u>. Make sure you include a <u>category</u> to fit <u>every possible</u> data label or value.

3) Tables for <u>grouped</u> data are covered on page 169.

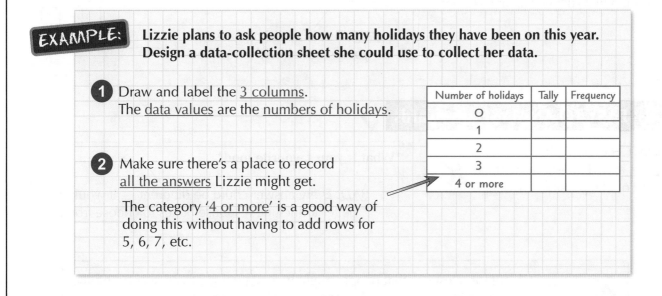

EXAMPLE: Lizzie plans to ask people how many holidays they have been on this year. Design a data-collection sheet she could use to collect her data.

1 Draw and label the <u>3 columns</u>. The <u>data values</u> are the <u>numbers of holidays</u>.

2 Make sure there's a place to record <u>all the answers</u> Lizzie might get.

The category '<u>4 or more</u>' is a good way of doing this without having to add rows for 5, 6, 7, etc.

Number of holidays	Tally	Frequency
0		
1		
2		
3		
4 or more		

Tables are a really good way to record data

Just make sure you clearly label each of the columns in your table. Once you've recorded your data, you can then process it to figure out what it all means. But don't worry — that's still to come.

Collecting Data

Questionnaires are good for collecting data — but they need to be written carefully.

Design Your Questionnaire Carefully

You need to be able to say what's wrong with questionnaire questions and write your own good questions.

A GOOD question is:

1) CLEAR and EASY TO UNDERSTAND ✓

WATCH OUT FOR:
confusing wording or no time frame ✗

How much do you spend on food? ☐ a little ☐ average amount ☐ a lot

BAD: Wording is vague and no time frame is specified (e.g. each week or month).

BAD: Response boxes might be interpreted differently by different people.

2) EASY TO ANSWER ✓

WATCH OUT FOR:
response boxes that overlap, or don't allow for all possible answers ✗

How many pieces of fruit do you eat a day on average? ☐ 1-2 ☐ 2-3 ☐ 3-4 ☐ 4-5 ☐ > 5

BAD: Response boxes overlap and don't allow an answer of zero.

3) FAIR — NOT LEADING or BIASED ✓

WATCH OUT FOR:
wording that suggests an answer ✗

Do you agree that potatoes taste better than cabbage? ☐ Yes ☐ No

BAD: This is a leading question — you're more likely to say 'Yes'.

4) EASY TO ANALYSE afterwards ✓

WATCH OUT FOR:
open-ended questions, with no limit on the possible answers ✗

What is your favourite food? ....

BAD: Every answer could be different — it would be better to include response boxes to choose from.

✗ Also watch out for questions that people might be embarrassed to answer truthfully. E.g. ones asking for personal information, like someone's exact age.

Remember the four important points when designing a questionnaire

Make sure you learn the four key points for writing good questions. You may be asked to criticise a questionnaire in the exam — if so, just think about what makes a good or bad question.

Mean, Median, Mode and Range

Mean, median, mode and range pop up all the time in statistics questions — make sure you know them.

MODE = MOST common (F)

MEDIAN = MIDDLE value (when values are in order of size)

MEAN = TOTAL of items ÷ NUMBER of items

RANGE = Difference between highest and lowest

REMEMBER:

Mode = most (emphasise the 'mo' in each when you say them)

Median = mid (emphasise the m*d in each when you say them)

Mean is just the average, but it's mean 'cos you have to work it out.

The Golden Rule

There's one vital step for finding the median that lots of people forget:

Always REARRANGE the data in ASCENDING ORDER (and check you have the same number of entries!)

You must do this when finding the median, but it's also really useful for working out the mode too.

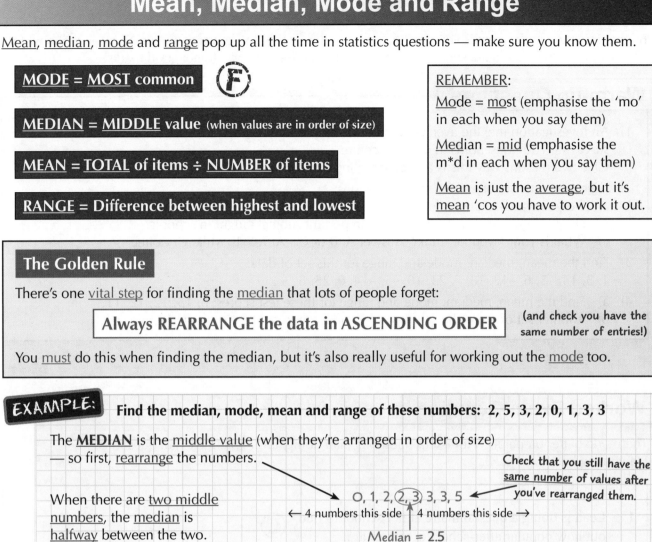

EXAMPLE: Find the median, mode, mean and range of these numbers: 2, 5, 3, 2, 0, 1, 3, 3

The **MEDIAN** is the middle value (when they're arranged in order of size) — so first, rearrange the numbers.

Check that you still have the same number of values after you've rearranged them.

When there are two middle numbers, the median is halfway between the two.

0, 1, 2, 2, 3, 3, 3, 5

← 4 numbers this side ↑ 4 numbers this side →

Median = 2.5

MODE (or modal value) is the most common value. ⟶ Mode = 3

Some data sets have more than one mode, or no mode at all.

$$\textbf{MEAN} = \frac{\text{total of items}}{\text{number of items}} \longrightarrow \frac{0+1+2+2+3+3+3+5}{8} = \frac{19}{8} = 2.375$$

RANGE = distance from lowest to highest value, i.e. from 0 up to 5. ⟶ 5 − 0 = 5

You might be asked to compare two data sets using the mean, median or mode and the range. Say which data set has the higher/lower value and what that means in the context of the data. (E)

EXAMPLE: Some children take part in a 'guess the weight of the baby hippo' competition. Here is some information about the weights they guess. Compare the weights guessed by the boys and the girls.

Boys: Mean = 40 kg, Range = 42 kg
Girls: Mean = 34 kg, Range = 40 kg

1) Compare the means:
   The boys' mean is higher than the girls' mean, so the boys generally guessed heavier weights.

2) Compare the ranges:
   The boys' guesses have a bigger range, so the weights guessed by the boys are more spread out.

Mean, median, mode, range — easy marks for learning 4 definitions

The maths involved in working these out is simple, so you'd be mad not to learn the definitions.

Warm-up and Worked Exam Questions

By the time the big day comes you need to know all the facts in these warm-up questions — and how to use them to answer exam questions. It's not easy, but it's the only way to get good marks.

Warm-up Questions

1) An investigation into the average number of people in households in Britain was done by surveying 100 households in one city centre.
   Give two reasons why this is a poor sampling technique.

2) Give one criticism of each of these questions:
   a) Do you watch a lot of television?
   b) Do you agree that maths is the most important subject taught in schools?
   c) What is your favourite drink? Answer A, B or C. A) Tea B) Milk C) Coffee

3) Find the mean, median, mode and range for this set of data:
   1, 3, 14, –5, 6, –12, 18, 7, 23, 10, –5, –14, 0, 25, 8

4) a) Find the mean, median, mode and range for these test scores: 6, 15, 12, 12, 11.
   b) Another set of scores has a mean of 9 and a range of 12. Compare the two sets.

Worked Exam Question

There's no better preparation for exam questions than doing, err... practice exam questions. Hang on, what's this I see...

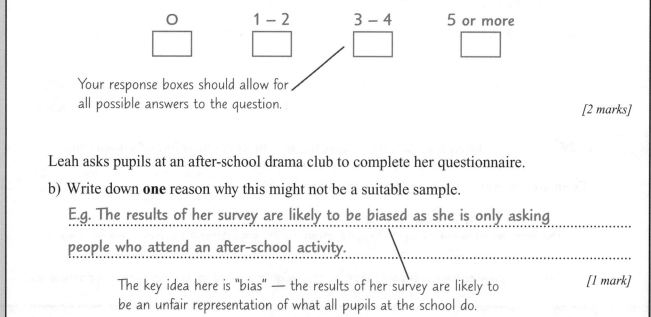

1   Leah is doing a questionnaire at her school to find out how popular after-school activities are.

a) Design a question for Leah to include in her questionnaire.
   You should include suitable response boxes.

   E.g. How many different after-school activities do you attend each week?

   O   1 – 2   3 – 4   5 or more
   ☐   ☐   ☐   ☐

   Your response boxes should allow for
   all possible answers to the question.

   [2 marks]

Leah asks pupils at an after-school drama club to complete her questionnaire.

b) Write down **one** reason why this might not be a suitable sample.

   E.g. The results of her survey are likely to be biased as she is only asking
   people who attend an after-school activity.

   The key idea here is "bias" — the results of her survey are likely to
   be an unfair representation of what all pupils at the school do.

   [1 mark]

Exam Questions

2 Sam thinks of three different whole numbers. (F)

The numbers have a range of 6 and a mean of 4.
What are the three numbers?

....................., ..., ...
[2 marks]

3 Company A employs 5 people.
Their annual salaries are listed below.

£18 000   £38 500   £18 000   £25 200   £18 000

a) Write down the mode. (F)

£ ....
[1 mark]

b) What is the median annual salary? (F)

£ ....
[1 mark]

Company B has a mean annual salary of £24 150.

c) Compare the mean annual salary of Company A and Company B. (E)

....

....
[3 marks]

4 Jane wants to find out how often people are visiting the leisure centre.

a) Design a suitable data collection sheet she could use to record this information. (D)

[3 marks]

Jane wants to improve membership numbers. She designs a questionnaire.
Here is one of her questions:

> Do you agree that the facilities need improving?

b) Give one reason why this is not a good question. (C)

....
[1 mark]

c) Rewrite the question so that it is more suitable. (C)

[2 marks]

160

Pictograms and Bar Charts

Pictograms and bar charts both show <u>frequencies</u>. (Remember... frequency = 'how many of something'.)

Pictograms Show Frequencies Using **Symbols**

Every pictogram has a <u>key</u> telling you what one symbol represents.

With pictograms, you <u>MUST</u> use the <u>KEY</u>.

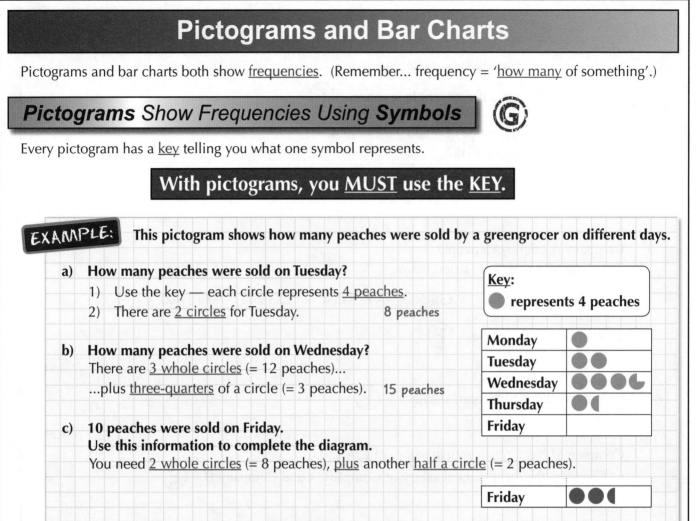

EXAMPLE: This pictogram shows how many peaches were sold by a greengrocer on different days.

a) **How many peaches were sold on Tuesday?**
   1) Use the key — each circle represents <u>4 peaches</u>.
   2) There are <u>2 circles</u> for Tuesday.   *8 peaches*

b) **How many peaches were sold on Wednesday?**
   There are <u>3 whole circles</u> (= 12 peaches)...
   ...plus <u>three-quarters</u> of a circle (= 3 peaches).   *15 peaches*

c) **10 peaches were sold on Friday.**
   Use this information to complete the diagram.
   You need <u>2 whole circles</u> (= 8 peaches), <u>plus</u> another <u>half a circle</u> (= 2 peaches).

Bar Charts Show Frequencies Using **Bars**

1) <u>Bar charts</u> are very similar to pictograms.
   Frequencies are shown by the <u>heights</u> of the different bars.

2) <u>Dual bar charts</u> show two things at once — they're good for <u>comparing</u> different sets of data.

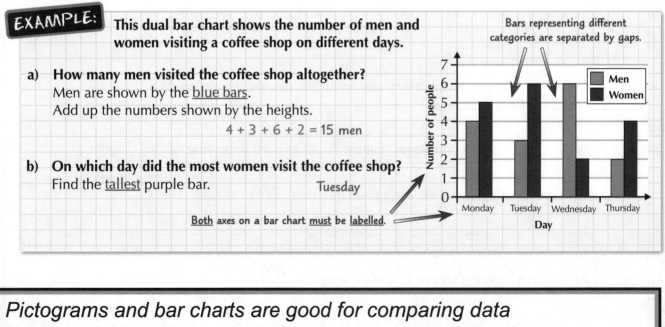

EXAMPLE: This dual bar chart shows the number of men and women visiting a coffee shop on different days.

a) **How many men visited the coffee shop altogether?**
   Men are shown by the <u>blue bars</u>.
   Add up the numbers shown by the heights.
   4 + 3 + 6 + 2 = 15 men

b) **On which day did the most women visit the coffee shop?**
   Find the <u>tallest</u> purple bar.   Tuesday

Both axes on a bar chart <u>must</u> be <u>labelled</u>.

Pictograms and bar charts are good for comparing data

Don't be scared if you're asked to draw a histogram — it's just a bar chart that you draw using a frequency table. And don't forget to give a key with every pictogram you draw.

Two-Way Tables

Two-way tables are another way to represent data.
They show how many things or people fall into different categories.

Two-Way Tables *Show Frequencies* (E)

Fill in any information you're <u>not</u> given in the question by <u>adding</u> or <u>subtracting</u>.

EXAMPLE:

This two-way table shows the number of cakes and loaves of bread a bakery sells on Friday and Saturday one week.

	Cakes	Loaves of bread	Total
Friday		10	22
Saturday	4	14	
Total	16	24	40

a) **Work out how many items in total were sold on Saturday.**

<u>Either</u>:(i) <u>add</u> the number of <u>cakes</u> for Saturday to the number of <u>loaves of bread</u>.

<u>Or</u>: (ii) <u>take away</u> the total items sold on <u>Friday</u> from the total sold over <u>both days</u>.

	Cakes	Loaves of bread	Total
Friday		10	22
Saturday	4	14	18
Total	16	24	40

b) **Work out how many cakes were sold on Friday.**

<u>Either</u>:(i) <u>take away</u> the <u>loaves of bread</u> for Friday from the <u>total</u> number of items for Friday.

<u>Or</u>: (ii) <u>take away</u> the number of cakes for <u>Saturday</u> from the total number of cakes over <u>both days</u>.

	Cakes	Loaves of bread	Total
Friday	12	10	22
Saturday	4	14	18
Total	16	24	40

You Might Have to *Draw Your Own* Table (E)

Sometimes they don't even give you a table — just <u>a few bits</u> of information.

EXAMPLE:

200 men and 200 women are asked whether they are left-handed or right-handed.
- **63 people altogether are left-handed.**
- **164 of the women are right-handed.**

How many of the men are right-handed?

When there's only <u>one</u> thing in a row or column that you don't know, you can always work it out.

1. Draw a table to show the information from the <u>question</u> — this is in the <u>yellow cells</u>.
2. Then fill in the gaps by <u>adding</u> and <u>subtracting</u>.

	Women	Men	Total
Left-handed	200 − 164 = 36	63 − 36 = 27	63
Right-handed	164	200 − 27 = 173	400 − 63 = 337
Total	200	200	200 + 200 = 400

So there are 173 right-handed men.

You'll need your adding and subtracting skills for two-way tables

If you're given a two-way table with only a few bits of data in, you'll need to use that information to fill in the rest of the table — but it's just a case of adding and subtracting. Easy peasy.

Pie Charts

Pie charts can be tricky exam questions, but not if you've learnt this Golden Rule:

The TOTAL of Everything = 360°

1) Fraction of the Total = Angle ÷ 360°   (E)

EXAMPLE: This pie chart shows the colour of all the cars sold by a dealer. **What fraction of the cars were red?**

Just remember that 'everything = 360°'.

$$\text{Fraction of red cars} = \frac{\text{angle of red cars}}{\text{angle of everything}} = \frac{72°}{360°} = \frac{1}{5}$$

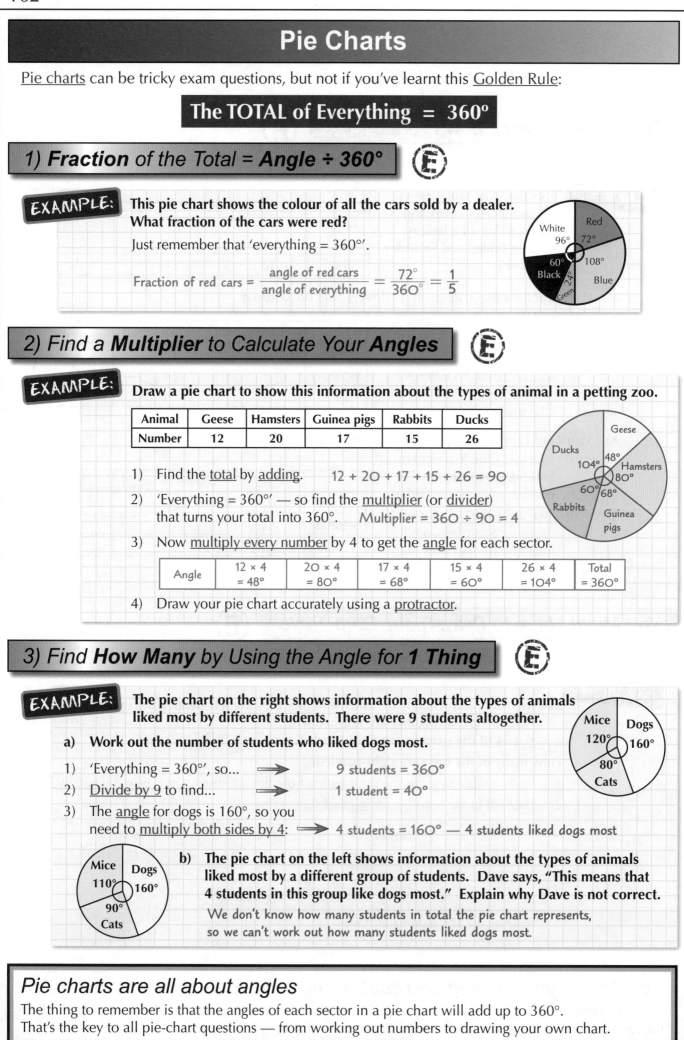

2) Find a Multiplier to Calculate Your Angles   (E)

EXAMPLE: Draw a pie chart to show this information about the types of animal in a petting zoo.

Animal	Geese	Hamsters	Guinea pigs	Rabbits	Ducks
Number	12	20	17	15	26

1) Find the total by adding.   12 + 20 + 17 + 15 + 26 = 90

2) 'Everything = 360°' — so find the multiplier (or divider) that turns your total into 360°.   Multiplier = 360 ÷ 90 = 4

3) Now multiply every number by 4 to get the angle for each sector.

Angle	12 × 4 = 48°	20 × 4 = 80°	17 × 4 = 68°	15 × 4 = 60°	26 × 4 = 104°	Total = 360°

4) Draw your pie chart accurately using a protractor.

3) Find How Many by Using the Angle for 1 Thing   (E)

EXAMPLE: The pie chart on the right shows information about the types of animals liked most by different students. There were 9 students altogether.

a) Work out the number of students who liked dogs most.

1) 'Everything = 360°', so... ⟹ 9 students = 360°
2) Divide by 9 to find... ⟹ 1 student = 40°
3) The angle for dogs is 160°, so you need to multiply both sides by 4: ⟹ 4 students = 160° — 4 students liked dogs most

b) The pie chart on the left shows information about the types of animals liked most by a different group of students. Dave says, "This means that 4 students in this group like dogs most." Explain why Dave is not correct.

We don't know how many students in total the pie chart represents, so we can't work out how many students liked dogs most.

Pie charts are all about angles

The thing to remember is that the angles of each sector in a pie chart will add up to 360°. That's the key to all pie-chart questions — from working out numbers to drawing your own chart.

Scatter Graphs

Scatter graphs are really useful — they show you if there's a <u>link</u> between two things.

Scatter Graphs Show *Correlations* (D)

1) A <u>scatter graph</u> shows how closely two things are <u>related</u>. The fancy word for this is <u>CORRELATION</u>.

2) If the two things <u>are related</u>, then you'd be able to draw a <u>straight line</u> ← *Called a line of best fit.*
passing <u>pretty close</u> to <u>most</u> of the points on the scatter diagram.

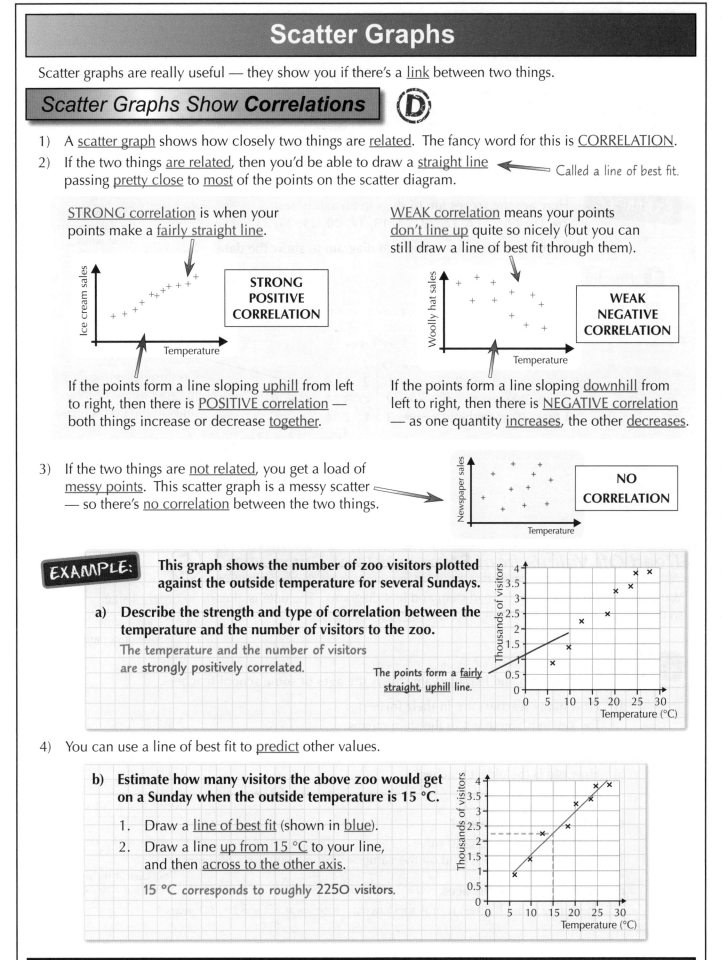

<u>STRONG correlation</u> is when your points make a <u>fairly straight line</u>.

STRONG POSITIVE CORRELATION

If the points form a line sloping <u>uphill</u> from left to right, then there is <u>POSITIVE</u> correlation — both things increase or decrease <u>together</u>.

<u>WEAK correlation</u> means your points <u>don't line up</u> quite so nicely (but you can still draw a line of best fit through them).

WEAK NEGATIVE CORRELATION

If the points form a line sloping <u>downhill</u> from left to right, then there is <u>NEGATIVE</u> correlation — as one quantity <u>increases</u>, the other <u>decreases</u>.

3) If the two things are <u>not related</u>, you get a load of <u>messy points</u>. This scatter graph is a messy scatter — so there's <u>no correlation</u> between the two things.

NO CORRELATION

EXAMPLE: This graph shows the number of zoo visitors plotted against the outside temperature for several Sundays.

a) **Describe the strength and type of correlation between the temperature and the number of visitors to the zoo.**

The temperature and the number of visitors are strongly positively correlated.

The points form a <u>fairly straight, uphill</u> line.

4) You can use a line of best fit to <u>predict</u> other values.

b) **Estimate how many visitors the above zoo would get on a Sunday when the outside temperature is 15 °C.**

1. Draw a <u>line of best fit</u> (shown in <u>blue</u>).

2. Draw a line <u>up from 15 °C</u> to your line, and then <u>across to the other axis</u>.

15 °C corresponds to roughly 2250 visitors.

Correlation means that there's a relationship between two things

Make sure you know all the terms in case you have to describe a correlation — positive correlation, negative correlation, and how to say how strong the correlation is.

Stem and Leaf Diagrams

Stem and leaf diagrams don't look as pretty as you might expect, but they're a useful way to show data.

Stem and Leaf Diagrams *Put the Data* **in Order** D

An ordered stem and leaf diagram shows a set of data in order of size.
To draw one, you need to decide how to make the stem, then the rest isn't too bad.

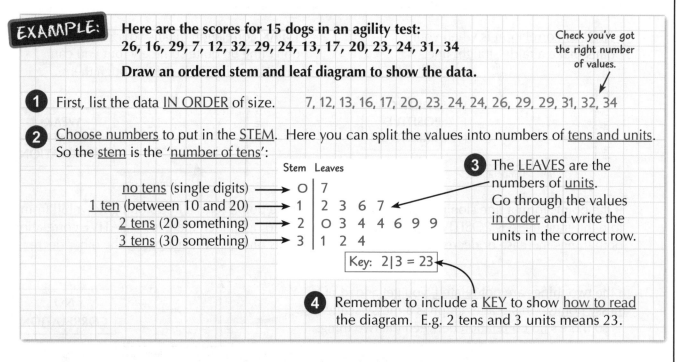

EXAMPLE: Here are the scores for 15 dogs in an agility test:
26, 16, 29, 7, 12, 32, 29, 24, 13, 17, 20, 23, 24, 31, 34

Draw an ordered stem and leaf diagram to show the data.

Check you've got the right number of values.

1. First, list the data IN ORDER of size.   7, 12, 13, 16, 17, 20, 23, 24, 24, 26, 29, 29, 31, 32, 34

2. Choose numbers to put in the STEM. Here you can split the values into numbers of tens and units.
So the stem is the 'number of tens':

Stem Leaves

no tens (single digits) ⟶ 0 | 7
1 ten (between 10 and 20) ⟶ 1 | 2 3 6 7
2 tens (20 something) ⟶ 2 | 0 3 4 4 6 9 9
3 tens (30 something) ⟶ 3 | 1 2 4

Key: 2|3 = 23

3. The LEAVES are the numbers of units. Go through the values in order and write the units in the correct row.

4. Remember to include a KEY to show how to read the diagram. E.g. 2 tens and 3 units means 23.

Read Off Values *from* Stem and Leaf Diagrams E

You also need to know how to read stem and leaf diagrams. The data is already in order of size, so that makes it easy to find things like the median and range.

See p157 for a reminder about median and range.

EXAMPLE:

This stem and leaf diagram shows the ages of some school teachers.

a) **How many teachers are in their forties?**
The key tells you that 4 in the stem means 40, so count the leaves in the second row.   **4 teachers**

b) **How old is the oldest teacher?**
Read off the last value.   6 | 3 = 63 years old

c) **What is the median age?**
The median is the middle value.   There are 11 values, so the median is the 6th value.
Find its position, then read off the value.   So median age = 48 years

d) **Find the range of the ages.**
The range is the highest minus the lowest.   Range = 63 − 33 = 30 years

3 | 3 5
4 | 0 5 7 8
5 | 1 4 9
6 | 1 3

Key: 5|4 = 54 years

Look at the stem and the leaf — it's 48, not 8.

Stem and leaf diagrams need a key

If you have to draw a stem and leaf diagram, follow these steps: order the data, decide how to split it into groups (e.g. tens), draw the stems to the left of a line, add the leaves, and write a key.

Warm-up and Worked Exam Questions

There's a nice selection of warm-up questions here covering tables, charts and graphs.
Now's the time to go back over any bits you're not sure of — in the exam it'll be too late.

Warm-up Questions

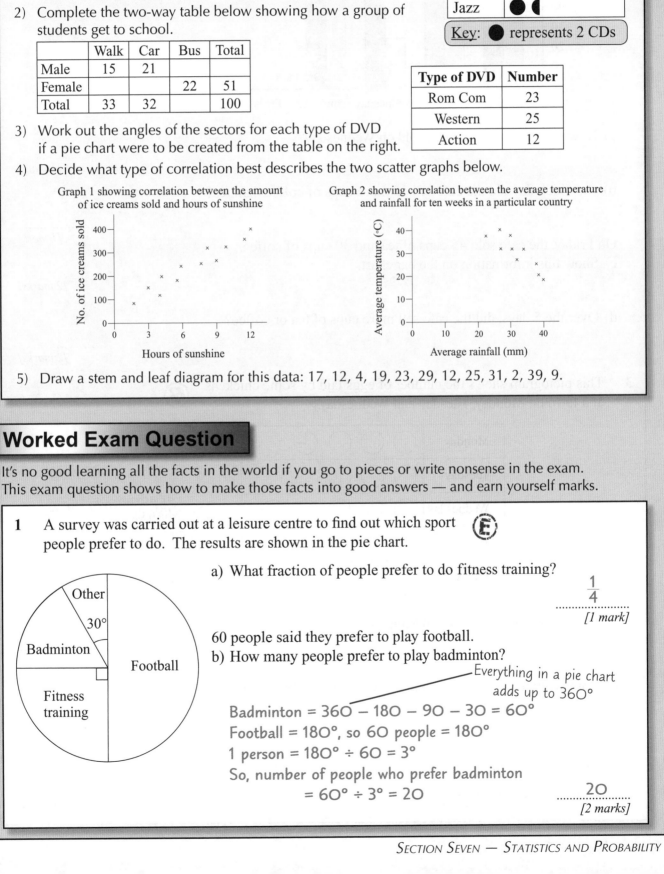

Rock	●	
Blues	●	●
Opera		
Jazz	●	◖

Key: ● represents 2 CDs

1) This pictogram shows the different types of CDs Javier owns.
   a) How many jazz CDs does Javier own?
   b) He owns 5 opera CDs. Complete the pictogram.

2) Complete the two-way table below showing how a group of students get to school.

	Walk	Car	Bus	Total
Male	15	21		
Female			22	51
Total	33	32		100

Type of DVD	Number
Rom Com	23
Western	25
Action	12

3) Work out the angles of the sectors for each type of DVD if a pie chart were to be created from the table on the right.

4) Decide what type of correlation best describes the two scatter graphs below.

Graph 1 showing correlation between the amount of ice creams sold and hours of sunshine

Graph 2 showing correlation between the average temperature and rainfall for ten weeks in a particular country

5) Draw a stem and leaf diagram for this data: 17, 12, 4, 19, 23, 29, 12, 25, 31, 2, 39, 9.

Worked Exam Question

It's no good learning all the facts in the world if you go to pieces or write nonsense in the exam.
This exam question shows how to make those facts into good answers — and earn yourself marks.

1 A survey was carried out at a leisure centre to find out which sport (E) people prefer to do. The results are shown in the pie chart.

a) What fraction of people prefer to do fitness training?

$\frac{1}{4}$

[1 mark]

60 people said they prefer to play football.
b) How many people prefer to play badminton?

Everything in a pie chart adds up to 360°

Badminton = 360 − 180 − 90 − 30 = 60°
Football = 180°, so 60 people = 180°
1 person = 180° ÷ 60 = 3°
So, number of people who prefer badminton
       = 60° ÷ 3° = 20

20

[2 marks]

Exam Questions

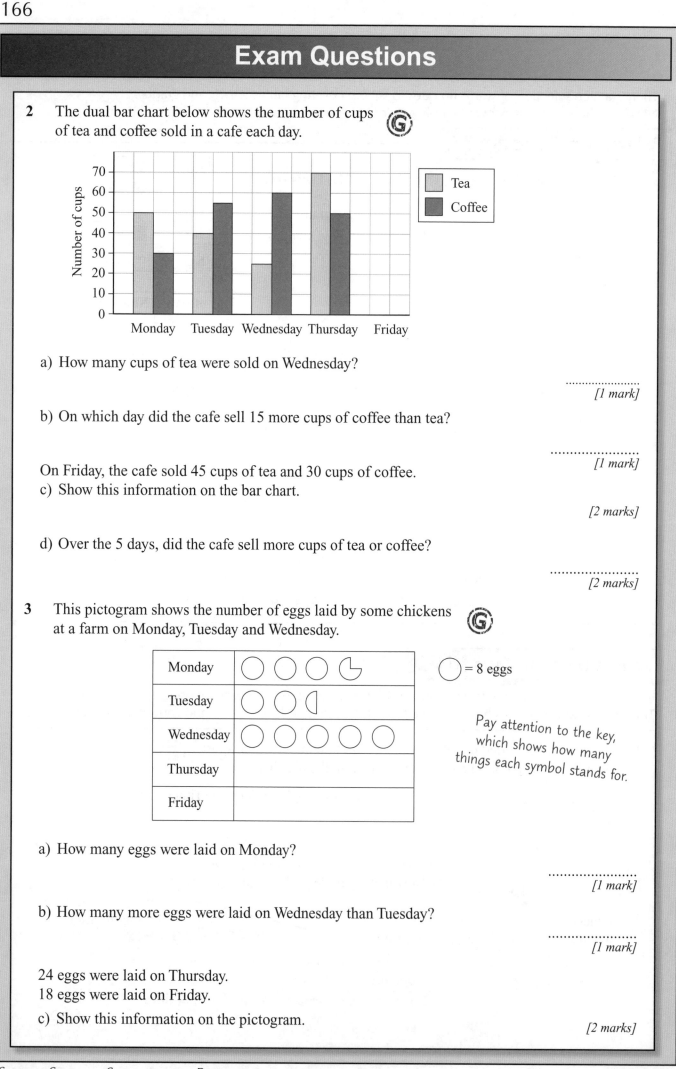

2   The dual bar chart below shows the number of cups
     of tea and coffee sold in a cafe each day.   (G)

a) How many cups of tea were sold on Wednesday?

....
[1 mark]

b) On which day did the cafe sell 15 more cups of coffee than tea?

....
[1 mark]

On Friday, the cafe sold 45 cups of tea and 30 cups of coffee.
c) Show this information on the bar chart.

[2 marks]

d) Over the 5 days, did the cafe sell more cups of tea or coffee?

....
[2 marks]

3   This pictogram shows the number of eggs laid by some chickens   (G)
     at a farm on Monday, Tuesday and Wednesday.

◯ = 8 eggs

Pay attention to the key,
which shows how many
things each symbol stands for.

a) How many eggs were laid on Monday?

....
[1 mark]

b) How many more eggs were laid on Wednesday than Tuesday?

....
[1 mark]

24 eggs were laid on Thursday.
18 eggs were laid on Friday.
c) Show this information on the pictogram.

[2 marks]

Exam Questions

4 115 athletes took part in a sports day. Some took part in swimming, some took part in athletics and the rest took part in football. 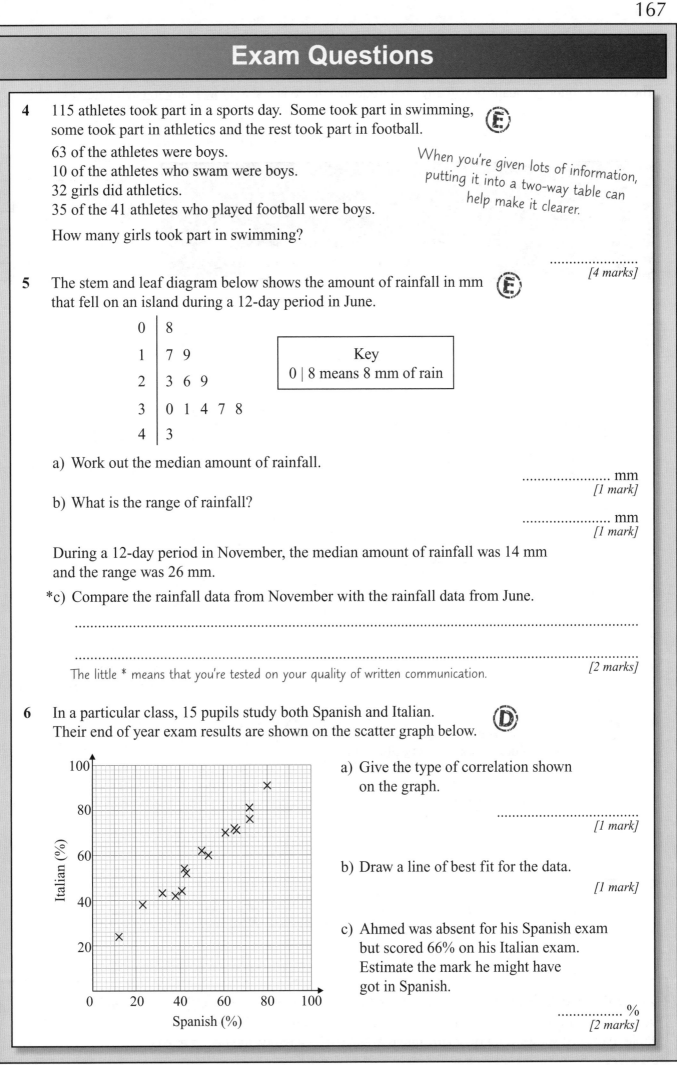 **E**

63 of the athletes were boys.
10 of the athletes who swam were boys.
32 girls did athletics.
35 of the 41 athletes who played football were boys.

How many girls took part in swimming?

When you're given lots of information, putting it into a two-way table can help make it clearer.

....
[4 marks]

5 The stem and leaf diagram below shows the amount of rainfall in mm that fell on an island during a 12-day period in June. **E**

```
0 | 8
1 | 7 9
2 | 3 6 9
3 | 0 1 4 7 8
4 | 3
```

Key
0	8 means 8 mm of rain

a) Work out the median amount of rainfall.

.... mm
[1 mark]

b) What is the range of rainfall?

.... mm
[1 mark]

During a 12-day period in November, the median amount of rainfall was 14 mm and the range was 26 mm.

*c) Compare the rainfall data from November with the rainfall data from June.

....

....

*The little * means that you're tested on your quality of written communication.*
[2 marks]

6 In a particular class, 15 pupils study both Spanish and Italian. Their end of year exam results are shown on the scatter graph below. **D**

a) Give the type of correlation shown on the graph.

....
[1 mark]

b) Draw a line of best fit for the data.

[1 mark]

c) Ahmed was absent for his Spanish exam but scored 66% on his Italian exam. Estimate the mark he might have got in Spanish.

.... %
[2 marks]

Frequency Tables and Averages

The word <u>frequency</u> means <u>how many</u>, so a frequency table is just a '<u>How many in each category</u>' table.
E.g. in the table below, 17 people don't have a cat, 22 have one cat, etc.

You found <u>averages</u> on page 157, and you can do the same sort of thing from tables.

D

1) The <u>MODE</u> is just the <u>CATEGORY with the MOST ENTRIES</u>.

2) The <u>MEDIAN</u> is the <u>CATEGORY</u> of the <u>middle value</u>.

3) To find the <u>MEAN</u>, you have to <u>WORK OUT A THIRD COLUMN</u> yourself.

The <u>MEAN</u> is then: | 3rd Column Total ÷ 2nd Column Total |

Categories   How many

Number of cats	Frequency	
0	17	
1	22	
2	15	
3	7	

Mysterious 3rd column...

EXAMPLE:

Some people were asked how many sisters they have.
The table opposite shows the results.

Find the <u>mode</u>, the <u>mean</u>, the <u>median</u> and the <u>range</u> of the data.

Number of sisters	Frequency
0	7
1	15
2	12
3	8
4	4

Here the categories are the values 0, 1, 2, 3 and 4 sisters.

1 The <u>MODE</u> is the <u>category</u> with the <u>most entries</u> — i.e. the one with the <u>highest frequency</u>:

1 The highest frequency is 15 for '1 sister', so <u>MODE</u> = 1

2 To find the <u>MEAN</u>, <u>add a 3rd column</u> to the table showing '<u>number of sisters × frequency</u>'.

Label the first column x, the second column f and your third column $(f \times x)$.

Number of sisters (x)	Frequency (f)	No. of sisters × Frequency $(f \times x)$
0	7	0
1	15	15
2	12	24
3	8	24
4	4	16
Total	46	79

2 $\underline{\text{MEAN}} = \dfrac{\text{total of } (f \times x) \text{ column}}{\text{total of } f \text{ column}} = \dfrac{79}{46} = 1.72 \text{ (3 s.f.)}$

total number of sisters

total number of people asked

3 The <u>MEDIAN</u> is the <u>category</u> of the <u>middle</u> value. Work out its <u>position</u>, then <u>count through</u> the 2nd column to find it.

It helps to imagine the data set out in an ordered list:
0000000111111111111111222222222222333333334444

↑
median

3 There are <u>46</u> values, so the middle value is <u>halfway</u> between the <u>23rd and 24th</u> values.

There are (7 + 15) = <u>22</u> values in the first two categories, and <u>12</u> more in the third category makes <u>34</u>. So the 23rd and 24th values must both be in category '2 sisters', which means the <u>MEDIAN</u> is 2.

4 The <u>RANGE</u> is the <u>difference</u> between the <u>highest and lowest</u> numbers of sisters. Looking at the <u>first</u> column, that's 4 sisters and no sisters, so:

4 <u>RANGE</u> = 4 – 0 = 4

A frequency table is just a "how many in each category" table

As so often happens in maths, the words can be really off-putting. Once you realise that a frequency table really just shows you how many things are in each category, life becomes easier.

Grouped Frequency Tables

Grouped frequency tables group together the data into classes.
They look like ordinary frequency tables, but they're a slightly trickier kettle of fish...

Choose the Classes Carefully Ⓓ

1) Numerical data is either discrete — it can only take certain exact values (e.g. number of points scored in a rugby match), or it's continuous — it can take any value in a range (e.g. height of a plant).

Don't worry about the terms 'discrete' and 'continuous'. Just learn the differences between the types of data.

2) When you're grouping data in a table, you need to make sure that none of the classes overlap and they cover all the possible values. The way you write the class intervals depends on the type of data.

1)   For **DISCRETE** data, you need '**GAPS**' between the classes...

For example, this table shows the number of points scored in some rugby matches.

Number of points	Frequency
0-10	5
11-20	20
21-30	31

The data can only take whole-number values.
So if the first class ends at 10 points, the second begins at 11 points because you don't have to fit 10.5 points in.

2)   When the data is **CONTINUOUS**, you **don't** want 'gaps'...

For example, this table shows the heights of some plants.

Height (h millimetres)	Frequency
$5 < h \leq 10$	12
$10 < h \leq 15$	15
$15 < h \leq 20$	11

See p60 for more on inequality symbols.

Use inequality symbols to cover all possible values.
$5 < h \leq 10$ means the height is greater than 5 mm and less than or equal to 10 mm. So, 10 would go in the 1st class, but 10.1 would go in the 2nd class.

Grouping data can make it simpler to analyse

Practise putting data into grouped tables so you get used to choosing the intervals. Turn over to see an example of how grouped frequency tables can be used in exam questions.

More Grouped Frequency Tables

Now you've learnt all about <u>grouped frequency tables</u>, have a look at the example below.
Then you can read all about <u>mid-interval values</u> too.

Using Grouped Frequency Tables Ⓓ

EXAMPLE:

Jonty wants to find out about the ages (in whole years) of people who use his local library.
Design a data-collection sheet he could use to collect his data.

1 <u>Choose NON-OVERLAPPING classes</u>.
These should have <u>gaps</u> because the
data can only be <u>whole numbers</u>.

> E.g. if the 2nd class ends at 39,
> the 3rd begins at 40.

Age (whole years)	Tally	Frequency
0-19		
20-39		
40-59		
60-79		
80 or over		

2 <u>Draw COLUMNS</u>
for <u>Age</u>, <u>Tally</u>
and <u>Frequency</u>
(see p155).

3 <u>Cover ALL POSSIBLE values</u>.

> E.g. group everything over 80 together so
> you don't have too many rows in the table.

Mid-Interval Values Ⓓ

The <u>middle</u> of a data <u>class</u> is called the <u>MID-INTERVAL VALUE</u>.

You need to <u>find mid-interval values</u> when <u>estimating the mean</u> of grouped data (see the next page).

Height (h millimetres)	Frequency
$5 < h \leq 10$	12
$10 < h \leq 15$	15
$15 < h \leq 20$	11
$20 < h \leq 25$	10
$25 < h \leq 30$	8

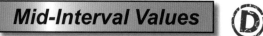

To find MID-INTERVAL VALUES:

Add together the <u>end values</u> of the
<u>class</u> and <u>divide by 2</u>.

E.g. for the first class: $\dfrac{5 + 10}{2} = \underline{7.5}$

Mid-interval values are easy to calculate

You'll often need to calculate mid-interval values, but you might not always be told to do it.
Just remember that you'll have to use them when estimating the mean of grouped data.

Grouped Frequency Tables — Averages

Another page on grouped frequency tables... lucky you. The methods for finding averages are similar to the ones used for normal frequency tables (see p168), except now you need to add two extra columns.

Find **Averages** from **Grouped Frequency Tables** Ⓒ

Unlike with ordinary frequency tables, you don't know the actual data values, only the classes they're in. So you have to ESTIMATE THE MEAN, rather than calculate it exactly. Again, you do this by adding columns:

> Add a 3RD COLUMN and enter the MID-INTERVAL VALUES for each class.
>
> *Finding mid-interval values is on the previous page.*
>
> Add a 4TH COLUMN to show 'FREQUENCY × MID-INTERVAL VALUE' for each class.
>
> The ESTIMATED MEAN is then: **4th Column Total ÷ 2nd Column Total**

And you'll be asked to find the MODAL CLASS and the CLASS CONTAINING THE MEDIAN, not exact values.

EXAMPLE: This table shows information about the weights, in kilograms, of 60 schoolchildren.

a) Write down the modal class.
b) Write down the class containing the median.
c) Calculate an estimate for the mean weight.

Weight (w kg)	Frequency
$30 < w \le 40$	8
$40 < w \le 50$	16
$50 < w \le 60$	18
$60 < w \le 70$	12
$70 < w \le 80$	6

a) The modal class is the one with the highest frequency. Modal class is $50 < w \le 60$

b) Work out the position of the median, then count through the 2nd column.

There are 60 values, so the median is halfway between the 30th and 31st values.

Both these values are in the third class, so the class containing the median is $50 < w \le 60$.

c) Add extra columns for 'mid-interval value' and 'frequency × mid-interval value'.

Label the frequency column f, your third column x and your fourth column $(f \times x)$.

Weight (w kg)	Frequency (f)	Mid-interval value (x)	($f \times x$)
$30 < w \le 40$	8	35	280
$40 < w \le 50$	16	45	720
$50 < w \le 60$	18	55	990
$60 < w \le 70$	12	65	780
$70 < w \le 80$	6	75	450
Total	60	—	3220

The mid-interval values are used to work out the 4th column. You don't need to add them up.

$$\text{Mean} = \frac{\text{total of } (f \times x) \text{ column}}{\text{total of } f \text{ column}} \quad \begin{matrix} \leftarrow \text{4th column total} \\ \leftarrow \text{2nd column total} \end{matrix}$$

$$= \frac{3220}{60}$$

$$= 53.7 \text{ kg (3 s.f.)}$$

Use the mid-interval values to estimate the mean

Learn all the details about the mean, median and modal class. When the question asks you to estimate the mean of grouped data, start by drawing two extra columns and labelling them clearly.

Frequency Polygons

Frequency polygons can be used to show data from grouped frequency tables.

Frequency Polygons Show Frequencies Ⓓ

Frequency polygons can be <u>fiddly</u> to draw. But remember this <u>rule</u> and you'll be okay.

Always plot your point at the <u>mid-interval value</u> of a class.

EXAMPLE:

Draw a frequency polygon to show the information in this table.

Add the endpoints of the class, then divide by 2.

Age (a) of people at concert	Frequency	mid-interval value
$10 < a \le 20$	20	$(10 + 20) \div 2 = 15$
$20 < a \le 30$	36	$(20 + 30) \div 2 = 25$
$30 < a \le 40$	32	$(30 + 40) \div 2 = 35$
$40 < a \le 50$	26	$(40 + 50) \div 2 = 45$
$50 < a \le 60$	24	$(50 + 60) \div 2 = 55$

1) Add a column to show the <u>mid-interval values</u>.

2) Plot the <u>mid-interval values</u> on the <u>horizontal axis</u> and the <u>frequencies</u> on the <u>vertical axis</u>.

   So plot the points (<u>15, 20</u>), (<u>25, 36</u>), (<u>35, 32</u>), (<u>45, 26</u>) and (<u>55, 24</u>).

3) <u>Join</u> the points of your frequency polygon using <u>straight lines</u> (not a curve).

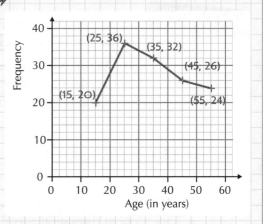

Use Frequency Polygons to **Interpret** *Sets of Data* Ⓓ

EXAMPLE:

The frequency polygon in blue represents the ages of people at a football game (while the red frequency polygon shows the ages of people at the above concert).

Make one comment to compare the two distributions.

Whatever you say must involve <u>both</u> distributions.

There were more people over 40 years old at the football game than at the concert.

There are loads of things you could say. For example, you could say there were far more people aged 20-30 at the concert than at the football game.

Practise these polygons frequently

If you learn the stuff above, then you should be able to cope with any question on frequency polygons.

Warm-up and Worked Exam Questions

Think of the exam as a big race. In order to do your best, you should really warm up first. So give these questions a go and get your brain all ready for the big day.

Warm-up Questions

1) 50 people were asked how many times a week they play sport. The table to the right shows the results.
   a) Find the median.   b) Find the mode.

No. of times sport played	Frequency
0	8
1	15
2	17
3	6
4	4
5 or more	0

2) Here are the heights of some adults to the nearest 0.1 cm. Design and fill in a frequency table to record the data.
   150.4   163.5   156.7   164.1
   182.8   175.4   171.2   169.0
   173.3   185.6   167.0   162.6

3) Find the mid-interval value for each data class in this table.

Length (l cm)	$15.5 \leq l < 16.5$	$16.5 \leq l < 17.5$	$17.5 \leq l < 18.5$	$18.5 \leq l < 19.5$
Frequency	12	18	23	8

4) Draw a frequency polygon to represent the information shown in this frequency table.

Height (h, in m) of student	Frequency
$1.0 < h \leq 1.2$	5
$1.2 < h \leq 1.4$	8
$1.4 < h \leq 1.6$	6
$1.6 < h \leq 1.8$	2

Worked Exam Question

This worked exam question is just like one that could come up in the exam. But the one in the exam won't have the answers filled in, complete with handy hints. Make the most of it now.

1   As part of their coursework, the students in a Year 11 maths class recorded Ⓒ their arm spans. The results are shown below.

Use the mid-interval value of each arm span group to find an estimate of the mean arm span.

Arm Span, x cm	Frequency	Mid-Interval Value	Frequency × Mid-Interval Value
$120 \leq x < 130$	13	(120 + 130) ÷ 2 = 125	13 × 125 = 1625
$130 \leq x < 140$	6	(130 + 140) ÷ 2 = 135	6 × 135 = 810
$140 \leq x < 150$	4	(140 + 150) ÷ 2 = 145	4 × 145 = 580
$150 \leq x < 160$	7	(150 + 160) ÷ 2 = 155	7 × 155 = 1085
Total	30		4100

So the estimate of the mean arm span is 4100 ÷ 30 = 136.6666...

Give your answer to a sensible level of accuracy

...........136.7..... cm
[4 marks]

SECTION SEVEN — STATISTICS AND PROBABILITY

Exam Questions

2 A traffic survey at a road junction recorded the following 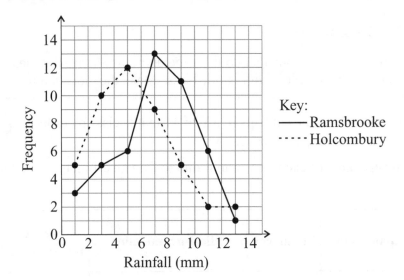 numbers of vehicles arriving per minute.

Vehicles per minute	0	1	2	3	4
Frequency	13	8	6	2	1

a) What is the median number of vehicles per minute?

....
[2 marks]

b) What is the mean number of vehicles per minute?

....
[3 marks]

3 The frequency polygons below show the amount of rainfall per day in two different towns over a period of 45 days.

Key:
—— Ramsbrooke
----- Holcombury

a) Write down one comparison of the amount of rainfall in the two towns.

....
[1 mark]

b) Use the frequency polygon to calculate an estimate of the mean amount of rainfall in Ramsbrooke over the 45 days.

.... mm
[3 marks]

Probability Basics

A lot of people think <u>probability</u> is tough. But learn the <u>basics</u> well, and it'll all make sense.

All *Probabilities* are *Between 0 and 1* (F)

- Probabilities are <u>always</u> between 0 and 1.

- The <u>higher</u> the probability of something, the <u>more likely</u> it is.

- A probability of <u>ZERO</u> means it will <u>NEVER HAPPEN</u>.

- A probability of <u>ONE</u> means it <u>DEFINITELY WILL HAPPEN</u>.

You can show the probability of something happening on a <u>scale</u> from 0 to 1.

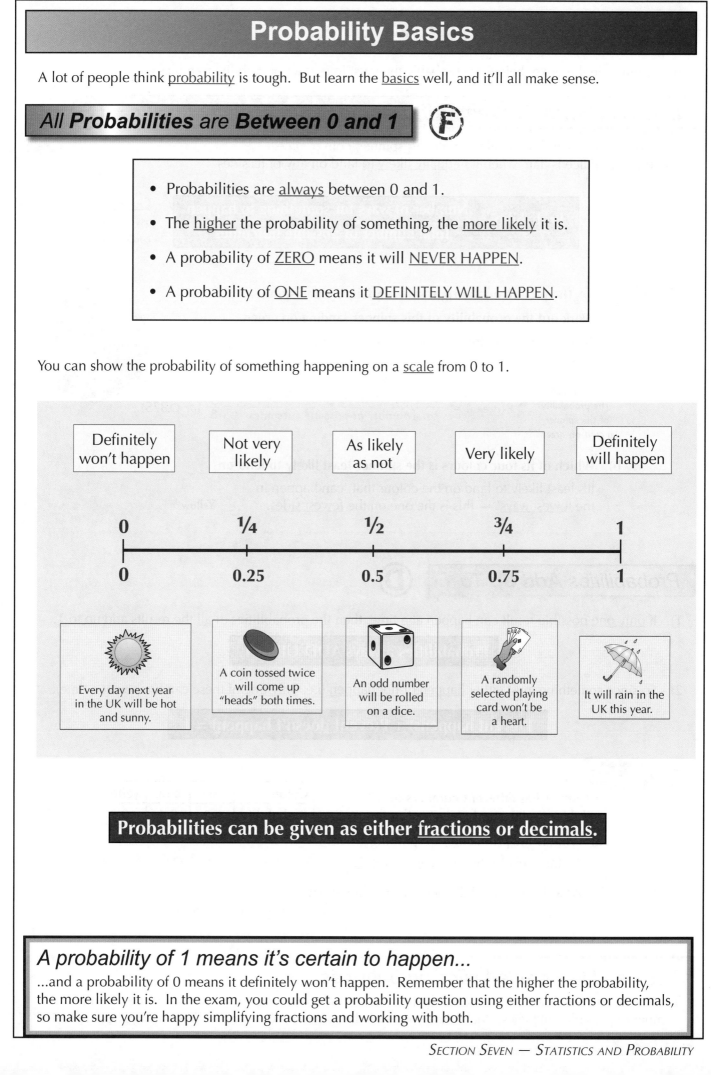

| Definitely won't happen | Not very likely | As likely as not | Very likely | Definitely will happen |

| 0 | ¼ | ½ | ¾ | 1 |
| 0 | 0.25 | 0.5 | 0.75 | 1 |

Every day next year in the UK will be hot and sunny.

A coin tossed twice will come up "heads" both times.

An odd number will be rolled on a dice.

A randomly selected playing card won't be a heart.

It will rain in the UK this year.

Probabilities can be given as either <u>fractions</u> or <u>decimals</u>.

A probability of 1 means it's certain to happen...

...and a probability of 0 means it definitely won't happen. Remember that the higher the probability, the more likely it is. In the exam, you could get a probability question using either fractions or decimals, so make sure you're happy simplifying fractions and working with both.

Probability

The formulas in the boxes below are probably the <u>most important</u> probability formulas ever.

Use This **Formula** *When* **All Outcomes** *are* **Equally Likely**

Use this formula to find probabilities for a <u>fair</u> spinner, coin or dice.
A spinner/coin/dice is 'fair' when it's <u>equally likely</u> to land on <u>any</u> of its sides.

$$\text{Probability} = \frac{\text{Number of ways for something to happen}}{\text{Total number of possible outcomes}}$$

<u>Outcomes</u> are just 'things that could happen'.

EXAMPLE: **The picture on the right shows a fair, 8-sided spinner.**

a) Work out the probability of this spinner landing on green.

1) Each side is a possible outcome — so there are <u>8 possible outcomes</u>.

2) And there are <u>3 ways</u> for it to land on green.

P(green) means 'The probability of the spinner landing on green'.

$$P(\text{green}) = \frac{\text{number of ways for 'green' to happen}}{\text{total number of possible outcomes}} = \frac{3}{8} \text{ (or } 0.375)$$

b) Which of its four colours is the spinner least likely to land on?

It's least likely to land on the colour that 'can happen in the <u>fewest ways</u>' — this is the one on the <u>fewest sides</u>. Yellow

Probabilities **Add Up To 1**

1) If <u>only one</u> possible result can happen at a time, then the probabilities of <u>all</u> the results <u>add up to 1</u>.

Probabilities always ADD UP to 1.

2) So since something must either <u>happen</u> or <u>not happen</u> (i.e. <u>only one</u> of these can happen at a time):

P(event happens) + P(event doesn't happen) = 1

EXAMPLE:

A spinner has different numbers of red, blue, yellow and green sections.

Colour	red	blue	yellow	green
Probability	0.1	0.4	0.3	

a) What is the probability of spinning green?

All the probabilities must <u>add up to 1</u>.

$$P(\text{green}) = 1 - (0.1 + 0.4 + 0.3) = 0.2$$

b) What is the probability of not spinning green?

$$P(\text{green}) + P(\text{not green}) = 1$$

$$P(\text{not green}) = 1 - P(\text{green}) = 1 - 0.2 = 0.8$$

You need to be able to find probabilities

To work out probabilities, just work out the number of ways of getting the result and divide it by the number of possible outcomes. Make sure you learn the formula too — probabilities always add up to 1.

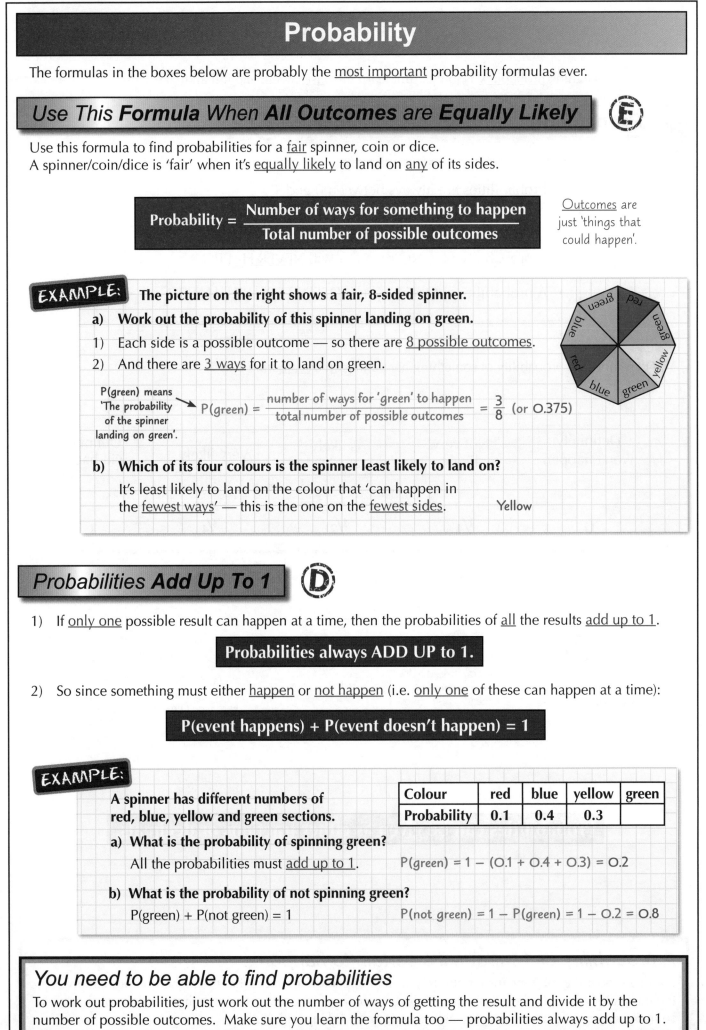

More Probability

Believe me, probability's not as bad as you think, but <u>you must learn the basic facts</u> on the previous two pages. Then you're on to sample space diagrams — these can help make probability questions less daunting.

Sample Space Diagrams *Show All Possible Outcomes*   Ⓓ

When there are <u>two things</u> happening (e.g. two spinners being spun),
you can use a <u>table</u> as a <u>sample space diagram</u>.

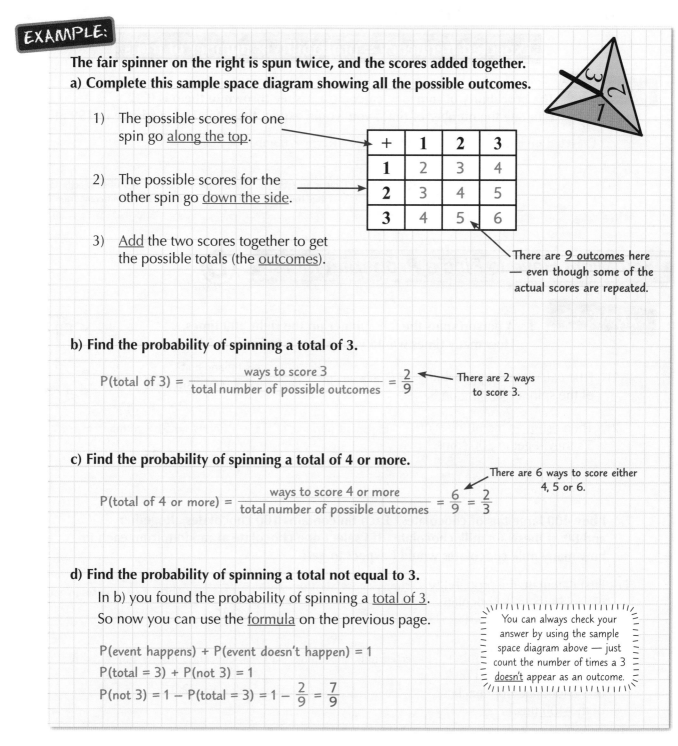

EXAMPLE:

The fair spinner on the right is spun twice, and the scores added together.
a) Complete this sample space diagram showing all the possible outcomes.

1) The possible scores for one spin go <u>along the top</u>.

2) The possible scores for the other spin go <u>down the side</u>.

+	1	2	3
1	2	3	4
2	3	4	5
3	4	5	6

3) <u>Add</u> the two scores together to get the possible totals (the <u>outcomes</u>).

There are <u>9 outcomes</u> here — even though some of the actual scores are repeated.

b) Find the probability of spinning a total of 3.

$$P(\text{total of }3) = \frac{\text{ways to score }3}{\text{total number of possible outcomes}} = \frac{2}{9}$$

There are 2 ways to score 3.

c) Find the probability of spinning a total of 4 or more.

There are 6 ways to score either 4, 5 or 6.

$$P(\text{total of 4 or more}) = \frac{\text{ways to score 4 or more}}{\text{total number of possible outcomes}} = \frac{6}{9} = \frac{2}{3}$$

d) Find the probability of spinning a total not equal to 3.

In b) you found the probability of spinning a <u>total of 3</u>.
So now you can use the <u>formula</u> on the previous page.

$$P(\text{event happens}) + P(\text{event doesn't happen}) = 1$$
$$P(\text{total} = 3) + P(\text{not }3) = 1$$
$$P(\text{not }3) = 1 - P(\text{total} = 3) = 1 - \frac{2}{9} = \frac{7}{9}$$

You can always check your answer by using the sample space diagram above — just count the number of times a 3 <u>doesn't</u> appear as an outcome.

List all possible outcomes in a methodical way

It's a good idea to draw a sample space diagram when you're working out probability questions,
even if it's not asked for — it makes sure that you don't miss any of the possible combinations.

Expected Frequency

You can use probabilities to work out how often you'd <u>expect</u> something to happen.

Use Probability to Find an "Expected Frequency" Ⓓ

1) Once you know the <u>probability</u> of something, you can <u>predict</u> how many times it will happen in a certain number of trials.

A 'trial' could be any activity — e.g. rolling a dice.

2) For example, you can predict the number of sixes you could expect if you rolled a fair dice 20 times. This prediction is called the <u>expected frequency</u>.

Expected frequency = probability × number of trials

EXAMPLE:

The probability of someone catching a frisbee thrown to them is 0.92.
Estimate the number of times you would expect them to catch a frisbee in 150 attempts.

Expected number of catches = probability of a catch × number of trials

= 0.92 × 150

= 138   *This is an <u>estimate.</u> They might not catch the frisbee <u>exactly</u> 138 times, but the number of catches shouldn't be too different from this.*

You Might Have to Find a **Probability** First Ⓓ

EXAMPLES:

1. A person spins the fair spinner on the right 200 times.
   How many times would you expect it to land on 5?

1) First calculate the probability of the spinner <u>landing on 5</u>.

$$P(\text{lands on 5}) = \frac{\text{ways to land on 5}}{\text{number of possible outcomes}} = \frac{1}{8}$$

2) Then <u>estimate</u> the number of 5's they'll get in <u>200</u> spins.

Expected number of 5's = P(lands on 5) × number of trials

$$= \frac{1}{8} \times 200 = 25$$

2. I buy 400 large tins of chocolates. Each tin contains 100 chocolates altogether, and 80 of these are milk chocolate. If I select one chocolate at random from each tin, how many milk chocolates would I expect to get?

1) First calculate the <u>probability</u> of <u>picking a milk chocolate</u> from <u>one tin</u>.

$$P(\text{milk chocolate from 1 tin}) = \frac{\text{number of ways to get a milk chocolate}}{\text{total number of chocolates in each tin}}$$

$$= \frac{80}{100} = \frac{4}{5}$$

2) Then <u>estimate</u> the number of milk chocolates if I pick one chocolate from each of the <u>400 tins</u>.

Expected milk chocolates = P(milk chocolate from 1 tin) × number of tins

$$= \frac{4}{5} \times 400 = 320$$

The expected frequency is just what's likely to happen

The number of trials is just the number of times you are going to do something.
Once you've understood everything here, move to the next page to learn all about relative frequency.

Relative Frequency

Relative frequency is nothing too difficult — it's just a way of estimating <u>probabilities</u>.

Fair or Biased?

'<u>Fair</u>' just means that all the possible scores are <u>equally likely</u>.

1) You can use the formula on page 176 to work out that the probability of rolling a 3 on a dice is $\frac{1}{6}$.

2) BUT this only works if it's a <u>fair dice</u>. If the dice is <u>wonky</u> (the technical term is '<u>biased</u>') then each number <u>won't</u> have an equal chance of being rolled.

3) This is where <u>relative frequency</u> is useful. You can use it to <u>estimate</u> probabilities when things are wonky.

Do the Experiment Again and Again and Again...

You need to do an experiment <u>over and over again</u> and count how often a result happens (its <u>frequency</u>). Then you can find its <u>relative frequency</u>.

$$\text{Relative frequency} = \frac{\text{Frequency}}{\text{Number of times you tried the experiment}}$$

An experiment could just mean rolling a dice.

You can use the <u>relative frequency</u> of a result as an <u>estimate</u> of its <u>probability</u>.

EXAMPLE:

The spinner on the right was spun 100 times. The results are in the table below. Estimate the probability of getting each of the scores.

Score	1	2	3	4	5	6
Frequency	10	14	36	20	11	9

The spinner was spun <u>100 times</u>.
So <u>divide</u> each of the frequencies by 100 to find the <u>relative frequencies</u>.

Score	1	2	3	4	5	6
Relative Frequency	$\frac{10}{100} = 0.1$	$\frac{14}{100} = 0.14$	$\frac{36}{100} = 0.36$	$\frac{20}{100} = 0.2$	$\frac{11}{100} = 0.11$	$\frac{9}{100} = 0.09$

The <u>MORE TIMES</u> you do the experiment, the <u>MORE ACCURATE</u> your estimate of the probability is likely to be. If you spun the above spinner <u>1000 times</u>, chances are you'd get a <u>better</u> estimate of the probability of each score.

For a <u>fair</u> dice or spinner, the relative frequencies should all be <u>roughly the same</u> after a large number of trials. If some of them are very <u>different</u>, the dice or spinner is probably <u>biased</u>.

Do the above results suggest that the spinner is biased?
Yes, because the relative frequency of 3 is much higher than the relative frequencies of 1, 5 and 6.

For a fair 6-sided spinner, you'd expect all the relative frequencies to be about 1 ÷ 6 = 0.17(ish).

Use decimals for relative frequency — they're easier to compare

If a coin/dice/spinner is fair, you can tell the probability of each of the results 'just by looking at it'. But if it's biased, then you have to use relative frequencies to estimate the probabilities.

Warm-up and Worked Exam Questions

Probability is really not that difficult, but it's easy to throw away marks by being a little slap-dash with your calculations. So it's important to get loads of practice. Try these questions.

Warm-up Questions

1) What is the probability of picking a white ball at random from a bag containing 3 black balls, 4 brown balls, 2 white balls and one purple ball?

2) The probability of rolling a double from two dice rolls is 1/6. What is the probability of not rolling a double?

3) A 4-sided spinner is spun twice and the scores added together. Complete the sample space diagram showing all the possible outcomes.

4) A 3-sided spinner is spun 100 times — it lands on red 43 times, blue 24 times and green the other times. Calculate the relative frequency of each outcome.

+	1	2	3	4
1				
2				
3				
4				

Worked Exam Questions

Take a look at these worked exam questions. They should give you a good idea of how to answer similar questions. You'll usually get at least one probability question in the exam.

---

1   Sarah has stripy, spotty and plain socks in her drawer.   **(D)**
She picks out a sock from the drawer at random.

The probability that she will pick a plain sock is 0.4.
The probability that she will pick a spotty sock from the drawer is x.
The probability that she will pick a stripy sock from the drawer is $2x$.

What is the probability that the sock she picks is stripy?
Give your answer as a decimal.

P(spotty sock) = x, P(stripy sock) = $2x$, P(plain sock) = 0.4

0.4 + x + $2x$ = 1 ⟶ Probabilities always add up to 1
$3x$ = 1 − 0.4 = 0.6
x = 0.6 ÷ 3 = 0.2

P(stripy sock) = $2x$, so P(stripy sock) = 2 × 0.2 = 0.4

.......**0.4**.......
[3 marks]

---

2   The probability of a train arriving in Udderston on time is 0.64.   **(D)**

Hester gets the train to Udderston 200 times a year.
How many times a year can Hester expect to arrive at the station on time?

200 × 0.64 = 128 times

.......**128**.......
[2 marks]

Exam Questions

3 Alecia has an 8-sided spinner labelled as shown on the right. It has an equal probability of landing on each of the 8 segments.

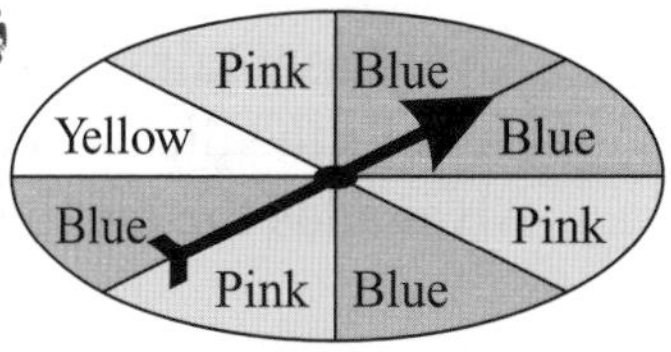

a) What colour segment is the spinner most likely to land on?

....
[1 mark]

b) What is the probability of the spinner landing on a pink segment?

....
[1 mark]

4 Steven asks all the members of his football team whether their favourite position is in attack, midfield, defence or goal. The table below shows his results.

Position	Frequency
Attack	6
Midfield	9
Defence	4
Goal	1

A member of the team is chosen at random.

What is the probability that this person's favourite position is in midfield?

....
[2 marks]

5 Katie decides to attend two new after-school activities. She can do one on Monday and one on Thursday. Below are lists of the activities she could do on these days.

Monday	**Thursday**
Hockey	Netball
Orchestra	Choir
Drama	Orienteering

List all the possible combinations of two clubs Katie could try in one week.

[2 marks]

Exam Questions

6 Alvar has a fair 6-sided dice and a set of five cards numbered 2, 4, 6, 8 and 10.
He rolls the dice and chooses a card at random.
Alvar adds the number on the dice to the number on the card to calculate his total score.

a) Complete the table below to show all of the possible scores.

Cards

		2	4	6	8	10
Dice	1					
	2					12
	3				11	13
	4			10	12	14
	5		9	11	13	15
	6	8	10	12	14	16

[2 marks]

b) Find the probability that Alvar will score exactly 9.
Give your answer as a fraction in its simplest form.

....
[2 marks]

7 Cameron has a bag containing a large number of counters. **©**
They are each numbered with a number from 1 to 5.

He selects one counter at random from the bag, records its number, and then puts it back in the bag. Cameron does this 100 times. He records his results in the table below.

Number on counter	1	2	3	4	5
Frequency	23	25	22	21	9
Relative frequency					

a) Complete the table, giving the relative frequencies of selecting each number.

[2 marks]

Cameron thinks there is the same number of counters with each number on in the bag.

b) Is he right? Give a reason for your answer.

....
....
[1 mark]

Revision Questions for Section Seven

Here's the inevitable list of straight-down-the-middle questions to test how much you know.
- Have a go at each question... but only tick it off when you can get it right without cheating.
- And when you think you could handle pretty much any statistics question, tick off the whole topic.

Collecting Data and Finding Averages (p154-157) ☑

Pet	Tally	Frequency

1) What is a sample and why does it need to be representative?

2) Complete this frequency table for the data below. ⟶
    Cat, Cat, Dog, Dog, Dog, Rabbit, Fish, Cat, Rabbit, Rabbit, Dog, Dog, Cat, Cat, Dog, Rabbit, Cat, Fish, Cat, Cat

3) List the four key things you should bear in mind when writing questionnaire questions.

4) Find the mode, median, mean and range of this data: 2, 8, 11, 15, 22, 24, 27, 30, 31, 31, 41

Statistics Graphs and Charts (p160-164) ☑

5) As well as counting the number of symbols on a pictogram, you need to check one other thing before you can find a frequency. What's the other thing?

6) The numbers of students in different years at a village school are shown in this table. Draw a bar chart to show this data.

School Year	7	8	9	10	11
Number of students	40	30	40	45	25

7) 125 boys and 125 girls were asked if they prefer Maths or Science. 74 of the boys said they prefer Maths, while 138 students altogether said they prefer Science. How many girls said they prefer Science?

8) Draw a pie chart to represent the data in question 6.

9) Sketch graphs to show:
    a) weak positive correlation, b) strong negative correlation, c) no correlation

10) This stem and leaf diagram shows the speeds of the fastest serves of some tennis players.
    Find the: a) fastest speed recorded b) median speed c) range of speeds

```
10| 2  5
11| 1  4  6
12| 0  2  2
13| 6  8
```
Key: 10|2 = 102 mph

Frequency Tables and Frequency Polygons (p168-172) ☑

11) Explain how you would find the mode, median and mean of the data in a frequency table.

12) For this grouped frequency table showing the lengths of some pet alligators:
    a) find the modal class,
    b) find the class containing the median,
    c) estimate the mean.

Length (y, in m)	Frequency
$1.4 \leq y < 1.5$	4
$1.5 \leq y < 1.6$	8
$1.6 \leq y < 1.7$	5
$1.7 \leq y < 1.8$	2

13) The frequency polygons on the right show the times (t, in seconds) taken to run 60 m by groups of Year 7 and Year 10 students. Make one comment to compare the two sets of data.

Probability (p175-179) ☑

14) What does a probability of 0 mean? What about a probability of ½?

15) I pick a random number between 1 and 50. Find the probability that my number is a multiple of 6.

16) If the probability of a spinner landing on red is 0.3, what is the probability that it doesn't land on red?

HT means Heads on the first flip and Tails on the second.

First flip	Second flip	
	Heads	Tails
Heads		HT
Tails		

17) I flip a fair coin twice.
    a) Complete this sample space diagram showing all the possible results.
    b) Use your diagram to find the probability of getting 2 Heads.

18) Write down the formula for estimating how many times you'd expect something to happen in n trials.

19) I flip a fair coin 100 times. How many times would you expect it to land on Tails?

20) When might you need to use relative frequency to find a probability?

Practice Exam 1: Non-calculator

As final preparation for the exams, we've included two full practice papers to really put your Maths skills to the test. Paper 1 is a non-calculator paper — Paper 2 (on page 198) requires a calculator. There's a formula sheet for both papers on page 214. Good luck...

Watch step-by-step solutions on video

Your free Online Edition of this book includes videos of our expert Maths tutors explaining the answers to this whole Exam Paper. (If you haven't accessed your Online Edition yet, you can find out how to get it at the front of this book.)

Candidate Surname		Candidate Forename(s)
Patel		Kailen

Centre Number	Candidate Number	Candidate Signature
I D k	I D k	test

GCSE

Mathematics   **Foundation Tier**

Paper 1 (Non-Calculator)

Practice Paper
Time allowed: 1 hour 45 minutes

You must have:
Pen, pencil, eraser, ruler, protractor, pair of compasses.
You may use tracing paper.

You are **not allowed** to use a calculator.

Instructions to candidates
- Use **black** ink to write your answers.
- Write your name and other details in the spaces provided above.
- Answer **all** questions in the spaces provided.
- In calculations, show clearly how you worked out your answers.
- Do all rough work on the paper.

Information for candidates
- The marks available are given in brackets at the end of each question.
- You may get marks for method, even if your answer is incorrect.
- There are 27 questions in this paper. There are no blank pages.
- There are 100 marks available for this paper.
- In questions labelled with an asterisk (*), you will be assessed on the quality of your written communication — take particular care here with spelling, punctuation and the quality of explanations.

Answer ALL the questions.

Write your answers in the spaces provided.

You must show all of your working.

1   (a)   What is the value of the 7 in the number 13 973?

.... tens ....

[1]

(b)   Round the number 331 to the nearest 10

.... 330 ....

[1]

(c)   Write out the number 1758 in words.

.... One thousand, seven hundred and fifty eight. ....

[1]

(d)   Write 85.7 to the nearest whole number.

.... 86 ....

[1]

[Total 4 marks]

2   (a)   Reflect the shaded shape in the mirror line.

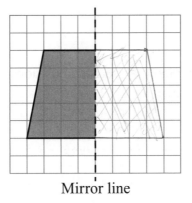

Mirror line

[1]

(b)   Here is a shape with rotational symmetry.

Write down the order of rotational symmetry of this shape.

.... 4 ....

[1]

[Total 2 marks]

1

3 (a) Work out 7.55×10

75.5

[1]

(b) Work out $427.3 \div 100$

4.273

[1]

(c) Work out 0.302×1000

302

[1]

[Total 3 marks]

---

4 Sam needs to buy grammar books and dictionaries for a class of students.
She needs 15 copies of each. She sees them on sale at a local book shop.

Grammar Book £3

Dictionary £2

(a) Calculate how much it would cost Sam to buy all the books for her class from the shop.

15
×3
45

15×2 = 30

45+30 = 75

£ *75*

[2]

(b) A website sells a set of the same grammar book and dictionary together for £4.
The website has a £5 delivery charge.
Should Sam buy the books from the shop or the website? Explain your answer.

4×15=60
60+5=65

Yes she should because it is ten pounds cheaper than from the local book shop.

[2]

[Total 4 marks]

5   A group of people were asked what they had eaten for breakfast that morning.
     The incomplete bar chart below shows some of the results.

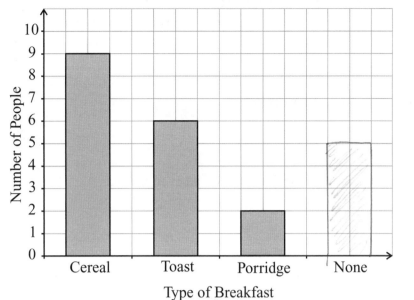

(a) How many people had porridge for breakfast?

                                                                    2
     ....

                                                                         [1]

(b) Write down how many more people had cereal for breakfast than toast.

              9-6= 3   3
     ....

                                                                         [2]

(c) Five people said that they didn't have any breakfast.
     Use this information to complete the bar chart.

                                                                         [1]

(d) Which type of breakfast was the mode?

                                                    Toast
     ....

                                                                         [1]

                                                              [Total 5 marks]

6   A clothes shop has 112 shirts in stock.
     On Monday it sells 17 shirts, and on Tuesday morning it gets a delivery of 38 shirts.
     How many shirts does it have after the delivery arrives on Tuesday?

       112   95
     - 17   + 38
       95   133   133
                                          ....

                                                              [Total 2 marks]

3

7 Use the words below to describe the likelihood of each of the events.

Impossible   Unlikely   Even   Likely   Certain

(a) A card drawn at random from a pack of playing cards is a two of diamonds.

........................*unlikely*........................

[1]

(b) Christmas will fall in November next year.

........................*Impossible*........................

[1]

(c) When you roll a fair six-sided dice, it lands on an even number.

........................*even*........................

[1]

[Total 3 marks]

---

8 The shape is drawn on a grid of centimetre squares.

(a) What is the perimeter of the shape?

18

.... cm

[1]

(b) What is the area of the shape? Make sure you give the units of your answer.

........................*12cm²*........................

[2]

[Total 3 marks]

---

9 The picture shows a man standing next to an elephant.

Estimate the height of the elephant.

........................*60*........................ m

[Total 3 marks]

4

10 The table shows the temperatures in Glasgow and Barcelona at different times on the same day.

Time	Temperature in Glasgow (°C)	Temperature in Barcelona (°C)
04 00	−8	6
08 00	−2	9
12 00	3	15
16 00	1	14
20 00	−3	11

(a) Write down the maximum temperature across the two cities.

.................15................. °C

[1]

(b) How much colder was Glasgow than Barcelona at 08 00?

2 + 9 = 11

.................11................. °C

[2]

(c) What was the difference between the highest temperature in Glasgow and the lowest temperature in Barcelona?

6 − 3 = 3

.................3................. °C

[3]

[Total 6 marks]

11 Look at the triangles below.

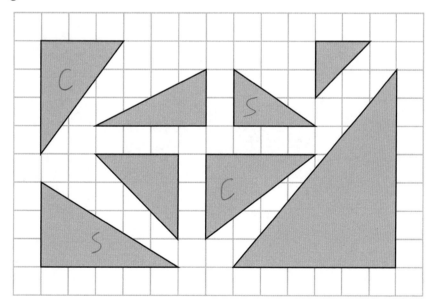

(a) Write C in the two triangles that are **congruent**.

[1]

(b) Write S in the two triangles that are **similar**.

[1]

[Total 2 marks]

5

190

12 Arthur thinks of three different numbers. The numbers have a range of 6 and a mean of 5. What are the three numbers?

$6 \times 3 = 18$

.........5......., ...5......., ...8......

[Total 2 marks]

13 For each of the statements below, choose a number from the list which matches the description.

19   125   36   18   84   32   124   48

(a) An even number less than 29.

.........18.........
[1]

(b) A power of 2.

.........48.........
[1]

(c) A square number.

.........36.........
[1]

(d) A cube number.

.........48.........
[1]

[Total 4 marks]

6

PRACTICE PAPER 1

14 Below is part of a bus timetable.

Windsor	08 12	08 46	09 08	09 33
Datchet	08 17	08 51	09 13	09 38
Old Windsor	08 23	08 57	09 19	09 44
Egham	08 33	09 07	09 29	09 54
Staines-upon-Thames	08 40	09 14	09 36	10 01
Chertsey	08 52	09 26	09 48	10 13
Weybridge	09 01	09 35	09 57	10 22

(a) (i) A bus leaves Datchet at 08 51.
        What time should it arrive in Weybridge?

        ....9:35..............

    (ii) A later bus leaves Old Windsor at 09 44.
        How long should it take to travel to Chertsey?

        .... minutes
        [3]

(b) Jake has a bus pass that gives him $\frac{1}{3}$ off the price of bus tickets.

    He buys a return ticket from Windsor to Egham.
    The normal price of the ticket is £3.60.

    How much does Jake pay?

    £ ....
    [2]

[Total 5 marks]

15 Work out the value of:
(a) $8 \times 2 + 7$

    ....
    [1]

(b) 2^3

    ....
    [1]

(c) $\sqrt{49}$

    ....
    [1]

[Total 3 marks]

192

*16 For work one week, Hassan drives 75 miles each day for 4 days.
He is able to claim in expenses 20p for every mile he drives.

Calculate the amount, in pounds, that he can claim in expenses for his working week.

£ ....

[Total 3 marks]

17 A quadrilateral has been drawn on the grid.

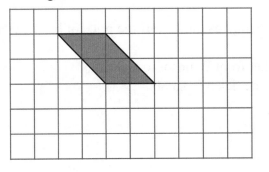

(a) Write down the name of this quadrilateral.

....
[1]

(b) Show how this quadrilateral can tessellate on the grid.
You should draw at least 6 quadrilaterals.

[2]

[Total 3 marks]

8

18 (a) Complete the table of values for $y = 2 + 2x$.

x	−1	0	1	2	3	4
y	0				8	10

[2]

(b) On the grid below, draw the graph of $y = 2 + 2x$.

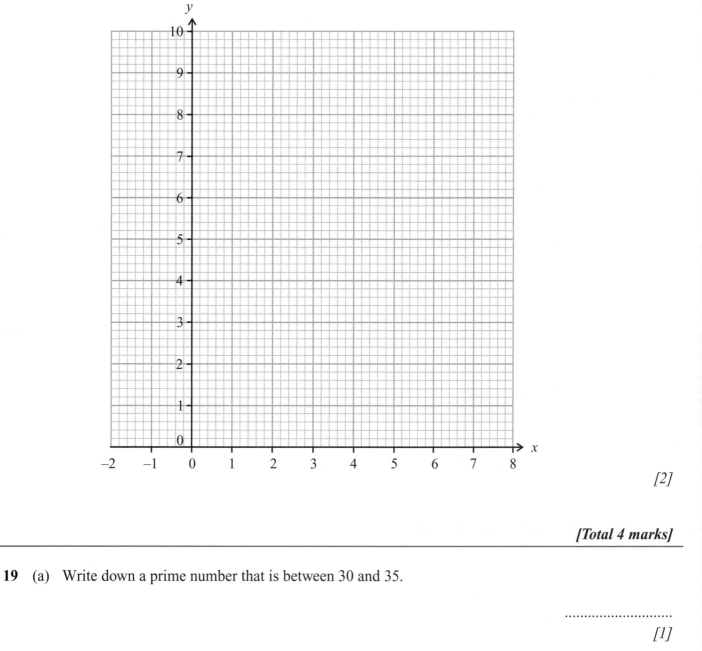

[2]

[Total 4 marks]

19 (a) Write down a prime number that is between 30 and 35.

....

[1]

(b) Write down all the factors of 28.

....

[2]

[Total 3 marks]

9

20 (a) Here are the first five terms of a number sequence.

57   46   35   24   13

(i) Write down the next number in the sequence.

....

(ii) Explain how you found your answer.

....

....

[2]

(b) The *n*th term of another sequence is given by $2n + 2$.
(i) Find the 4th term of this sequence.

....

*(ii) Rita says that the number 65 will be in the sequence.
Is she correct? Explain your answer.

....

....

[3]

[Total 5 marks]

21 Here is a right-angled triangle.

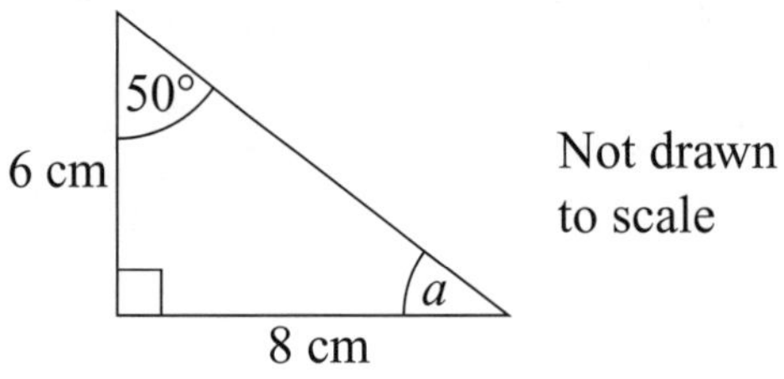

50°

6 cm

Not drawn
to scale

a

8 cm

*(a) (i) Work out the size of angle *a*.

.... °

(ii) Give a reason for your answer.

....

[2]

(b) Work out the area of the triangle.

.... cm²

[2]

[Total 4 marks]

22 (a) Solve $3x = 21$

$x =$ ....

[1]

(b) Solve $2m + 5 = -9$

$m =$ ....

[2]

(c) Solve $5n - 11 = 2n + 4$

$n =$ ....

[2]

[Total 5 marks]

23 Robert is the conductor of a choir that is performing a concert for charity.

Robert needs to buy sheet music for each of the 40 singers in the choir.
A website offers the sheet music for £7.00 per copy
with a 10% discount for an order of 10 or more copies.
Robert also has to pay £318 to rent the venue for the concert.

Robert sells 95 tickets for the concert. A ticket for the concert costs £10.

Robert has agreed to donate 50% of the concert profits to charity.
How much money should he donate to charity?

£ ....

[Total 7 marks]

196

24 (a) Expand and simplify $3(2 + b) + 5(3 - b)$

....

[2]

(b) Factorise $6c - 8$

$3x - 9 \cdot 3$

....

[1]

(c) Factorise fully $2d^2 + 10d$

....

[2]

[Total 5 marks]

*25 A Youth Centre decides to organise a trip to a theme park.
They plan to hire a coach that costs £100 for the day.
The cost to get into the theme park is £15 per person.

The Youth Centre will charge £23 per person for the trip,
which includes the coach journey and entry to the theme park.

The trip can only go ahead if the Youth Centre makes enough money to cover its costs.
Work out how many people need to go on the trip for it to go ahead.

....

[Total 4 marks]

12

26 A large supermarket chain is planning to open a new store in Digton.
Residents of Digton are asked to complete a questionnaire to give their views on the idea.

This is one of the questions from the questionnaire.

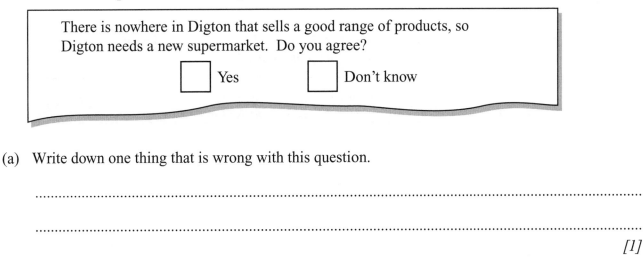

There is nowhere in Digton that sells a good range of products, so
Digton needs a new supermarket. Do you agree?

☐ Yes   ☐ Don't know

(a) Write down one thing that is wrong with this question.

....

....

[1]

(b) Design a more suitable question that could be used instead. Include some response boxes.

[2]

[Total 3 marks]

27 (a) Solve the inequality $6x < 3x + 9$.

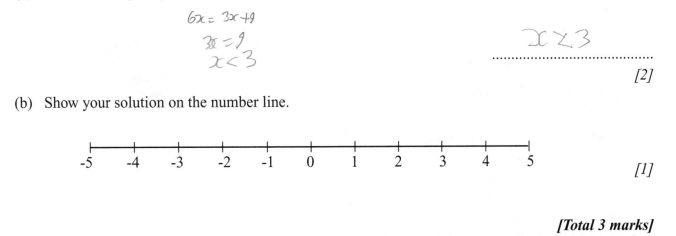

$6x = 3x + 9$

$3x = 9$

$x < 3$

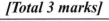

$x < 3$

[2]

(b) Show your solution on the number line.

-5 -4 -3 -2 -1 0 1 2 3 4 5

[1]

[Total 3 marks]

[TOTAL FOR PAPER = 100 MARKS]

13

Practice Exam 2: Calculator
Right, here's Exam Paper 2 — you'll need a calculator for this one. Don't forget there's
a formula sheet on page 214 if you need it (you'll get one of these in the real exam too).

Watch step-by-step solutions on video
Your free Online Edition of this book includes videos of our expert Maths tutors
explaining the answers to this whole Exam Paper. (If you haven't accessed your
Online Edition yet, you can find out how to get it at the front of this book.)

Candidate Surname	Candidate Forename(s)

Centre Number	Candidate Number	Candidate Signature

GCSE

Mathematics

Paper 2 (Calculator)

Foundation Tier

Practice Paper
Time allowed: 1 hour 45 minutes

You must have:
Pen, pencil, eraser, ruler, protractor, pair of compasses.
You may use tracing paper.

You **may use** a calculator.

Instructions to candidates
* Use **black** ink to write your answers.
* Write your name and other details in the spaces provided above.
* Answer **all** questions in the spaces provided.
* In calculations, show clearly how you worked out your answers.
* Do all rough work on the paper.
* Unless a question tells you otherwise, take the value of π to be 3.142,
  or use the π button on your calculator.

Information for candidates
* The marks available are given in brackets at the end of each question.
* You may get marks for method, even if your answer is incorrect.
* There are 25 questions in this paper. There are no blank pages.
* There are 100 marks available for this paper.
* In questions labelled with an asterisk (*), you will be assessed
  on the quality of your written communication — take particular
  care here with spelling, punctuation and the quality of explanations.

Answer ALL the questions.

Write your answers in the spaces provided.

You must show all of your working.

1 (a) Write twenty four thousand and twelve in digits.

....

[1]

(b) Put the following numbers in order of size, from smallest to largest.

       85.3   95.3   85.03   90.9   87.2

.... , ... , ... , ... , ...

[1]

Jill has 4 cards, each with a number written on.

| 5 | | 9 | | 6 | | 2 |

She lines all of the cards up to make a 4 digit number.

(c) (i) What is the largest number she can make?

....

(ii) What is the smallest number she can make?

....

[2]

[Total 4 marks]

2 The diagram below shows a quadrilateral.

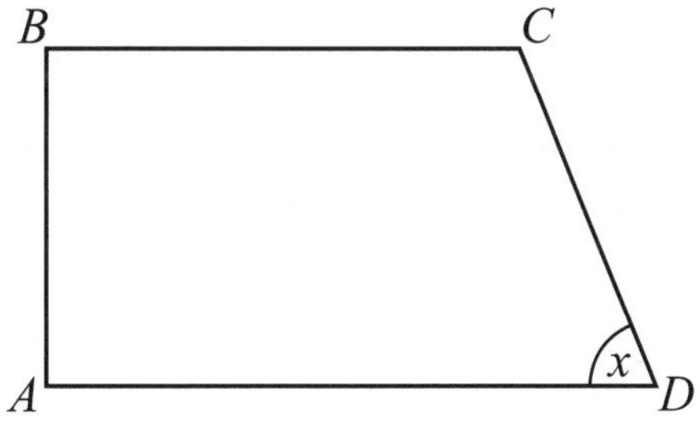

(a) Measure the length of *AB*.

.... cm

[1]

(b) On the diagram, mark the midpoint of the line *AB* with a cross.

[1]

(c) Mark a right angle with the letter R.

[1]

(d) Give the mathematical name for the type of angle marked *x*.

....

[1]

[Total 4 marks]

1

200

3 (a) Write down the fraction of this shape that is shaded.

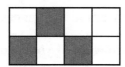

....
[1]

(b) Write down the fraction of the shape that is shaded.
Give your answer in its simplest form.

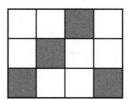

....
[2]

(c) (i) Shade 10% of this shape.

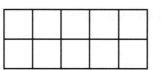

(ii) Write down the proportion of the shape that is **not** shaded.
Give your answer as a decimal.

....
[2]

[Total 5 marks]

2

4 The pictogram shows the number of pies eaten by customers at a restaurant one week.

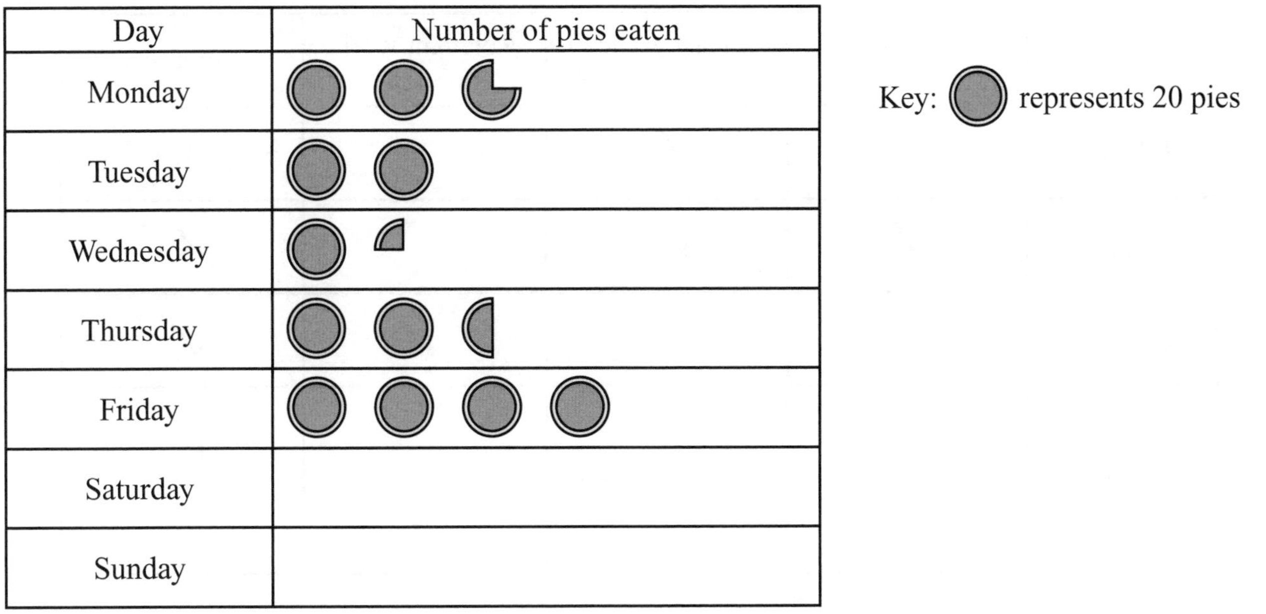

Day	Number of pies eaten
Monday	
Tuesday	
Wednesday	
Thursday	
Friday	
Saturday	
Sunday	

Key: ⬤ represents 20 pies

(a) How many pies were eaten on Monday?

....

[1]

(b) How many more pies were eaten on Thursday than on Wednesday?

....

[1]

90 pies were eaten on Saturday.
35 pies were eaten on Sunday.

(c) Complete the pictogram using this information.

[2]

[Total 4 marks]

3

5   The table below gives some information about five candidates who have applied for a job as a postal worker.

	Driver's licence	At least grade G in GCSE Maths	Relevant work experience	Reference
Edward		✓		
Fiona	✓	✓		✓
Graham	✓		✓	✓
Nadia		✓		✓
Isaac		✓	✓	

(a)   Which candidates have at least a grade G in GCSE Maths?

....

[1]

(b)   Write down all the information from the table about Graham.

....

....

[1]

(c)   Write down the fraction of candidates who have relevant work experience.

....

[1]

[Total 3 marks]

6   Daisy is monitoring the amount of rainfall (in cm) each month in the towns of Duluth and Norcross. Her results for a six-month period are shown in the table below.

Month	April	May	June	July	August	September
Duluth (cm of rain)	6	3	5	6	7	10
Norcross (cm of rain)	5	4	4	8	7	8

Draw a suitable chart or diagram that could be used to compare the numbers of centimetres of rain each month in the two towns.

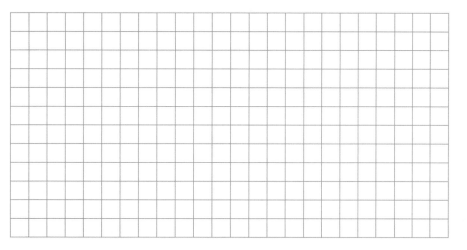

[Total 4 marks]

7   (a)   Write down the mathematical name of each shape

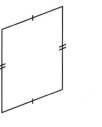   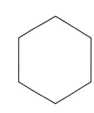

(i) ....   (ii) ....   *[2]*

(b)   Here is a 3D shape.

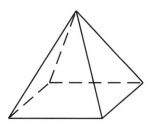

(i)   Write down the number of faces of the shape.

....

(ii)   Write down the number of edges of the shape.

....

[2]

[Total 4 marks]

8   Complete the pattern so that it has rotational symmetry of order 4.

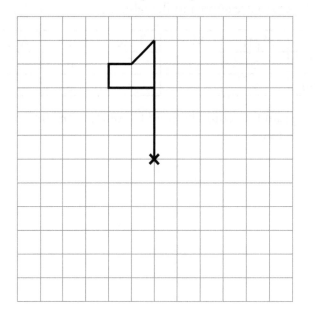

[Total 2 marks]

---

9   Look at the coordinate grid.

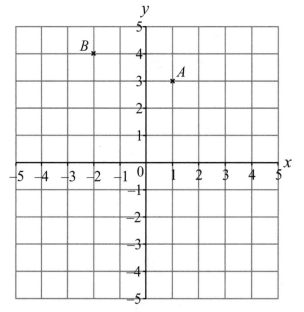

(a)   Write down the coordinates of point *A*.

(.................., ...)

[1]

(b)   Write down the coordinates of point *B*.

(.................., ...)

[1]

(c)   Plot the point (−3, −4) on the grid. Label this point *C*.

[1]

[Total 3 marks]

6

10 Jemma is the manager of a shoe shop.
One week Jemma recorded the size of each pair of a particular type of shoe that was sold.

Here are her results.

6   4   7   6   4   8   4

(a) Write down the modal size sold.

....

[1]

(b) Work out the median size sold.

....

[2]

(c) Explain why Jemma might be more interested in the modal size sold than the median size.

....

....

[1]

[Total 4 marks]

---

11 Andrew runs a snack shop. These are the prices for some of the snacks he sells.

Price list	
Chocolate bar	£0.70
Crisps	£0.35
Flapjack	£0.95
Cereal bar	£1.10
Apple	£0.50

(a) How much does it cost to buy a chocolate bar, a cereal bar and a flapjack?

£ ....

[1]

(b) A customer buys two packets of crisps and an apple. They pay with a £5 note.
How much change should Andrew give them?

£ ....

[2]

(c) Next week, Andrew is going to put all of his prices up by 20%.
How much will it cost to buy a cereal bar and an apple next week?

£ ....

[3]

[Total 6 marks]

12 (a) Write down the number shown by the arrow.

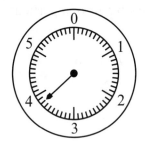

.... [1]

(b) Find the number 84 on the number line.
Mark it with an arrow.

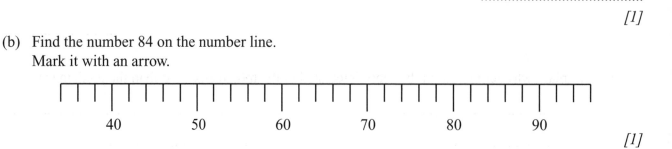

[1]

[Total 2 marks]

13 Mark is following a recipe for roasting a chicken.
The recipe gives the following information on the cooking time for a chicken:

> *"Cook the chicken for 50 minutes per kg, plus an extra 20 minutes."*

(a) Mark has bought a chicken that weighs 2.4 kg.
Calculate how long he should cook the chicken for.

.... hours .... minutes

[2]

Mark is also making a salad to go with his chicken.

> **Salad Dressing: Serves 4**
> 3 tablespoons olive oil
> 1 tablespoon vinegar
> Seasoning

(b) Mark wants to make a salad dressing for six people using the recipe above.
How much olive oil should he use?

.... tablespoons

[1]

[Total 3 marks]

14

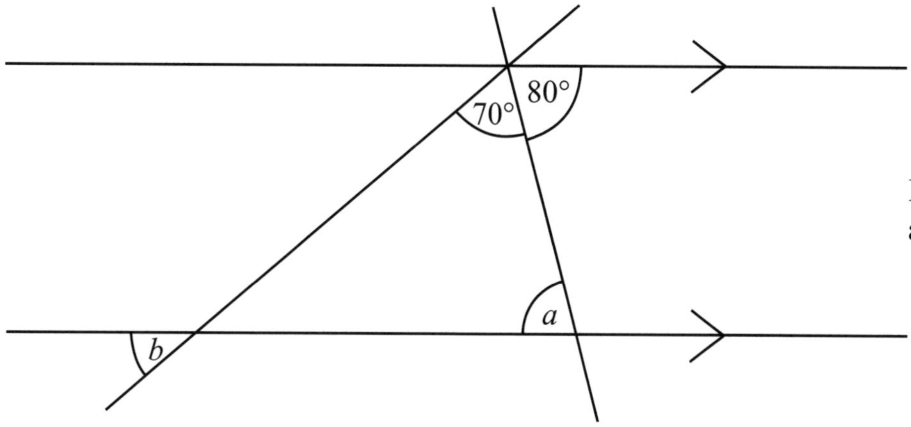

Diagram not accurately drawn

(a) Write down the size of angle *a* and give a reason for your answer.

....

[2]

(b) Work out the size of angle *b*.

....°

[2]

[Total 4 marks]

15 In a probability experiment, Amanda flips an ordinary coin and notes the result.
Then she rolls an ordinary unbiased 6-sided dice and notes the result.

(a) List all the possible combinations she could get.

....

....

[2]

Amanda flips the coin and rolls the dice.

(b) Work out the probability that Amanda gets heads on the coin **and** a 3 on the dice.

....

[1]

(c) Work out the probability that she gets heads on the coin **and** a 3 or higher on the dice.
Give your answer as a fraction in its simplest form.

....

[2]

[Total 5 marks]

208

16 $e = 3, f = 5$ and $g = -7$

Find the value of

(a) $e^2 + g^2$

....
[2]

(b) $\dfrac{efg}{2}$

....
[2]

[Total 4 marks]

---

17 Below is a scale drawing of Sharon's kitchen.

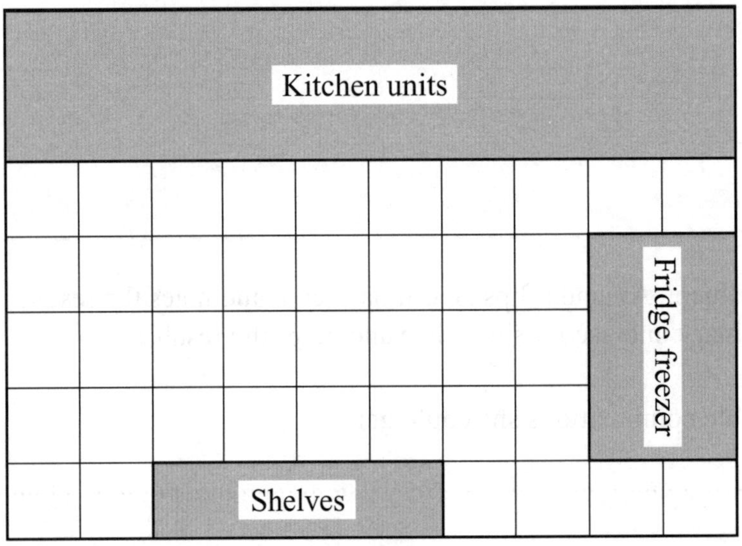

(a) The shelves are 2 m long and 0.5 m wide. What is the scale of the drawing?

1 cm to .... m
[1]

(b) What is the area of the real kitchen?

.... m²
[3]

[Total 4 marks]

18 The graph below can be used to convert between pounds (£) and euros (€).

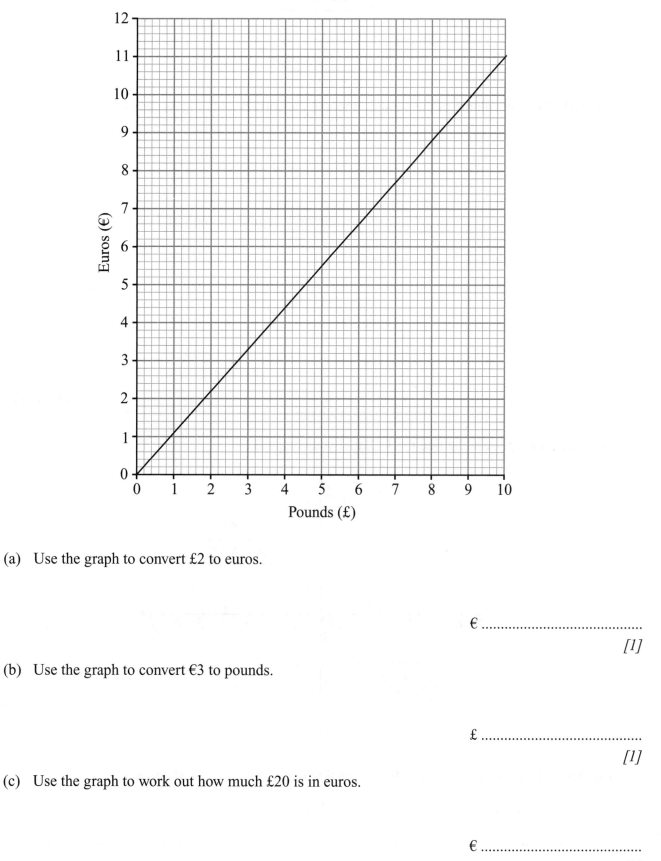

(a) Use the graph to convert £2 to euros.

€ ....

[1]

(b) Use the graph to convert €3 to pounds.

£ ....

[1]

(c) Use the graph to work out how much £20 is in euros.

€ ....

[2]

[Total 4 marks]

11

19 The diagram shows a cuboid.

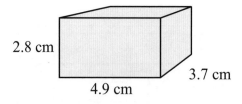

2.8 cm

3.7 cm

4.9 cm

Diagram not
accurately drawn

(a) Calculate the volume of the cuboid. Include the correct units in your answer.

$3.7 \times 4.9 \times 2.8 = 50.764$

$50.764 \, cm^2$
....

[3]

Another cuboid has a volume of 56 cm³, a width of 4 cm and a length of 7 cm.

(b) Calculate h, the height of the cuboid.

.... cm

[2]

[Total 5 marks]

20 (a) Reflect shape **A** in the line $x = 1$. Label your shape **B**.

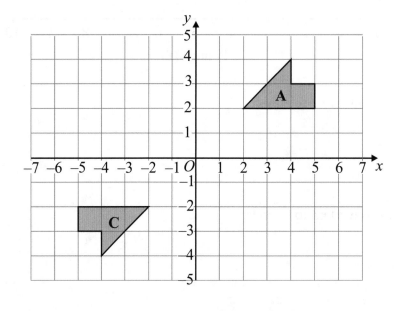

[2]

(b) Describe fully the single transformation that maps shape **C** onto shape **A**.

....

....

....

[3]

[Total 5 marks]

21 A painter wants to calculate the cost of painting the four outside walls of a warehouse.
The diagram below gives the dimensions of the warehouse.

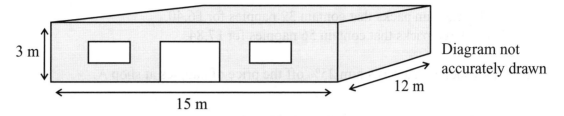

3 m

15 m

12 m

Diagram not
accurately drawn

There are 5 windows in the warehouse that each measure 2 m by 1 m and
a door that measures 3 m by 2.5 m.

The paint covers 13 m² per litre.
The paint can be bought in tins that contain 5 litres or 2.5 litres.
5 litre tins cost £20.99
2.5 litre tins cost £12.99

Calculate the cheapest price for painting the warehouse. You must show all your working.

£ ....

[Total 6 marks]

22 The equation $x^3 - x - 4 = 0$ has a solution between 1 and 2.

Use trial and improvement to find this solution.
Give your answer correct to one decimal place.

You must show **ALL** of your working.

$x = $ ....

[Total 4 marks]

*23 Tom is buying nappies for his son.
He can buy nappies from two different shops.

Shop A sells medium packs that contain 32 nappies for £6.40,
Shop B sells large packs that contain 56 nappies for £7.84.

Tom has a voucher that gives him 25% off the price of nappies at shop A.

Should Tom buy the nappies at shop A or shop B?
You must show all your working.

[Total 4 marks]

---

24 Simplify:
(a) $a^4 \times a^5$

....

[1]

(b) $b^9 \div b^3$

....

[1]

[Total 2 marks]

---

25 (a) The diagram shows a right-angled triangle.

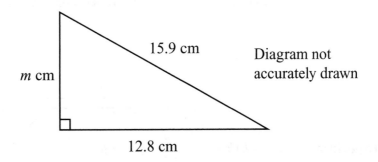

15.9 cm

m cm

Diagram not accurately drawn

12.8 cm

Calculate the value of m. Give your answer correct to 1 decimal place.

$15.9^2 - 12.8^2 = 88.97$

$\sqrt{88.97} = 9.43$

9.4

....

[2]

*(b) Explain why triangle ABC below cannot be a right-angled triangle.

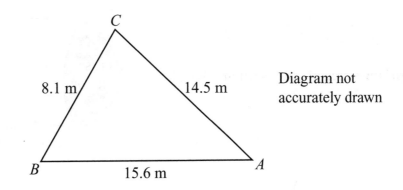

C

8.1 m

14.5 m

Diagram not accurately drawn

B

15.6 m

A

....

....

....

....

....

[3]

[Total 5 marks]

[TOTAL FOR PAPER = 100 MARKS]

PRACTICE PAPER 2

Formula Sheet: Foundation Tier

Area of trapezium $= \dfrac{1}{2}(a + b)h$

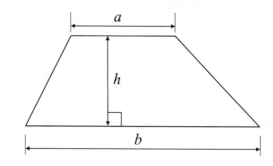

Volume of prism = area of cross-section × length

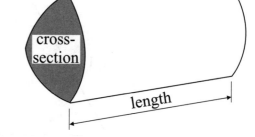

Section One — Numbers

Page 10 (Warm-up Questions)

1. a) One million, two hundred and thirty-four thousand, five hundred and thirty-one.
   b) Twenty-three thousand, four hundred and fifty-six.
   c) Three thousand, four hundred and two.
2. 56 421
3. 9, 23, 87, 345, 493, 1029, 3004
4. 0.008, 0.09, 0.1, 0.2, 0.307, 0.37
5. 171 cm
6. 1.72 litres
7. a) 1230
   b) 48 *(E.g. work out 2.4 × 2, then × by 10)*
8. a) 0.245
   b) 5 *(You can divide both numbers by 100 to get: 40 ÷ 8 = 5)*
9. a) 336
   b) 832
   c) 179.2
   d) 6.12
10. a) 12
    b) 121
    c) 56
    d) 30
    These multiplications and divisions involving decimals are much easier if you turn them into whole-number calculations.

Pages 11-12 (Exam Questions)

2. a) Five thousand and seventy-nine *[1 mark]*
   b) 6105 *[1 mark]*
   c) 90/ninety *[1 mark]*
3. 522 – (197 + 24) = 301
   [2 marks available — 1 mark for subtracting the two numbers from 522, 1 mark for the correct answer]
4. Total cost = £2.15 + £2.40 + £2.40 = £6.95
   Change = £10 – £6.95 = £3.05
   [2 marks available — 1 mark for adding amounts and subtracting from £10, 1 mark for the correct answer]
5. 53.3, 52.91, 35.6, 35.54, 35.06 *[1 mark]*
6. 98 649, 98 653, 100 003, 100 010 *[1 mark]*
7. a)
$$\begin{array}{r} 113 \\ \times\ 76 \\ \hline 678 \\ +\ 7910 \\ \hline 8588 \end{array}$$
   [2 marks available — 1 mark for a correct method, 1 mark for the correct answer]
   "A correct method" here can be any non-calculator multiplication method.
   b)
$$\begin{array}{r} 376 \\ \times\ 48 \\ \hline 3008 \\ +\ 15040 \\ \hline 18048 \end{array}$$
   [2 marks available — 1 mark for a correct method, 1 mark for the correct answer]
8. a) 19 + 26 ÷ 2 = 19 + 13 = 32
   [2 marks available — 1 mark for doing the calculation steps in the correct order, 1 mark for the correct answer]
   b) (22 – 18) × (3 + 8) = 4 × 11 = 44
   [2 marks available — 1 mark for doing the calculation steps in the correct order, 1 mark for the correct answer]
9.
$$54\overline{)75^{21}6}\quad =\ 14$$
   So 756 ÷ 54 = 14 cushions
   [2 marks available — 1 mark for any division method, 1 mark for the correct answer]

10. a) 5.6 × 4.27 = (23 912 ÷ 10) ÷ 100 = 23.912 *[1 mark]*
    b) 0.56 × 4 270 000 = (23 912 ÷ 100) × 10 000 = 2 391 200 *[1 mark]*
    c) 2391.2 ÷ 4.27 = (56 ÷ 10) × 100 = 560 *[1 mark]*
11. a) $14 \div 0.7 = \frac{14}{0.7} = \frac{140}{7} = 20$
    [2 marks available — 1 mark for a correct method, 1 mark for the correct answer]
    b) $23 \div 0.46 = \frac{23}{0.46} = \frac{2300}{46} = 50$
    [2 marks available — 1 mark for a correct method, 1 mark for the correct answer]

Page 19 (Warm-up Questions)

1. a) 12
   b) –6
   c) 9
   d) –3
2. a) 2, 4, 6, 8, 10
   b) 1, 3, 5, 7, 9
   c) 1, 4, 9
   d) 1, 8
3. 31, 37
4. 27 ÷ 3 = 9. So 27 is not a prime number because it divides by 3 and 9.
5. 1, 2, 4, 5, 8, 10, 20, 40
6. 2 and 5
7. 20 (multiples of 4 are: 4, 8, 12, 16, 20, ..., multiples of 5 are: 5, 10, 15, 20, ...)
8. 12 (factors of 36 are: 1, 2, 3, 4, 6, 9, 12, 18, 36, factors of 96 are: 1, 2, 3, 4, 6, 8, 12, 16, 24, 32, 48, 96)

Page 20 (Exam Questions)

2. a) 41 *[1 mark]*
   b) 100 *[1 mark]*
   c) 27 *[1 mark]*
3. a) –11 × 7 = –77 *[1 mark]*
   b) –72 ÷ –8 = 9 *[1 mark]*
4. a) E.g. 21, 42 *[1 mark for any two multiples of 21]*
   b) 47 *[1 mark]*
5. Multiples of 35 are: 35, 70, 105, 140, 175, 210, 245, 280, 315, 350, (385), 420, ...
   Multiples of 55 are: 55, 110, 165, 220, 275, 330, (385), 440, ...
   So the LCM is 385, which is the minimum number of jars he needs.
   So the minimum number of packs he needs is 385 ÷ 35 = 11
   [3 marks available — 1 mark for a correct method to find LCM, 1 mark for LCM correct, 1 mark for correct number of packs]

Page 26 (Warm-up Questions)

1. 0.7
2. 66.$\dot{6}$% or $66\frac{2}{3}$%
3. $\frac{2}{5}$
4. $\frac{3}{4}$
5. $\frac{2}{6}$ and $\frac{5}{15}$
6. a) $\frac{4}{15}$
   b) $\frac{2}{5} \div \frac{2}{3} = \frac{2}{5} \times \frac{3}{2} = \frac{6}{10} = \frac{3}{5}$
   c) $\frac{2}{5} + \frac{2}{3} = \frac{6}{15} + \frac{10}{15} = \frac{16}{15} = 1\frac{1}{15}$
   d) $\frac{2}{3} - \frac{2}{5} = \frac{10}{15} - \frac{6}{15} = \frac{4}{15}$
7. 0.$\dot{2}$8571$\dot{4}$

Page 27 (Exam Questions)

2 a) $\frac{3}{4} = 3 \div 4 = 0.75$ *[1 mark]*

b) $0.06 \times 100 = 6\%$ *[1 mark]*

c) $35\% = \frac{35}{100}$ *[1 mark]*

$= \frac{35 \div 5}{100 \div 5} = \frac{7}{20}$ *[1 mark]*

[2 marks available in total — as above]

3 a) $(60 \div 5) \times 3 = 12 \times 3 = 36$

[2 marks available — 1 mark for dividing by 5, 1 mark for the correct answer]

You could also multiply 60 by 0.6.

b) $\frac{15}{40} = \frac{3}{8}$

[2 marks available — 1 mark for putting the numbers into a fraction, 1 mark for the correct final answer]

4 a) $\frac{1}{2} \times \frac{1}{6} = \frac{1 \times 1}{2 \times 6} = \frac{1}{12}$ *[1 mark]*

b) $\frac{2}{3} \div \frac{3}{5} = \frac{2}{3} \times \frac{5}{3} = \frac{2 \times 5}{3 \times 3} = \frac{10}{9}$ or $1\frac{1}{9}$

[2 marks available — 1 mark for changing to the reciprocal fraction and multiplying, 1 mark for the correct answer]

5 a) $\frac{18}{7} = 2\frac{4}{7}$ *[1 mark]*

b) $1\frac{3}{4} = \frac{4+3}{4} = \frac{7}{4}$ *[1 mark]*

6 a) $\frac{1}{6} + \frac{2}{3} = \frac{1}{6} + \frac{4}{6} = \frac{1+4}{6} = \frac{5}{6}$

[2 marks available — 1 mark for finding a common denominator, 1 mark for the correct answer]

b) $\frac{7}{8} - \frac{3}{4} = \frac{7}{8} - \frac{6}{8} = \frac{7-6}{8} = \frac{1}{8}$

[2 marks available — 1 mark for finding a common denominator, 1 mark for the correct answer]

Page 33 (Warm-up Questions)

1 £1.28 *(Cost per bar = £0.96 ÷ 3 = £0.32 Cost for 4 bars = £0.32 × 4 = £1.28)*

2 525 g *(In the 250 g jar you get 250 g ÷ £1.25 = 200 g per £, in the 350 g jar you get 350 g ÷ £2.10 = 166.67 g per £, in the 525 g jar you get 525 g ÷ £2.50 = 210 g per £)*

3 £17 *(34 ÷ 100 × 50 = 17)*

4 £138 *(15% of £120 is = 0.15 × £120 = £18, which you add to the £120)*

5 74% *(37 ÷ 50 × 100)*
If you didn't have your calculator you could do this by doubling both numbers to make it out of 100, which is a percentage.

6 9 *(10% of 60 = 6, so 5% = 3. Add together to get 15%.)*

7 £205 *(2.5% of £200 = £5, which you add to the £200.)*

8 a) 1:2   b) 4:3

9 5:9

10 £1000:£1400 *(5 + 7 = 12 parts, so £2400 ÷ 12 = £200 per part.)*

Page 34 (Exam Questions)

2 Price per ml:
250 ml bottle: £2.30 ÷ 250 = £0.0092
330 ml bottle: £2.97 ÷ 330 = £0.009
500 ml bottle: £4.10 ÷ 500 = £0.0082
Therefore the 500 ml bottle is the best value for money.
[2 marks available — 1 mark for finding the price per ml (or amount per penny/pound), 1 mark for the correct answer]

3 $100\% - (60\% + 30\%) = 10\%$ *[1 mark]*
10% of the total number of elephants = 3 *[1 mark]*
Total number of elephants = $(3 \div 10) \times 100$ *[1 mark]*
$= 30$ *[1 mark]*
[4 marks available in total — as above]

4 a) boys : girls
= 12 : 14
= 6 : 7 *[1 mark]*

b) $25 \div (2 + 3) = 5$ *[1 mark]*
Number of girls is $5 \times 3 = 15$ *[1 mark]*
[2 marks available in total — as above]

5 $£160 \div (3 + 6 + 7) = £160 \div 16$ *[1 mark]*
$= £10$
So Christine's share = $£10 \times 7 = £70$ *[1 mark]*
[2 marks available in total — as above]

Page 40 (Warm-up Questions)

1 a) 3.2
b) 1.8
c) 2.3
d) 0.5
e) 9.8

2 a) 3
b) 5
c) 2
d) 7
e) 3

3 a) 350 *(the decider is 2, so keep the 5, and fill the missing place with zero)*
b) 500 *(the decider is 6, so round the 4 up to 5, and fill the missing places with zeros)*
c) 12.4 *(the decider is 8, so round the 3 up to 4)*
d) 0.036 *(the decider is 6, so round the 5 up to 6)*

4 a) 2900
b) 500
c) 100

5 a) 14.14 (other value is –14.14)
b) 20.00

6 4 *(This is approximately (30 – 10) ÷ 5)*

7 a) 3^8
b) 4
c) 8^{12}
d) 1
e) 7^6

8 a) 5^{12}
b) 36 or 6^2
c) 2^5

Page 41 (Exam Questions)

2 a) $8.7^3 = 658.503$ *[1 mark]*
b) $\sqrt{2025} = 45$ *[1 mark]*

3 Since $6^2 = 36$ and $7^2 = 49$, $6 < \sqrt{42} < 7$
So, $\sqrt{42} \approx 6.5$ *[1 mark]*
You would be given the mark here for any answer which was greater than 6 and less than 7.

4 a) 428.6 light years *[1 mark]*
b) 430 light years *[1 mark]*

5 E.g. $\frac{12.2 \times 1.86}{0.19} \approx \frac{10 \times 2}{0.2} = \frac{20}{0.2} = 100$
[3 marks available — 1 mark for rounding to suitable values, 1 mark for next calculation step, 1 mark for the correct final answer using your values]

6 $\frac{3^4 \times 3^7}{3^6} = \frac{3^{(4+7)}}{3^6} = \frac{3^{11}}{3^6} = 3^{(11-6)} = 3^5$
[2 marks available — 1 mark for a correct attempt at adding or subtracting powers, 1 mark for the correct final answer]

Page 42 (Revision Questions)

1 Twenty-one million, three hundred and six thousand, five hundred and fifteen.

2 2.09, 2.2, 3.51, 3.8, 3.91, 4.7

3 a) 882
b) 446
c) £4.17

ANSWERS

4 a) £120
   b) £0.50 = 50p
   c) 26
   d) 62.7
   e) 0.35
5 a) −16
   b) 7
   c) 20
6 A square number is a whole number multiplied by itself.
   The first ten are: 1, 4, 9, 16, 25, 36, 49, 64, 81 and 100.
7 41, 43, 47, 53, 59
8 The multiples of a number are its times table.
   a) 10, 20, 30, 40, 50, 60
   b) 4, 8, 12, 16, 20, 24
9 a) $210 = 2 \times 3 \times 5 \times 7$
   b) $1050 = 2 \times 3 \times 5 \times 5 \times 7$
10 a) 14
   b) 40
11 a) i) $\dfrac{4}{100} = \dfrac{1}{25}$
       ii) 4%
    b) i) $\dfrac{65}{100} = \dfrac{13}{20}$
        ii) 0.65
12 To simplify a fraction you divide the top and bottom by the same
   number. To simplify as far as possible you keep dividing until
   they won't go any further.
13 a) 320
   b) £60
14 a) $\dfrac{23}{8} = 2\dfrac{7}{8}$
   b) $\dfrac{11}{21}$
   c) $\dfrac{25}{16} = 1\dfrac{9}{16}$
   d) $\dfrac{44}{15} = 2\dfrac{14}{15}$
15 Recurring decimals have a pattern of numbers which repeats
   forever. You show a decimal is recurring by putting a dot above
   the digits at the start and end of the repeating part of the number
   (if only one digit is repeated then just put one dot above that
   digit).
16 £1.41
17 The 250 g tin is the best buy.
18 To find x as a percentage of y, divide x by y and then multiply by
   100.
19 £60
20 The top costs £38.25, so Carl can't afford it.
21 5 : 8
22 600, 960, 1440
23 a) 17.7
   b) 6700
   c) 4 000 000
24 a) E.g. 100
   b) E.g. 1400
   These estimates were found by rounding all the numbers to 1 s.f.
25 a) 11
   b) 4
   c) 56
   d) 10^4

26 a) 421.875
   b) 4.8
   c) 8
27 a) 1) When multiplying, add the powers.
        2) When dividing, subtract the powers.
        3) When raising one power to another, multiply the powers.
    b) 7^5

Section Two — Algebra

Page 47 (Warm-up Questions)

1 a) $4x + y - 4$
   b) $9x + 5xy - 5$
   c) $5x + 3x^2 + 5y^2$
   d) $6y - 4xy$
2 a) e^5
   b) $18fg$
3 a) h^9
   b) s^3
4 a) $2x - 4$
   b) $5x + x^2$
   c) $y^2 + xy$
   d) $6xy - 18y$
5 a) $5x(y + 3)$
   b) $a(5 - 7b)$
   c) $6y(2x + 1 - 6y)$

Page 48 (Exam Questions)

2 a) $10ab$ *[1 mark]*
   b) $4pq$ *[1 mark]*
   c) $x^2 + 4x$
       [2 marks available — 1 mark for x^2 and 1 mark for $4x$]
3 a) $3(x - 2)$
       $= (3 \times x) + (3 \times -2)$
       $= 3x - 6$ *[1 mark]*
   b) $x(x + 4)$
       $= (x \times x) + (x \times 4)$
       $= x^2 + 4x$ *[1 mark]*
4 a) $6x + 3 = (3 \times 2x) + (3 \times 1) = 3(2x + 1)$ *[1 mark]*
   b) $x(x + 7)$ *[1 mark]*
5 a) $4x^2 + 6xy = 2(2x^2 + 3xy)$
          $= 2x(2x + 3y)$
       *[2 marks available — 2 marks for the correct final answer,
       otherwise 1 mark if the expression is only partly factorised]*
   b) $2vw + 8v^2 = 2(vw + 4v^2)$
          $= 2v(w + 4v)$
       *[2 marks available — 2 marks for the correct final answer,
       otherwise 1 mark if the expression is only partly factorised]*

Page 54 (Warm-up Questions)

1 $x = 8$ *(Add 12 to both sides. Then divide both sides by 4.)*
2 $x = 7$ *(Subtract $3x$ from both sides and add 9 to both sides to give
   $14 = 2x$. Then divide both sides by 2.)*
3 a) $x = 4$ *(Subtract 1 from both sides. Then divide both sides by
       3.)*
   b) $q = 32$ *(Multiply both sides by 4.)*
   c) $y = -2$ *(Subtract $2y$ from both sides and subtract 4 from both
       sides to give $3y = -6$. Then divide both sides by 3.)*
4 $a = 156$
5 $y = 5x - 3$
6 $C = 95n$
7 $b = \dfrac{a}{2} + 3$ *(Divide both sides by 2. Then add 3 to both sides.)*

Pages 55-56 (Exam Questions)

2 a) $x + 3 = 12$
      $x = 12 - 3 = 9$ *[1 mark]*

   b) $6x = 24$
      $x = 24 \div 6 = 4$ *[1 mark]*

   c) $\dfrac{x}{5} = 4$
      $x = 4 \times 5 = 20$ *[1 mark]*

3 $Q = 7x - 3y$
   $Q = (7 \times 8) - (3 \times 7)$
   $Q = 56 - 21 = 35$
   [2 marks available — 1 mark for correct substitution of x and y, 1 mark for correct final answer]

4 $S = 4m^2 + 2.5n$
   $S = (4 \times 6.5 \times 6.5) + (2.5 \times 4)$
   $S = 169 + 10$
   $S = 179$
   [2 marks available — 1 mark for correct substitution of m and n, 1 mark for correct final answer]

5 $F = \dfrac{9C}{5} + 32$
   $F = (9 \times 35 \div 5) + 32$
   $F = 95\ °F$
   [2 marks available — 1 mark for correct substitution of C, 1 mark for correct final answer]

6 a) Total cost = flat fee + (cost per day × number of days)
      $C = 300 + (50 \times d)$
      $C = 300 + 50d$
      [2 marks available — 2 marks for correct formula, otherwise 1 mark for just 300 + 50d]

   b) Substitute $d = 3$ into formula:
      $C = 300 + (50 \times 3)$
      $C = 300 + 150 = 450$
      Therefore Alex would have to pay £450.
      [2 marks available — 1 mark for substitution of d = 3 into formula, 1 mark for correct final answer]

7 a) $40 - 3x = 17x$
      $40 = 17x + 3x$
      $40 = 20x$ *[1 mark]*
      $x = 40 \div 20 = 2$ *[1 mark]*
      [2 marks available in total — as above]

   b) $2y - 5 = 3y - 12$
      $-5 + 12 = 3y - 2y$ *[1 mark]*
      $y = 7$ *[1 mark]*
      [2 marks available in total — as above]

8 $7x - 12 = 4x$ *[1 mark]*
   $3x = 12$
   $x = 4$
   So Alexa's original number was 4. *[1 mark]*
   [2 marks available in total — as above]

9 a) $9(e - 2) = 3e + 6$
      $9e - 18 = 3e + 6$ *[1 mark]*
      $9e - 3e = 6 + 18$
      $6e = 24$ *[1 mark]*
      $e = 24 \div 6 = 4$ *[1 mark]*
      [3 marks available in total — as above]

   b) $5(2c - 1) = 4(3c - 2)$
      $10c - 5 = 12c - 8$ *[1 mark]*
      $-5 + 8 = 12c - 10c$
      $3 = 2c$ *[1 mark]*
      $c = 3 \div 2 = 1.5$ *[1 mark]*
      [3 marks available in total — as above]

10 a) $u = v - at$ *[1 mark]*

   b) $v - u = at$ *[1 mark]*

      $t = \dfrac{v - u}{a}$ *[1 mark]*
      [2 marks available in total — as above]

Page 61 (Warm-up Questions)

1 a) 20, 27 "Add one extra each time to the previous term"
   b) 2000, 20 000 "Multiply the previous term by 10"
   c) 4, 2 "Divide the previous term by 2"

2 $2n + 5$ *(The common difference is 2, then add 5 to 2n)*

3 $x = 1.6$ *(E.g. try 1.6: $1.6^3 - (2 \times 1.6) = 0.896...$ too low. Try 1.7: $1.7^3 - (2 \times 1.7) = 1.513...$ too high. Try 1.65: $1.65^3 - (2 \times 1.65) = 1.192...$ too high. So x must lie between 1.6 and 1.65, and rounding to 1 d.p. x = 1.6.)*

4
```
    ○———┼———┼———┼———┼———┼———┼———●
   -2  -1  0   1   2   3   4   5   6
```

5 $x \geq -4, x < 2$, *(First inequality: subtract 9 from both sides, then divide both sides by 2. Second inequality: subtract x from both sides, then divide both sides by 3.)*
   $x = -4, -3, -2, -1, 0, 1$ *(As x is greater than or equal to –4, and less than 2, it could be –4, or any whole number between –3 and 1, but not 2.)*

Page 62 (Exam Questions)

2 a)

      [1 mark]

   b) 81, because the number of triangles in the nth pattern is equal to n^2.
      [2 marks available — 1 mark for 81 and 1 mark for the correct reason]

3

x	$x^3 - 2x$	
1	$1^3 - (2 \times 1) = 1 - 2 = -1$	Too small
2	$2^3 - (2 \times 2) = 8 - 4 = 4$	Too big
1.5	$1.5^3 - (2 \times 1.5) = 0.375$	Too big
1.3	$1.3^3 - (2 \times 1.3) = -0.403$	Too small
1.4	$1.4^3 - (2 \times 1.4) = -0.056$	Too small
1.45	$1.45^3 - (2 \times 1.45) = 0.148...$	Too big

The solution is between $x = 1.4$ and $x = 1.45$,
so to 1 d.p. the solution is $x = 1.4$
[4 marks available — 1 mark for any trial between 1 and 2, 1 mark for any trial between 1.4 and 1.5 inclusive, 1 mark for a different trial between 1.42 and 1.45 inclusive, 1 mark for the correct final answer]

4 $x \geq -2$ *[1 mark]*
   It's ≥ because the circle above the number line is coloured in, so –2 is included.

5 a) $2p > 4$
      $p > 4 \div 2$
      $p > 2$ *[1 mark]*

   b) $4q - 5 < 23$
      $4q < 23 + 5$
      $4q < 28$ *[1 mark]*
      $q < 28 \div 4$
      $q < 7$ *[1 mark]*
      [2 marks available in total — as above]

   c) $4r - 2 \geq 2r + 5$
      $4r - 2r \geq 5 + 2$
      $2r \geq 7$ *[1 mark]*
      $r \geq 7 \div 2$
      $r \geq 3.5$ *[1 mark]*
      [2 marks available in total — as above]

Page 63 (Revision Questions)

1 a) 3e   b) 8f

2 a) $7x - y$ b) $3a + 9$

3 a) m^3 b) $7pq$ c) $18xy$

4 a) g^{11} b) c^3

5 a) $6x + 18$ b) $-9x + 12$ c) $5x - x^2$

6 $6x$

7 Putting in brackets (the opposite of multiplying out brackets).

8 a) $8(x + 3)$ b) $9(2x + 3y)$
  c) $5x(x + 3)$

9 a) $x = 7$ b) $x = 16$ c) $x = 3$

10 a) $x = 4$ b) $x = 2$ c) $x = 3$

11 $Q = 8$

12 $P = 7d + 5c$

13 2 hours

14 $v = \dfrac{W - 5}{4}$

15 a) 31, rule is add 7
   b) 256, rule is multiply by 4
   c) 19, rule is add two previous terms.

16 $6n - 2$

17 Yes, it's the 5th term.

18 A way of finding an approximate solution to an equation by trying different values in the equation.

19 $x = 4.1$

20 $x = 3.7$

21 a) x is greater than minus seven.
   b) x is less than or equal to 6.

22 $k = 1, 2, 3, 4, 5, 6, 7$

23 a) $x < 10$ b) $x > 14$ c) $x \geq 3$

24 $x \leq 7$

Section Three — Graphs

Page 70 (Warm-up Questions)

1 a) $(3,5)$
  b) $(5,1)$

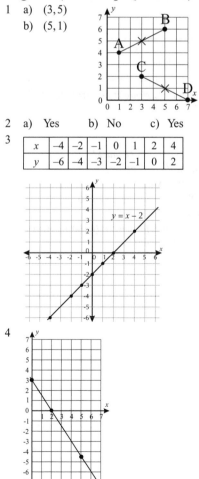

2 a) Yes   b) No   c) Yes

3

x	-4	-2	-1	0	1	2	4
y	-6	-4	-3	-2	-1	0	2

$y = x - 2$

4

Gradient $= -1.5$ *(3 ÷ 2 = 1.5 and it's downhill so it's negative)*

5 a) $m = 1, c = 5$ b) $m = -2, c = -3$ c) $m = -5, c = 4$
  Get each equation into the form $y = mx + c$ first.

Pages 71-72 (Exam Questions)

2 a) $(1, 3)$ *[1 mark]*
  b) $(1, 0)$ *[2 marks available — 2 marks for the correct answer, otherwise 1 mark for correctly drawing point S on graph]*
  c) $(0, 4)$ *[2 marks available — 2 marks for the correct answer, otherwise 1 mark for correctly drawing point T on graph]*

3 a)

x	-2	-1	0	1	2
y	-8	-5	-2	1	**4**

[2 marks available — 2 marks for all values correct, otherwise 1 mark for 2 correct values]

b)

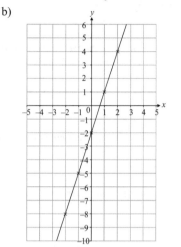

[2 marks available — 2 marks for all points plotted correctly and a straight line drawn from (–2, –8) to (2, 4), otherwise 1 mark for a correct straight line that passes through at least 3 correct points, or a straight line with the correct gradient, or a straight line with a positive gradient passing through (0, –2)]

c)

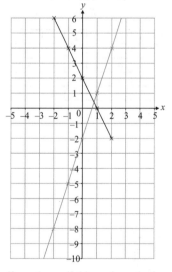

[3 marks available — 3 marks for a correct line drawn from (–2, 6) to (2, –2), otherwise 2 marks for a line that passes through (0, 2) and has a gradient of –2, or 1 mark for a line passing through (0, 2), or a line with a gradient of –2]

220

4 a)
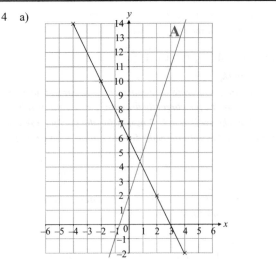

*[3 marks available — 3 marks for a correct line drawn from
(–4, 14) to (4, –2), otherwise 2 marks for a line that passes
through (0, 6) and has a gradient of –2, or 1 mark for a line
that passes through (0, 6), or a line with a gradient of –2]*

b) Gradient = $\dfrac{\text{change in } y}{\text{change in } x}$ = e.g. $\dfrac{14-2}{4-0} = \dfrac{12}{4} = 3$

*[2 marks available — 1 mark for a correct method, 1 mark
for correct final answer]*

5 a) $\left(\dfrac{2+4}{2}, \dfrac{1+3}{2}\right) = (3, 2)$

*[2 marks available — 1 mark for correct method and
1 mark for correct final answer]*

b) $\left(\dfrac{2+6}{2}, \dfrac{1+(-3)}{2}\right) = (4, -1)$

*[2 marks available — 1 mark for correct method and
1 mark for correct final answer]*

Page 77 (Warm-up Questions)

1 a) 36 litres (allow 35 – 37 litres)
   b) 4.5 gallons (allow 4.3 – 4.7 gallons)
2 a) 25 litres *(read up from 60 secs and across to y-axis)*
   b) 0.25 litres per second *(rate = gradient = e.g. 10 ÷ 40)*
3 a)

x	-2	-1	0	1	2	3	4	5
x^2	4	1	0	1	4	9	16	25
$-2x$	4	2	0	-2	-4	-6	-8	-10
-1	-1	-1	-1	-1	-1	-1	-1	-1
$y = x^2 - 2x - 1$	7	2	-1	-2	-1	2	7	14

b)
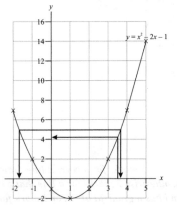

c) 4.25 (a value between 4.2 and 4.3 is acceptable).
d) $x = -1.65$ and 3.65 (values between –1.7 and –1.6 and
between 3.6 and 3.7 are acceptable).

Pages 78-79 (Exam Questions)

2 Holiday spending = €50 + €70 + €100 = €220
   Home spending = £30 + £50 + £80 = £160
   Convert £160 to euros:
   From graph £5 = €7
   And £160 ÷ £5 = 32
   So, £160 = €7 × 32 = €224
   Since €224 is more than €220, Edwige is correct — she spent
   less than she normally would while on holiday.
   *[5 marks available — 1 mark for totalling the home spending
   values, 1 mark for totalling the holiday spending values, 1 mark
   for a correct method to convert to a common currency, 1 mark
   for a correct conversion, 1 mark for a correct comparison]*
   *You could also answer this question by converting the holiday
   spending into pounds. Either way, as long as you convert them into
   the same units, you can make a comparison.*

3 a) $\dfrac{15-0}{1-0} = \dfrac{15}{1} = 15$ km/h
   *[2 marks available — 1 mark for a correct method, 1 mark
   for correct final answer]*
   b) The speed at which Selby was travelling. *[1 mark]*
   c) 3 hours *[1 mark]*
   *As he was at point A at O hours, all you have to do is read off
   the x-value at point C to see how long Selby's journey was.*
   d) 2.5 hours *[1 mark]*
   e)
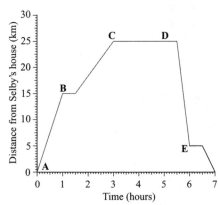

   *[2 marks available — 1 mark for a flat line from point E for
   30 minutes, and 1 mark for a straight line from the end of
   the flat line to (7, 0)]*

4 a) (i) £18 *[1 mark]*
       (ii) 40 ÷ 100 *[1 mark]*
            = 0.4, so it costs 40p per unit *[1 mark]*
       [2 marks available in total — as above]
   b) Mr Barker should use Plan A because it is cheaper.
       Using 85 units with Plan A would cost £26.50.
       85 units with Plan B would cost £34.
       *[2 marks available — 1 mark for correctly stating which
       plan, 1 mark for giving a reason]*

ANSWERS

5 a)

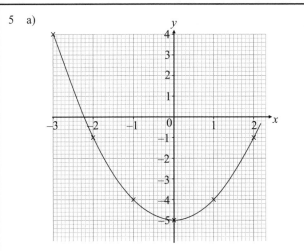

[2 marks available — 1 mark if all points are plotted correctly, 1 mark for a smooth curve joining the correctly plotted points]

b) –2.2 (allow –2.3 or –2.1) *[1 mark]*

Page 80 (Revision Questions)

1 A(5, –3), B(4, 0), C(0, 3), D(–4, 5), E(–2, –3)

2 (2, 1.5)

3

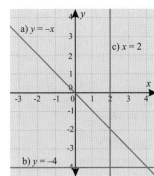

4 Straight-line equations just contain 'something x, something y and a number'. They don't contain any powers of x or y, xy, $1/x$ or $1/y$.

5 E.g.

x	0	1	2
y	3	5	7

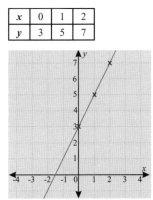

6 A line with a negative gradient slopes 'downhill' from left to right.

7 2

8 'm' is the gradient and 'c' is the y-intercept.

9 The object has stopped.

10 a) On his way home.
   b) 15 minutes

11 a) £10
   b) 20 minutes
   c) £20
   d) Answer in the range 25-26 minutes.

---

e) 67p (allow between 65p and 69p)

12 They are both "bucket shaped" graphs. $y = x^2 - 8$ is like a "u" whereas $y = -x^2 + 2$ is like an "n" (or an upturned bucket).

13

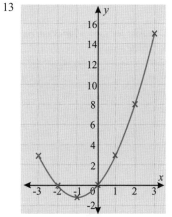

$x = -2.7$ (allow between –2.9 and –2.6) and $x = 0.7$ (allow between 0.6 and 0.9).

Section Four — Shapes and Area

Page 88 (Warm-up Questions)

1 and 2
   C 1 line of symmetry, rotational symmetry order 1
   W 1 line of symmetry, rotational symmetry order 1
   + 2 lines of symmetry, rotational symmetry order 2
   D 1 line of symmetry, rotational symmetry order 1
   Q 0 lines of symmetry, rotational symmetry order 1

3 E.g.

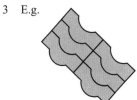

4 An equilateral triangle has 3 equal sides, 3 equal angles of 60°, 3 lines of symmetry and rotational symmetry of order 3.

5 A kite has 1 line of symmetry.

6 a) **B** and **E** are similar.
   b) **A** and **D** are congruent.

7 A cuboid has 6 faces, 8 vertices and 12 edges.

8

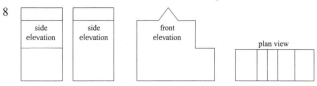

Pages 89-90 (Exam Questions)

2 A = Sphere *[1 mark]*
   B = Cone *[1 mark]*
   [2 marks available in total — as above]

3 a) A and B
      [2 marks available — 1 mark for each correct letter]
   b) E *[1 mark]*
   c)

      [1 mark]

4

a) [1 mark for shapes correctly labelled 'C' — as above]
b) [1 mark for shapes correctly labelled 'S' — as above]

5 E.g.

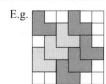

[2 marks available in total — 2 marks for four shapes correctly tessellated, otherwise 1 mark for three shapes correctly tessellated]

It's also possible to tessellate these shapes by rotating some of them.

6 a) Isosceles triangle *[1 mark]*
You need to say "isosceles triangle" to get the mark, not just "triangle".

   b) C *[1 mark]*

7 E.g.

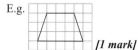 *[1 mark]*

There isn't just one right answer here — as long as your shape has four sides, with two of them parallel, you'll get the mark.

8

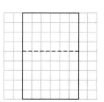

[3 marks available — 1 mark for a width of 6 squares, 1 mark for a height of 9 squares, 1 mark for a correct dotted or solid line marking the edge of the roof]

Page 97 (Warm-up Questions)

1 42 cm
2 a) area = length × width, A = l × w
   b) circumference = π × diameter, C = π × D (or C = 2πr)
   c) area = base × vertical height, A = b × h
3 10.5 m² *(Area = ½ × base × vertical height = 0.5 × 3 × 7)*
4 201.06 cm² to 2 d.p. (or 201.09 cm² to 2 d.p. using π = 3.142) *(Area = πr² = π × 8²)*
5 a) A straight line that just touches the outside of a circle.
   b) A line drawn across the inside of a circle.

Page 98 (Exam Questions)

2 a) 18 cm *[1 mark]*
   b) 14 cm² *[1 mark]*
3 a) 6.4 × 10 = 64 cm²
      [2 marks available — 1 mark for correct calculation, 1 mark for correct answer]
   b) √64 = 8 cm
      [2 marks available — 1 mark for correct calculation, 1 mark for correct answer]
4 a) Circumference = 2 × π × 0.25 *[1 mark]* = 1.57 m *[1 mark]*
      [2 marks available in total — as above]
   b) 500 ÷ 1.57 *[1 mark]* = 318.47... *[1 mark]*
      So the wheel makes 318 full turns. *[1 mark]*
      [3 marks available in total — as above]

Page 102 (Warm-up Questions)

1 8 cm³ *(8 × 1 cm³, or v = l × w × h = 2 cm × 2 cm × 2 cm)*
2 672 cm³ *(area of triangle × length = ½ × 12 × 8 × 14)*
3 E.g.

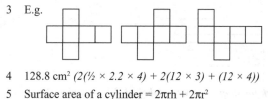

4 128.8 cm² *(2(½ × 2.2 × 4) + 2(12 × 3) + (12 × 4))*
5 Surface area of a cylinder = 2πrh + 2πr²

Page 103 (Exam Questions)

2 Area of cross-section is 7 squares and the length is 3 cubes, so volume = 7 × 3 = 21 cm³
   [2 marks available — 1 mark for correct method, 1 mark for correct answer]

3

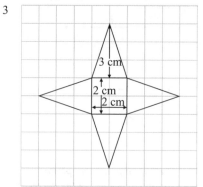

[3 marks available in total — 3 marks for a correct and accurately drawn diagram, otherwise 1 mark for a square and 4 triangles with the wrong measurements, or 2 marks for a correctly-sized square and 4 isosceles triangles with the wrong measurements]

There are other ways of drawing this net, but this is really the only sensible way of doing it.

4 Area of cross-section = π × 3² = 28.274... cm² *[1 mark]*
   Circumference = π × 6 = 18.849... cm
   Area of curved surface = circumference × length
                          = 18.849... × 11 = 207.345... cm²
   [1 mark]
   Total surface area = (2 × 28.274...) + 207.345... *[1 mark]*
   = 264 cm² (to 3 s.f.) *[1 mark]*
   [4 marks available in total — as above]

Page 104 (Revision Questions)

1 H: 2 lines of symmetry, rotational symmetry order 2
   Z: 0 lines of symmetry, rotational symmetry order 2
   T: 1 line of symmetry, rotational symmetry order 1
   N: 0 lines of symmetry, rotational symmetry order 2
   E: 1 line of symmetry, rotational symmetry order 1
   X: 4 lines of symmetry, rotational symmetry order 4
   S: 0 lines of symmetry, rotational symmetry order 2
2 E.g.

3 No
4 2 angles the same, 2 sides the same, 1 line of symmetry, no rotational symmetry.
5 2 lines of symmetry, rotational symmetry order 2
6 2 pairs of equal (parallel) sides, 2 pairs of equal angles, no lines of symmetry, rotational symmetry order 2.
7 Congruent shapes are exactly the same size and same shape. Similar shapes are the same shape but different sizes.
8 a) D and G
   b) C and F
9 a) faces = 5, edges = 8, vertices = 5
   b) faces = 2, edges = 1, vertices = 1
   c) faces = 5, edges = 9, vertices = 6
10 The view from directly above an object.

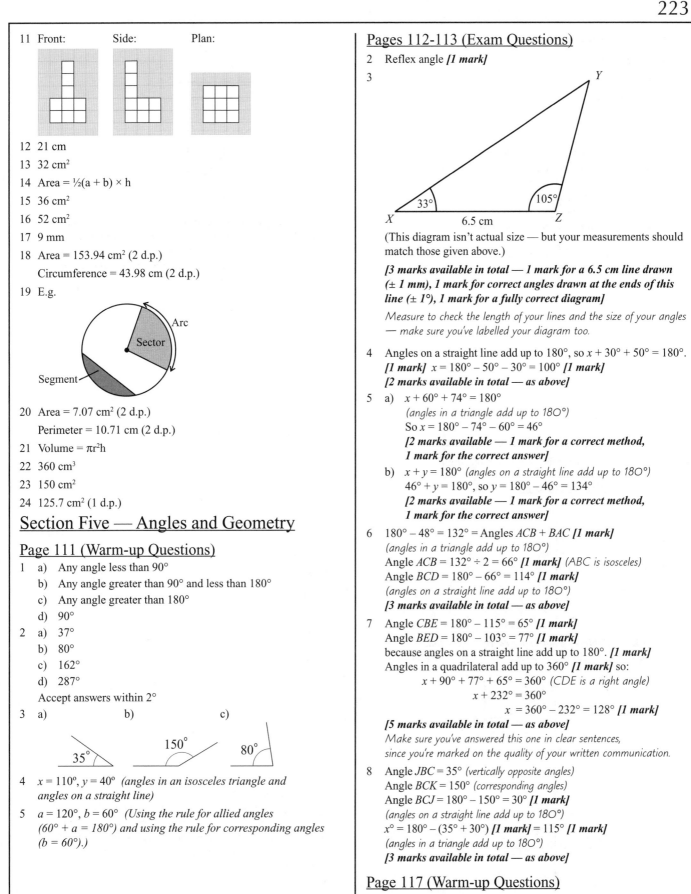

11 Front: Side: Plan:

12 21 cm

13 32 cm²

14 Area = ½(a + b) × h

15 36 cm²

16 52 cm²

17 9 mm

18 Area = 153.94 cm² (2 d.p.)

Circumference = 43.98 cm (2 d.p.)

19 E.g.

Arc

Sector

Segment

20 Area = 7.07 cm² (2 d.p.)

Perimeter = 10.71 cm (2 d.p.)

21 Volume = πr²h

22 360 cm³

23 150 cm²

24 125.7 cm² (1 d.p.)

Section Five — Angles and Geometry

Page 111 (Warm-up Questions)

1 a) Any angle less than 90°

b) Any angle greater than 90° and less than 180°

c) Any angle greater than 180°

d) 90°

2 a) 37°

b) 80°

c) 162°

d) 287°

Accept answers within 2°

3 a)   b)   c)

35°   150°   80°

4 $x = 110°$, $y = 40°$ (angles in an isosceles triangle and angles on a straight line)

5 $a = 120°$, $b = 60°$ (Using the rule for allied angles $(60° + a = 180°)$ and using the rule for corresponding angles $(b = 60°)$.)

Pages 112-113 (Exam Questions)

2 Reflex angle *[1 mark]*

3

Y

33°   105°

X   6.5 cm   Z

(This diagram isn't actual size — but your measurements should match those given above.)

[3 marks available in total — 1 mark for a 6.5 cm line drawn (± 1 mm), 1 mark for correct angles drawn at the ends of this line (± 1°), 1 mark for a fully correct diagram]

Measure to check the length of your lines and the size of your angles — make sure you've labelled your diagram too.

4 Angles on a straight line add up to 180°, so $x + 30° + 50° = 180°$. *[1 mark]* $x = 180° − 50° − 30° = 100°$ *[1 mark]*
[2 marks available in total — as above]

5 a) $x + 60° + 74° = 180°$
   (angles in a triangle add up to 180°)
   So $x = 180° − 74° − 60° = 46°$
   [2 marks available — 1 mark for a correct method, 1 mark for the correct answer]

b) $x + y = 180°$ *(angles on a straight line add up to 180°)*
   $46° + y = 180°$, so $y = 180° − 46° = 134°$
   [2 marks available — 1 mark for a correct method, 1 mark for the correct answer]

6 $180° − 48° = 132° =$ Angles $ACB + BAC$ *[1 mark]*
   (angles in a triangle add up to 180°)
   Angle $ACB = 132° ÷ 2 = 66°$ *[1 mark]* *(ABC is isosceles)*
   Angle $BCD = 180° − 66° = 114°$ *[1 mark]*
   (angles on a straight line add up to 180°)
   [3 marks available in total — as above]

7 Angle $CBE = 180° − 115° = 65°$ *[1 mark]*
   Angle $BED = 180° − 103° = 77°$ *[1 mark]*
   because angles on a straight line add up to 180°. *[1 mark]*
   Angles in a quadrilateral add up to 360° *[1 mark]* so:
   $x + 90° + 77° + 65° = 360°$ *(CDE is a right angle)*
   $x + 232° = 360°$
   $x = 360° − 232° = 128°$ *[1 mark]*
   [5 marks available in total — as above]
   Make sure you've answered this one in clear sentences, since you're marked on the quality of your written communication.

8 Angle $JBC = 35°$ *(vertically opposite angles)*
   Angle $BCK = 150°$ *(corresponding angles)*
   Angle $BCJ = 180° − 150° = 30°$ *[1 mark]*
   (angles on a straight line add up to 180°)
   $x° = 180° − (35° + 30°)$ *[1 mark]* $= 115°$ *[1 mark]*
   (angles in a triangle add up to 180°)
   [3 marks available in total — as above]

Page 117 (Warm-up Questions)

1 a) A regular polygon is a many-sided shape where all the sides and angles are the same.

b) Equilateral triangle, square, regular pentagon, regular hexagon, regular heptagon, regular octagon

2 Exterior angle = 45°, Interior angle = 135°

3 900° *(Sum of interior angles = (7 − 2) × 180°)*

Page 117 (Exam Questions)

2 a) 7 *[1 mark]*

   b) (Regular) octagon *[1 mark]*

3 a) Exterior angle = 360° ÷ 6 = 60° *[1 mark]*
      Interior angle = 180° − 60° = 120° *[1 mark]*
      [2 marks available in total — as above]

   b) Each angle in an equilateral triangle is 60°.
      The interior angle of a regular hexagon is 120°.
      Angles round a point add up to 360°. *[1 mark]*
      60° + 60° + 120° + 120° = 360°, *[1 mark]*
      so 2 regular hexagons and 2 equilateral triangles can meet at
      a point and leave no gap. *[1 mark]*
      [3 marks available in total — as above]

Page 123 (Warm-up Questions)

1 A → B — rotation of 90° clockwise about the origin.
   B → C — reflection in the line $y = x$.
   C → A — reflection in the y-axis.
   A → D — translation of 9 left and 7 down, or $\binom{-9}{-7}$.

2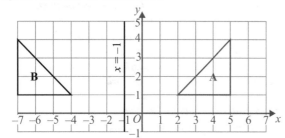

3 Scale factor 3 *(RS ÷ PQ = 4.5 cm ÷ 1.5 cm = 3)*

4 An enlargement of scale factor 2, centre the origin.

Page 124 (Exam Questions)

2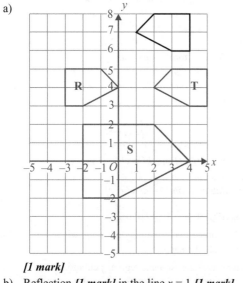

   *[2 marks available — 2 marks for correct reflection,
   otherwise 1 mark for triangle reflected but in wrong position]*

3 a)
      [1 mark]

   b) Reflection *[1 mark]* in the line $x = 1$ *[1 mark]*.
      [2 marks available in total — as above]

   c) Enlargement *[1 mark]* of scale factor 2 *[1 mark]*,
      centre (−4, 8) *[1 mark]*.
      [3 marks available in total — as above]

4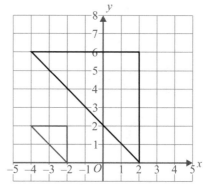

   *[3 marks available — 3 marks for correct enlargement,
   otherwise 2 marks for a correct triangle but in the wrong
   position or for an enlargement from the correct centre but of
   the wrong scale factor, or 1 mark for 2 lines enlarged by the
   correct scale factor anywhere on the grid]*

Page 132 (Warm-up Questions)

1
   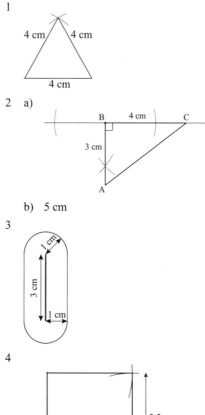

2 a)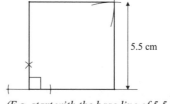

   b) 5 cm

3

4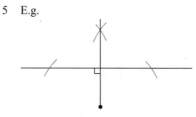

   *(E.g. start with the base line of 5.5 cm. Construct a 90° angle
   for the bottom left corner by extending the base line and using
   construction marks as shown, then extend the left side to
   5.5 cm. From the top left and bottom right corners, mark the
   final (top right) corner using compasses set to 5.5 cm, then
   join the corners.)*

5 E.g.

6 13.1 cm *(Add the squares of the two other sides,
   then take the square root)*

Pages 133-134 (Exam Questions)

2

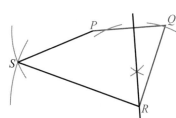

a) *[2 marks available — 1 mark for construction marks,
1 mark for point S drawn 4.2 cm from P and 6.7 cm
from R]*
*Be as accurate as you can with the measurements, but you'll still
get the marks if you are within 1 mm of the above lengths.*

b) *[2 marks available — 1 mark for construction arcs,
1 mark for the perpendicular]*

3   Scale: 1 cm represents 5 m

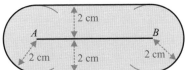

(diagram not actual size)
*[2 marks available — 2 marks for arcs with a radius of 2 cm
centred at A and B, lines 2 cm either side of AB and correct
area shaded, otherwise 1 mark for arcs with a radius of 2 cm
centred at A and B or for lines 2 cm either side of AB]*
*You'll still get the marks if you are within 1 mm of the
correct measurements.*

4

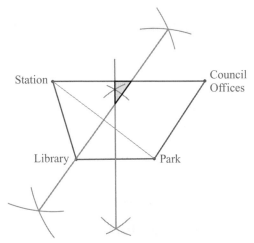

*[5 marks available — 1 mark for each pair of correct arcs
(centred at Library and Park and at Station and Park),
1 mark for each correct perpendicular bisector (of line between
Library and Park and line between Station and Park),
1 mark for correct shaded area]*

5   $2.1^2 + x^2 = 3.5^2$ *[1 mark]*
$x^2 = 3.5^2 - 2.1^2$
$x^2 = 12.25 - 4.41 = 7.84$
$x = \sqrt{7.84}$ *[1 mark]*
$x = 2.8$
2.8 m *[1 mark]*
[3 marks available in total — as above]

6

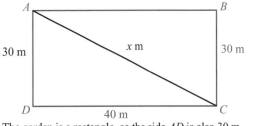

The garden is a rectangle, so the side AD is also 30 m.
The distance round the edge from A to C is 30 m + 40 m = 70 m.
[1 mark]
Let the distance across the diagonal from point A to point C be x m.
By Pythagoras' theorem $30^2 + 40^2 = x^2$ *[1 mark]*
$x = \sqrt{900 + 1600} = \sqrt{2500} = 50$ m *[1 mark]*
The price for laying the pipe across the diagonal is £8.35 per metre
plus the cost of digging the trench and replacing the grass:
£202.50 + (50 × £8.35) = £620.00 *[1 mark]*
The price for laying the pipe around the edge is £8.35 per metre:
70 × £8.35 = £584.50 *[1 mark]*
It is cheaper to lay the pipe round the edge — it will cost £620.00
to lay the pipe across the diagonal and only £584.50 to lay it around
the edge.
*[1 mark for concluding that it's cheaper to lay the pipe round the
edge, if the cost of each has been worked out above]*
[6 marks available in total — as above]
*Alternatively, instead of working out the cost of each option, you could
work out the cost of the extra 20 m of pipe needed to go around the
edge of the field, and show that this is less than the cost of digging
a trench.*

Page 135 (Revision Questions)

1   An obtuse angle

2   360°

3   a) 154°   b) 112°   c) 58°

4   60°

5   1800°

6   The interior angle of a hexagon is 120°, which divides into 360°
exactly (360° ÷ 120° = 3).

The interior angle of a pentagon is 180° − (360° ÷ 5) = 108°,
which doesn't divide into 360° exactly.

7   a)   Translation of $\begin{pmatrix} -2 \\ -4 \end{pmatrix}$.

b)   Reflection in $x = 0$ (the y-axis).

8

9   $b = 53$, $y = 5$

10   Not full size

226

11 A circle

12 First construct a 90° angle (construction marks shown in grey). Then construct the angle bisector of it (construction marks shown in black).

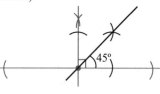

13

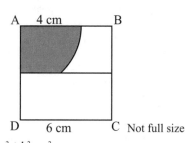

14 $a^2 + b^2 = c^2$
    You use Pythagoras' theorem to find the missing side of a right-angled triangle.

15 4.7 m

16 14.5 cm

17 5

Section Six — Measures

Page 140 (Warm-up Questions)

1 a) 200 cm
   b) 65 mm
2 a) 0.25 kg
   b) 1.5 l
3 160 kg
4 3 feet 10 inches
5 a) 320 km
   b) 4 feet
6 a) 230 000 cm²
   b) 3.45 m²
   To do these conversions, find the conversion factor, then multiply and divide by it. Then choose the most sensible answer.

Page 141 (Exam Questions)

2 2.5 litres × 1000 = 2500 ml
   2500 ÷ 250 = 10 cups
   [3 marks available — 1 mark for converting litres to ml, 1 mark for dividing by 250, and 1 mark for final answer]
   A correct method here could also be to convert 250 ml into litres (0.25 litres) and then divide 2.5 by 0.25.

3 2500 g ÷ 1000 = 2.5 kg *[1 mark]*
   Convert kg into lb: 2.5 × 2.2 = 5.5 lb *[1 mark]*
   5.5 ÷ 1.5 = 3.6666... *[1 mark]*
   Maximum number of books = 3 *[1 mark]*
   [4 marks available in total — as above]

4 a) 40 km into miles = $\frac{5 \times 40}{8}$ *[1 mark]*
          = 25 miles *[1 mark]*
       [2 marks available in total — as above]

   b) 60 km/h into mph = $\frac{5 \times 60}{8}$ = 37.5 mph *[1 mark]*
       Since 37.5 mph is less than 40 mph, the camel could outrun a giraffe. *[1 mark]*
       [2 marks available in total — as above]

5 39 200 ÷ 100 = 392
   392 ÷ 100 = 3.92 m²
   [2 marks available — 1 mark for correct method and 1 mark for correct final answer]
   When the unit is squared, you have to use the conversion factor twice (so you use 100, and then 100 again). You could also divide 39 200 by 10 000 instead.

Page 145 (Warm-up Questions)

1 a) 75 mph
   b) 3.2 °C
2 a) 4.2 cm
   b) M ———————×——————— N
3 Maximum weight = 3.5 kg, minimum weight = 2.5 kg
4 a) 02:36
   b) 21:52
   c) 11:32
   d) 12:16
   e) 00:05
5 a) 3.08 pm
   b) 4.40 am
   c) 5.30 pm
   d) 12.00 am
   e) 1.47 am
6 99 minutes *(9.37 to 10.37 = 1 hour (60 minutes), 10.37 to 11.00 = 23 minutes, 11.00 to 11.16 = 16 minutes)*
7 4.05 pm *(10:15 + 6 hours = 16:15, 16:15 – 10 minutes = 16:05)*

Page 146 (Exam Questions)

2 a) 40.6 m *[1 mark]*
   b)

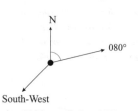

       [1 mark]

3 a) 17 04 *[1 mark]*
   b) 6 minutes *[1 mark]*
   c) 16 40 → 18 15 *[1 mark]*
       = 1 h 35 mins *[1 mark]*
       [2 marks available in total — as above]

4 Minimum weight = 56.5 kg *[1 mark]*
   Maximum weight = 57.5 kg *[1 mark]*
   [2 marks available in total — as above]

Page 151 (Warm-up Questions)

1

N

080°

South-West

2 a) 1100 m *(5.5 × 200)*
   b) 2.5 cm *(500 ÷ 200)*
3 67.5 km *(distance = speed × time = 45 × 1.5)*

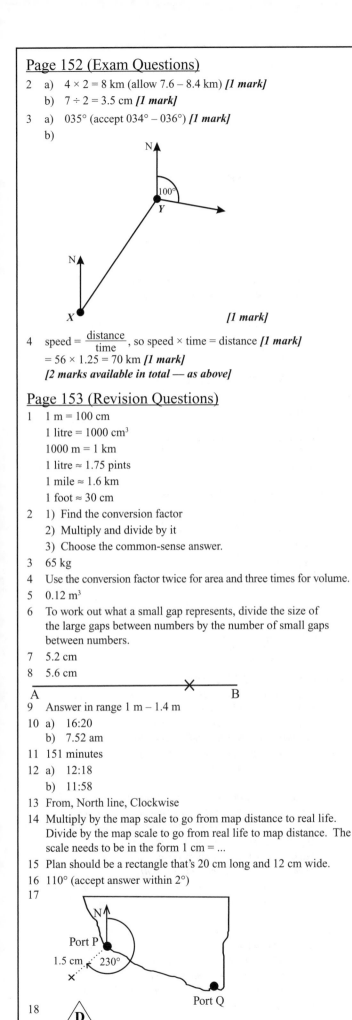

Page 152 (Exam Questions)

2 a) 4 × 2 = 8 km (allow 7.6 – 8.4 km) *[1 mark]*

   b) 7 ÷ 2 = 3.5 cm *[1 mark]*

3 a) 035° (accept 034° – 036°) *[1 mark]*

   b)

[1 mark]

4 speed = $\frac{\text{distance}}{\text{time}}$, so speed × time = distance *[1 mark]*
= 56 × 1.25 = 70 km *[1 mark]*
[2 marks available in total — as above]

Page 153 (Revision Questions)

1 1 m = 100 cm
1 litre = 1000 cm³
1000 m = 1 km
1 litre ≈ 1.75 pints
1 mile ≈ 1.6 km
1 foot ≈ 30 cm

2 1) Find the conversion factor
   2) Multiply and divide by it
   3) Choose the common-sense answer.

3 65 kg

4 Use the conversion factor twice for area and three times for volume.

5 0.12 m³

6 To work out what a small gap represents, divide the size of the large gaps between numbers by the number of small gaps between numbers.

7 5.2 cm

8 5.6 cm

A   B

9 Answer in range 1 m – 1.4 m

10 a) 16:20
    b) 7.52 am

11 151 minutes

12 a) 12:18
    b) 11:58

13 From, North line, Clockwise

14 Multiply by the map scale to go from map distance to real life. Divide by the map scale to go from real life to map distance. The scale needs to be in the form 1 cm = ...

15 Plan should be a rectangle that's 20 cm long and 12 cm wide.

16 110° (accept answer within 2°)

17

18

19 24 m

Section Seven — Statistics and Probability

Page 158 (Warm-up Questions)

1 Two from, e.g: sample too small, one city centre not representative of the whole of Britain, only done in one particular place.

2 a) This question is ambiguous. "A lot of television" can mean different things to different people.
   b) This is a leading question, inviting the person to agree.
   c) The answers to this question do not cover all possible options.

3 First, order the numbers:
–14, –12, –5, –5, 0, 1, 3, 6, 7, 8, 10, 14, 18, 23, 25
Mean = 5.27 (2 d.p.), Median = 6, Mode = –5, Range = 39

4 a) Mean = 11.2, median = 12,
mode = 12, range = 9
   b) The first set of scores has a higher mean, so those scores are generally higher. The second set has a bigger range, so those scores are more spread out.

Page 159 (Exam Questions)

2 1, 4, 7
[2 marks available — 2 marks for all three numbers correct, otherwise 1 mark for 3 numbers that have a range of 6 and a mean of 4 but aren't all different, or 3 different numbers that add up to 12 or that have a range of 6]

3 a) £18 000 *[1 mark]*
   b) £18 000 *[1 mark]*
   c) Mean for Company A
= (18 000 + 18 000 + 18 000 + 25 200 + 38 500) ÷ 5
= 117 700 ÷ 5 *[1 mark]* = £23 540 *[1 mark]*, so Company A has a lower mean annual salary than Company B. *[1 mark]*
[3 marks available in total — as above]

4 a) E.g.

How often?	Tally	Frequency
Never		
Less than once a month		
1 – 3 times a month		
4 – 6 times a month		
7 or more times a month		

[3 marks available — 1 mark for appropriate options, 1 mark for a tally column, 1 mark for a frequency or total column]

   b) E.g. It leads people to say the centre does need improving. *[1 mark]*

   c) E.g. Do you think any of the facilities need improving?
Yes   No   Don't know
☐   ☐   ☐
[2 marks available — 1 mark for an appropriate question and 1 mark for appropriate answer boxes]

Page 165 (Warm-up Questions)

1 a) 3
   b)

Rock	●
Blues	●●
Opera	●●◖
Jazz	●◖

2

	Walk	Car	Bus	Total
Male	15	21	13	49
Female	18	11	22	51
Total	33	32	35	100

3 Rom Com = 138° *(23 × 6)*
Western = 150° *(25 × 6)*
Action = 72° *(12 × 6)*
(23 + 25 + 12 = 60, so the multiplier is 360 ÷ 60 = 6)

4   Graph 1 — Strong positive correlation.
    Graph 2 — Moderate negative correlation.
5
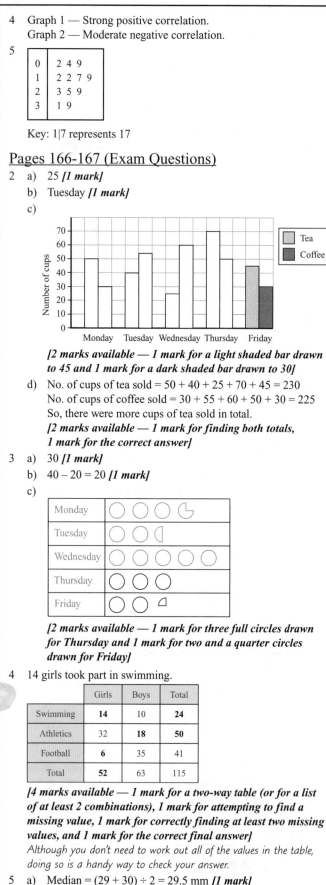

0	2 4 9
1	2 2 7 9
2	3 5 9
3	1 9

Key: 1|7 represents 17

Pages 166-167 (Exam Questions)

2 a) 25 *[1 mark]*
   b) Tuesday *[1 mark]*
   c)

[2 marks available — 1 mark for a light shaded bar drawn to 45 and 1 mark for a dark shaded bar drawn to 30]
   d) No. of cups of tea sold = 50 + 40 + 25 + 70 + 45 = 230
       No. of cups of coffee sold = 30 + 55 + 60 + 50 + 30 = 225
       So, there were more cups of tea sold in total.
       [2 marks available — 1 mark for finding both totals, 1 mark for the correct answer]
3 a) 30 *[1 mark]*
   b) 40 − 20 = 20 *[1 mark]*
   c)

[2 marks available — 1 mark for three full circles drawn for Thursday and 1 mark for two and a quarter circles drawn for Friday]
4   14 girls took part in swimming.

	Girls	Boys	Total
Swimming	14	10	24
Athletics	32	18	50
Football	6	35	41
Total	52	63	115

[4 marks available — 1 mark for a two-way table (or for a list of at least 2 combinations), 1 mark for attempting to find a missing value, 1 mark for correctly finding at least two missing values, and 1 mark for the correct final answer]
Although you don't need to work out all of the values in the table, doing so is a handy way to check your answer.
5 a) Median = (29 + 30) ÷ 2 = 29.5 mm *[1 mark]*
   b) Range = 43 − 8 = 35 mm *[1 mark]*
   c) E.g. The daily rainfall was generally much higher for the days in June, as the median was 15.5 mm higher. The daily rainfall for the days in June was more varied than for the days in November as the range was higher for the days in June.
       [2 marks available — 1 mark for a correct statement comparing the two medians, 1 mark for a correct statement comparing the two ranges]

6 a) positive *[1 mark]*
   b) E.g.
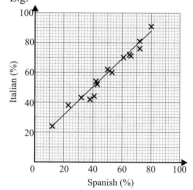
[1 mark for line of best fit that lies between (12, 16) and (12, 28) and also between (80, 82) and (80, 96)]
   c) 56% (depending on line of best fit)
       [2 marks available — 1 mark for indicating 66 on the y-axis, 1 mark for the x-coordinate of the line of best fit when y = 66]

Page 173 (Warm-up Questions)

1 a) Median = 2
   b) Mode = 2
2   E.g.

Height (h cm)	Tally	Frequency
$150 < h \leq 160$	||	2
$160 < h \leq 170$	ⵘ	5
$170 < h \leq 180$	|||	3
$180 < h \leq 190$	||	2

3   16 cm, 17 cm, 18 cm, 19 cm
4
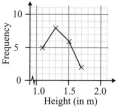

Page 174 (Exam Questions)

2 a) Total number of vehicles = 13 + 8 + 6 + 2 + 1 = 30, so the median is halfway between the 15th and 16th values *[1 mark]*, so the median = 1 *[1 mark]*
       [2 marks available in total — as above]
   b) ((13 × 0) + (8 × 1) + (6 × 2) + (2 × 3) + (1 × 4)) ÷ 30
       = (0 + 8 + 12 + 6 + 4) ÷ 30
       = 30 ÷ 30 = 1
       [3 marks available — 1 mark for multiplying the vehicles per minute by the frequency and adding, 1 mark for dividing by 30, 1 mark for the correct final answer]
3 a) E.g. There was between 4 and 6 mm of rainfall more often in Holcombury than in Ramsbrooke.
       [1 mark for any correct comparison]
       There are quite a few other comparisons you could make here — for example, you could look at the total rainfall over the 45 days.
   b) (3 × 1) + (5 × 3) + (6 × 5) + (13 × 7) + (11 × 9) + (6 × 11) + (1 × 13) = 3 + 15 + 30 + 91 + 99 + 66 + 13 = 317
       317 ÷ 45 = 7.0444.... = 7.0 mm (to 1 d.p.)
       [3 marks available — 1 mark for multiplying the frequencies by the midpoints, 1 mark for dividing by 45, 1 mark for the correct final answer]

Page 180 (Warm-up Questions)

1 1/5 or 0.2 *(2 out of 10, then cancel down)*

2 5/6 *(1 – 1/6)*

3

+	1	2	3	4
1	2	3	4	5
2	3	4	5	6
3	4	5	6	7
4	5	6	7	8

4 Landing on red: 0.43, landing on blue: 0.24, landing on green: 0.33

Pages 181-182 (Exam Questions)

3 a) Blue *[1 mark]*

   b) $\frac{3}{8}$ *[1 mark]*

4 Total number of people in the team = 6 + 9 + 4 + 1 = 20 *[1 mark]*
   So the probability that person's favourite position is midfield
   = $\frac{9}{20}$ *[1 mark]*
   [2 marks available in total — as above]

5 (hockey, netball), (hockey, choir), (hockey, orienteering),
   (orchestra, netball), (orchestra, choir), (orchestra, orienteering),
   (drama, netball), (drama, choir), (drama, orienteering)
   [2 marks available — 2 marks if all 9 possible outcomes are correct, otherwise 1 mark if at least 5 are correct]

6 a)

Cards

		2	4	6	8	10
	1	3	5	7	9	11
	2	4	6	8	10	12
Dice	**3**	5	7	9	11	13
	4	6	8	10	12	14
	5	7	9	11	13	15
	6	8	10	12	14	16

   [2 marks available — 2 marks if all entries are correct, otherwise 1 mark if at least 4 entries are correct]

   b) 3 ways of scoring exactly 9
   Total number of possible outcomes = 30
   Probability of scoring exactly 9 = $\frac{3}{30}$ *[1 mark]*
   = $\frac{1}{10}$ *[1 mark]*
   [2 marks available in total — as above]

7 a)

Number on counter	1	2	3	4	5
Frequency	23	25	22	21	9
Relative frequency	0.23	0.25	0.22	0.21	0.09

   [2 marks available — 2 marks for all correct answers, otherwise 1 mark for any frequency ÷ 100]

   b) E.g. No, because the relative frequency for selecting a counter numbered 5 is too low — they should each be around 0.2 if the number of counters is equal.
   [1 mark for saying that he's wrong (or probably wrong) because the relative frequency for 5 is too low]
   It's okay to say that he's probably wrong, because there is a (very) small possibility that he could have got these results even with the same number of each counter.

Page 183 (Revision Questions)

1 A sample is part of a population. Samples need to be representative so that conclusions drawn from sample data can be applied to the whole population.

2

Pet	Tally	Frequency
Cat	卌 |||	8
Dog	卌		6
Rabbit	||||	4
Fish	||	2

3 Questions need to be:
   (i) clear and easy to understand
   (ii) easy to answer
   (iii) fair (i.e. not leading or biased)
   (iv) easy to analyse afterwards

4 Mode = 31, Median = 24
   Mean = 22, Range = 39

5 You need to look at the key to see what each symbol represents.

6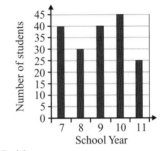

7 87 girls

8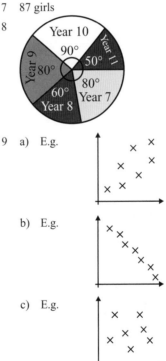

9 a) E.g.

   b) E.g.

   c) E.g.

10 a) Fastest speed = 138 mph
    b) Median = 118 mph
    c) Range = 36 mph

11 To find the mode: find the category with the highest frequency.
    To find the median: work out the position of the middle value, then count through the frequency column to find the category it's in.
    To find the mean: add a third column to the table showing the values in the first column (x) multiplied by the values in the frequency column (f). Work out the mean by dividing the total of the 3rd column (total of $f \times x$) by the total of the frequency column (total of f).

12 a) Modal class is: $1.5 \le y < 1.6$.
    b) Class containing median is: $1.5 \le y < 1.6$
    c) Estimated mean = 1.58 m (2 d.p.)

13 There are lots of things you could say — e.g. far more Year 10 students than Year 7 students took less than 10 seconds to run 60 m.

14 A probability of 0 means something will never happen.
A probability of ½ means something is as likely to happen as not.

15 $\frac{4}{25}$

16 0.7

17 a)

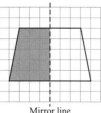

	Second flip	
	Heads	Tails
Heads	HH	HT
Tails	TH	TT

First flip

b) $\frac{1}{4}$

18 Expected frequency = probability × *n*

19 50 times

20 When you can't tell the probability of each result 'just by looking at it'. / When a dice, coin or spinner is biased.

Page 184 (Practice Paper 1)

1 a) 70 *[1 mark]*
  b) 330 *[1 mark]*
  c) One thousand, seven hundred and fifty eight *[1 mark]*
  d) 86 *[1 mark]*

2 a)

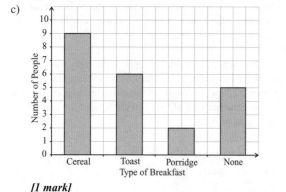

Mirror line
[1 mark]

  b) Order of rotational symmetry = 4 *[1 mark]*

3 a) 7.55 × 10 = 75.5 *[1 mark]*
  b) 427.3 ÷ 100 = 4.273 *[1 mark]*
  c) 0.302 × 1000 = 302 *[1 mark]*

4 a) 15 grammar books and 15 dictionaries:
     (15 × 3) + (15 × 2) = 45 + 30 = £75
     [2 marks available — 1 mark for correct method,
     1 mark for correct answer]
     You could have worked out the cost of one grammar book and
     one dictionary (3 + 2 = 5), then multiplied this by 15 to find the
     cost of 15 sets. The maths is a bit easier the first way though.
  b) Website: (15 × 4) + 5 = 60 + 5 = £65 *[1 mark]*. The books
     cost £75 from the bookshop or £65 from the website, so she
     should buy them from the website as it's cheaper *[1 mark]*.
     [2 marks available in total — as above]

5 a) 2 *[1 mark]*
  b) Cereal = 9, toast = 6, so difference = 9 – 6 = 3
     [2 marks available — 1 mark for correctly reading values
     off the graph, 1 mark for correct answer]
  c)

[1 mark]

---

  d) Mode = cereal *[1 mark]*

6 Monday: 112 – 17 = 95 *[1 mark]*
  Tuesday: 95 + 38 = 133 *[1 mark]*
  [2 marks available in total — as above]

7 a) Unlikely *[1 mark]*
  b) Impossible *[1 mark]*
  c) Even *[1 mark]*

8 a) 18 cm *[1 mark]*
  b) 12 cm²
     [2 marks available — 1 mark for correct answer,
     1 mark for correct units]

9 Height of man is about 1.8 m *[1 mark]*
  Elephant is approximately twice the height of the man, so height of
  elephant = height of man × 2 = 1.8 × 2 *[1 mark]* = 3.6 m *[1 mark]*
  [3 marks available in total — as above]

10 a) 15 °C *[1 mark]*
   b) Glasgow = –2 °C, Barcelona = 9 °C *[1 mark]*
      Difference = 9 – (–2) = 11 °C *[1 mark]*
      [2 marks available in total — as above]
   c) Highest temperature in Glasgow = 3 °C *[1 mark]*
      Lowest temperature in Barcelona = 6 °C *[1 mark]*
      Difference = 6 – 3 = 3 °C *[1 mark]*
      [3 marks available in total — as above]

11 a) and b)

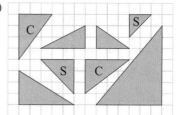

[2 marks available — 1 mark for correct triangles labelled
with a C, 1 mark for correct triangles labelled with an S]

12 Mean = 5, so total of the three numbers = 3 × 5 = 15. So you
   need three different numbers that add up to 15 and have a range
   of 6.
   By trial and error, these numbers are 2, 5 and 8.
   [2 marks available — 2 marks for all three numbers correct,
   otherwise 1 mark for two numbers correct OR 1 mark for
   working out that their total is 15]
   It's easy to check your answer — the range of these numbers is
   8 – 2 = 6, and the mean is (2 + 5 + 8) ÷ 3 = 15 ÷ 3 = 5,
   which is right.

13 a) 18 *[1 mark]*
   b) 32 *[1 mark]*
   c) 36 *[1 mark]*
   d) 125 *[1 mark]*

14 a) (i) 09 35 *[1 mark]*
      (ii) Bus arrives at Chertsey at 10 13 *[1 mark]*.
           So 09 44 to 10 13 is 16 + 13 = 29 minutes *[1 mark]*
           It's 16 minutes from 09 44 to 10 00, then 13 minutes from
           10 00 to 10 13.
           [3 marks available in total — as above]
   b) $\frac{1}{3}$ of £3.60 = £3.60 ÷ 3 = £1.20 *[1 mark]*
      £3.60 – £1.20 = £2.40 *[1 mark]*
      [2 marks available in total — as above]

15 a) 8 × 2 + 7 = 16 + 7 = 23 *[1 mark]*
   b) 2^3 = 2 × 2 × 2 = 8 *[1 mark]*
   c) 7^2 = 49, so $\sqrt{49}$ = 7 *[1 mark]*

16 Total number of miles driven = 75 × 4 = 300 miles *[1 mark]*
   Cost in p = 300 × 20 = 6000p *[1 mark]*
   Cost in £ = 6000 ÷ 100 = £60 *[1 mark]*
   [3 marks available in total — as above]

17 a) Parallelogram *[1 mark]*

b) E.g.

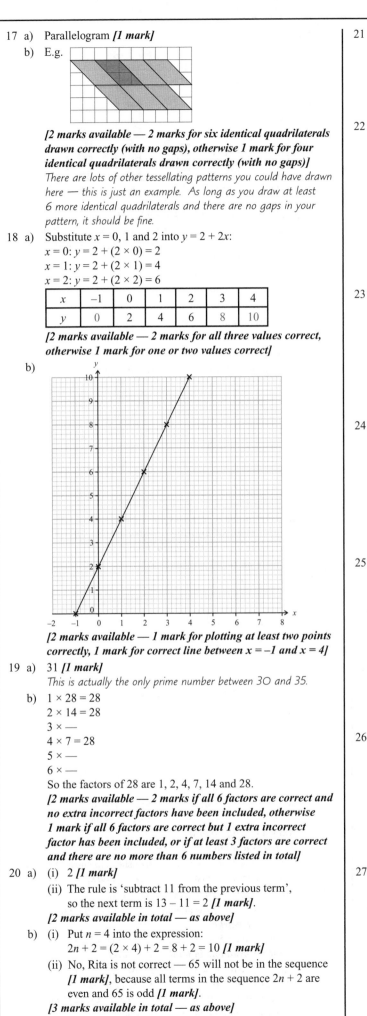

[2 marks available — 2 marks for six identical quadrilaterals drawn correctly (with no gaps), otherwise 1 mark for four identical quadrilaterals drawn correctly (with no gaps)]
There are lots of other tessellating patterns you could have drawn here — this is just an example. As long as you draw at least 6 more identical quadrilaterals and there are no gaps in your pattern, it should be fine.

18 a) Substitute $x = 0$, 1 and 2 into $y = 2 + 2x$:
$x = 0$: $y = 2 + (2 \times 0) = 2$
$x = 1$: $y = 2 + (2 \times 1) = 4$
$x = 2$: $y = 2 + (2 \times 2) = 6$

x	−1	0	1	2	3	4
y	0	2	4	6	8	10

[2 marks available — 2 marks for all three values correct, otherwise 1 mark for one or two values correct]

b)

[2 marks available — 1 mark for plotting at least two points correctly, 1 mark for correct line between $x = -1$ and $x = 4$]

19 a) 31 *[1 mark]*
This is actually the only prime number between 30 and 35.

b) $1 \times 28 = 28$
$2 \times 14 = 28$
$3 \times$ —
$4 \times 7 = 28$
$5 \times$ —
$6 \times$ —
So the factors of 28 are 1, 2, 4, 7, 14 and 28.
[2 marks available — 2 marks if all 6 factors are correct and no extra incorrect factors have been included, otherwise 1 mark if all 6 factors are correct but 1 extra incorrect factor has been included, or if at least 3 factors are correct and there are no more than 6 numbers listed in total]

20 a) (i) 2 *[1 mark]*
(ii) The rule is 'subtract 11 from the previous term', so the next term is $13 - 11 = 2$ *[1 mark]*.
[2 marks available in total — as above]

b) (i) Put $n = 4$ into the expression:
$2n + 2 = (2 \times 4) + 2 = 8 + 2 = 10$ *[1 mark]*
(ii) No, Rita is not correct — 65 will not be in the sequence *[1 mark]*, because all terms in the sequence $2n + 2$ are even and 65 is odd *[1 mark]*.
[3 marks available in total — as above]

21 a) (i) $a = 180° - 90° - 50° = 40°$ *[1 mark]*
(ii) Angles in a triangle add up to 180° *[1 mark]*
[2 marks available in total — as above]

b) Area of a triangle = ½ × base × height = ½ × 8 × 6 = 24 cm²
[2 marks available in total — 1 mark for using the correct formula, 1 mark for the correct answer]

22 a) $3x = 21$
$x = 7$ *[1 mark]*

b) $2m + 5 = -9$
$2m = -14$ *[1 mark]*
$m = -7$ *[1 mark]*
[2 marks available in total — as above]

c) $5n - 11 = 2n + 4$
$5n = 2n + 15$
$3n = 15$ *[1 mark]*
$n = 5$ *[1 mark]*
[2 marks available in total — as above]

23 Cost of music = 40 × 7 = £280 *[1 mark]*
10% discount: 10% of £280 = 280 ÷ 10 = £28
So cost of music after discount = 280 − 28 = £252 *[1 mark]*
Rent = £318, so total costs = 252 + 318 = £570 *[1 mark]*
Ticket sales = 95 × 10 = £950 *[1 mark]*
Profit = sales − cost = 950 − 570 = £380 *[1 mark]*
50% of £380 = 380 ÷ 2 *[1 mark]* = £190 *[1 mark]*
So Robert should donate £190 to charity.
[7 marks available in total — as above]

24 a) $3(2 + b) + 5(3 - b) = (3 \times 2) + (3 \times b) + (5 \times 3) + (5 \times -b)$
$= 6 + 3b + 15 - 5b = 21 - 2b$
[2 marks available — 1 mark for expanding the brackets, 1 mark for simplifying]

b) $6c - 8 = 2(3c - 4)$ *[1 mark]*

c) $2d^2 + 10d = 2d(d + 5)$
[2 marks available — 1 mark for taking out 2 as a factor, 1 mark for taking out d as a factor and filling in the brackets correctly]

25 Let the number of people going be n. Then the total cost of the trip will be $100 + 15n$ *[1 mark]*. The youth centre will charge £23 per person, which is $23n$ in total.
To cover the cost, $23n$ must equal $100 + 15n$.
So $23n = 100 + 15n$ *[1 mark]*
$8n = 100$
$n = 12.5$ *[1 mark]*
So there need to be 13 people on the trip for the Youth Centre to cover its costs. *[1 mark]*.
[4 marks available in total — as above]
You have to round up here — if only 12 people went, 23n would be less than 100 + 15n, so the Youth Centre would lose money.

26 a) E.g. one of: it is a leading question — 'do you agree' leads people to say 'yes' / there is no response box if people want to say 'no'.
[1 mark for any sensible criticism]

b) E.g. Do you think that Digton needs a new supermarket?

☐ Yes   ☐ No   ☐ Don't know

[2 marks available — 1 mark for a sensible (non-leading) question, 1 mark for suitable response boxes]

27 a) $6x < 3x + 9$
$3x < 9$ *[1 mark]*
$x < 3$ *[1 mark]*
[2 marks available in total — as above]

b)

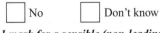

[1 mark]

Page 198 (Practice Paper 2)

1 a) 24 012 *[1 mark]*
   b) 85.03, 85.3, 87.2, 90.9, 95.3 *[1 mark]*
   c) (i) 9652 *[1 mark]*
       (ii) 2569 *[1 mark]*
       [2 marks available in total — as above]

2 a) 2.8 cm *[1 mark]*
   b), c)

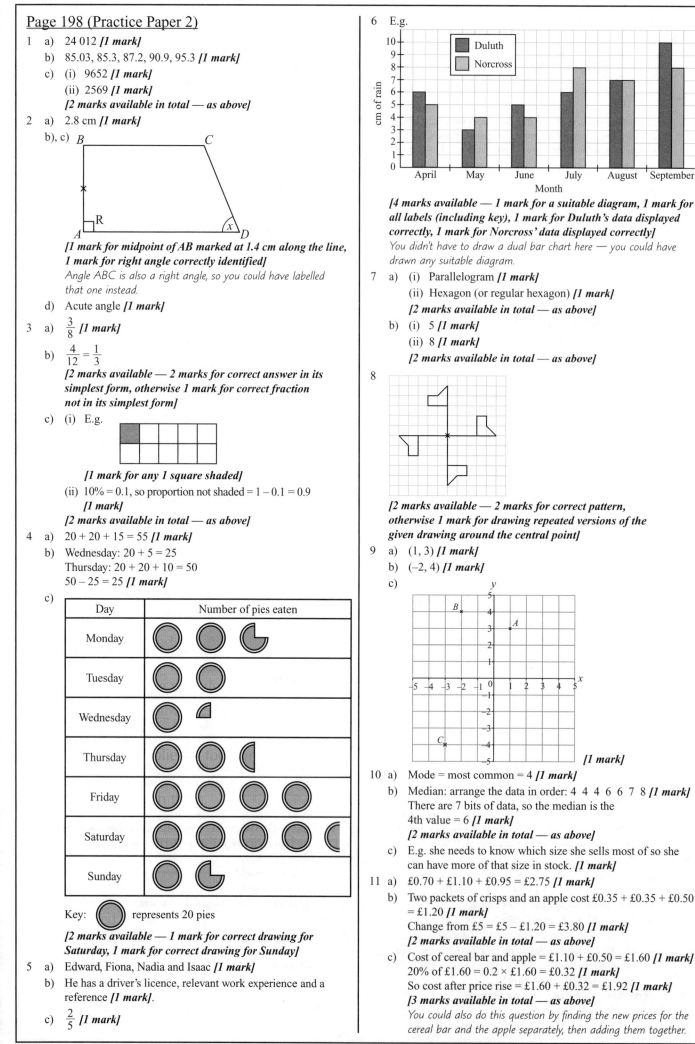

   [1 mark for midpoint of AB marked at 1.4 cm along the line, 1 mark for right angle correctly identified]
   Angle ABC is also a right angle, so you could have labelled that one instead.
   d) Acute angle *[1 mark]*

3 a) $\frac{3}{8}$ *[1 mark]*
   b) $\frac{4}{12} = \frac{1}{3}$
       [2 marks available — 2 marks for correct answer in its simplest form, otherwise 1 mark for correct fraction not in its simplest form]
   c) (i) E.g.

       [1 mark for any 1 square shaded]
       (ii) 10% = 0.1, so proportion not shaded = 1 – 0.1 = 0.9
       [1 mark]
       [2 marks available in total — as above]

4 a) 20 + 20 + 15 = 55 *[1 mark]*
   b) Wednesday: 20 + 5 = 25
       Thursday: 20 + 20 + 10 = 50
       50 – 25 = 25 *[1 mark]*
   c)

Day	Number of pies eaten
Monday	
Tuesday	
Wednesday	
Thursday	
Friday	
Saturday	
Sunday	

   Key: represents 20 pies
   [2 marks available — 1 mark for correct drawing for Saturday, 1 mark for correct drawing for Sunday]

5 a) Edward, Fiona, Nadia and Isaac *[1 mark]*
   b) He has a driver's licence, relevant work experience and a reference *[1 mark]*.
   c) $\frac{2}{5}$ *[1 mark]*

6 E.g.

   [4 marks available — 1 mark for a suitable diagram, 1 mark for all labels (including key), 1 mark for Duluth's data displayed correctly, 1 mark for Norcross' data displayed correctly]
   You didn't have to draw a dual bar chart here — you could have drawn any suitable diagram.

7 a) (i) Parallelogram *[1 mark]*
       (ii) Hexagon (or regular hexagon) *[1 mark]*
       [2 marks available in total — as above]
   b) (i) 5 *[1 mark]*
       (ii) 8 *[1 mark]*
       [2 marks available in total — as above]

8

   [2 marks available — 2 marks for correct pattern, otherwise 1 mark for drawing repeated versions of the given drawing around the central point]

9 a) (1, 3) *[1 mark]*
   b) (–2, 4) *[1 mark]*
   c)

   [1 mark]

10 a) Mode = most common = 4 *[1 mark]*
   b) Median: arrange the data in order: 4 4 4 6 6 7 8 *[1 mark]*
       There are 7 bits of data, so the median is the 4th value = 6 *[1 mark]*
       [2 marks available in total — as above]
   c) E.g. she needs to know which size she sells most of so she can have more of that size in stock. *[1 mark]*

11 a) £0.70 + £1.10 + £0.95 = £2.75 *[1 mark]*
   b) Two packets of crisps and an apple cost £0.35 + £0.35 + £0.50 = £1.20 *[1 mark]*
       Change from £5 = £5 – £1.20 = £3.80 *[1 mark]*
       [2 marks available in total — as above]
   c) Cost of cereal bar and apple = £1.10 + £0.50 = £1.60 *[1 mark]*
       20% of £1.60 = 0.2 × £1.60 = £0.32 *[1 mark]*
       So cost after price rise = £1.60 + £0.32 = £1.92 *[1 mark]*
       [3 marks available in total — as above]
       You could also do this question by finding the new prices for the cereal bar and the apple separately, then adding them together.

12 a) 3.8 *[1 mark]*

b)

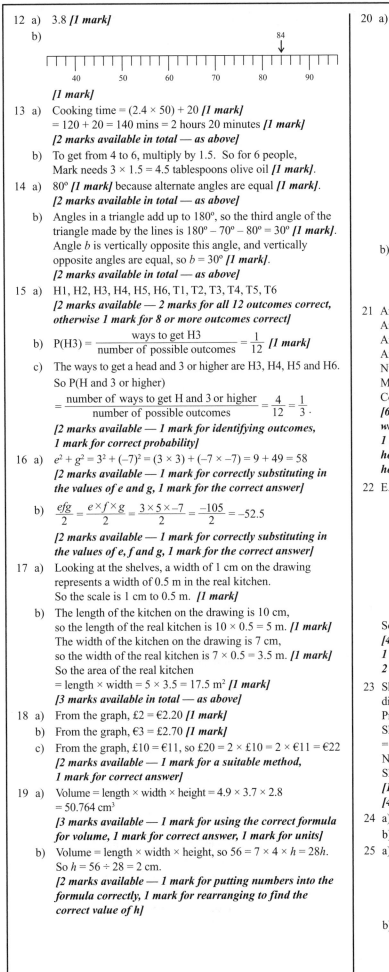

[1 mark]

13 a) Cooking time = (2.4 × 50) + 20 *[1 mark]*
= 120 + 20 = 140 mins = 2 hours 20 minutes *[1 mark]*
[2 marks available in total — as above]

b) To get from 4 to 6, multiply by 1.5. So for 6 people,
Mark needs 3 × 1.5 = 4.5 tablespoons olive oil *[1 mark]*.

14 a) 80° *[1 mark]* because alternate angles are equal *[1 mark]*.
[2 marks available in total — as above]

b) Angles in a triangle add up to 180°, so the third angle of the
triangle made by the lines is 180° – 70° – 80° = 30° *[1 mark]*.
Angle b is vertically opposite this angle, and vertically
opposite angles are equal, so b = 30° *[1 mark]*.
[2 marks available in total — as above]

15 a) H1, H2, H3, H4, H5, H6, T1, T2, T3, T4, T5, T6
*[2 marks available — 2 marks for all 12 outcomes correct,
otherwise 1 mark for 8 or more outcomes correct]*

b) P(H3) = $\frac{\text{ways to get H3}}{\text{number of possible outcomes}}$ = $\frac{1}{12}$ *[1 mark]*

c) The ways to get a head and 3 or higher are H3, H4, H5 and H6.
So P(H and 3 or higher)
= $\frac{\text{number of ways to get H and 3 or higher}}{\text{number of possible outcomes}}$ = $\frac{4}{12}$ = $\frac{1}{3}$.
*[2 marks available — 1 mark for identifying outcomes,
1 mark for correct probability]*

16 a) $e^2 + g^2 = 3^2 + (-7)^2 = (3 \times 3) + (-7 \times -7) = 9 + 49 = 58$
*[2 marks available — 1 mark for correctly substituting in
the values of e and g, 1 mark for the correct answer]*

b) $\frac{efg}{2} = \frac{e \times f \times g}{2} = \frac{3 \times 5 \times -7}{2} = \frac{-105}{2} = -52.5$
*[2 marks available — 1 mark for correctly substituting in
the values of e, f and g, 1 mark for the correct answer]*

17 a) Looking at the shelves, a width of 1 cm on the drawing
represents a width of 0.5 m in the real kitchen.
So the scale is 1 cm to 0.5 m. *[1 mark]*

b) The length of the kitchen on the drawing is 10 cm,
so the length of the real kitchen is 10 × 0.5 = 5 m. *[1 mark]*
The width of the kitchen on the drawing is 7 cm,
so the width of the real kitchen is 7 × 0.5 = 3.5 m. *[1 mark]*
So the area of the real kitchen
= length × width = 5 × 3.5 = 17.5 m² *[1 mark]*
[3 marks available in total — as above]

18 a) From the graph, £2 = €2.20 *[1 mark]*

b) From the graph, €3 = £2.70 *[1 mark]*

c) From the graph, £10 = €11, so £20 = 2 × £10 = 2 × €11 = €22
*[2 marks available — 1 mark for a suitable method,
1 mark for correct answer]*

19 a) Volume = length × width × height = 4.9 × 3.7 × 2.8
= 50.764 cm³
*[3 marks available — 1 mark for using the correct formula
for volume, 1 mark for correct answer, 1 mark for units]*

b) Volume = length × width × height, so 56 = 7 × 4 × h = 28h.
So h = 56 ÷ 28 = 2 cm.
*[2 marks available — 1 mark for putting numbers into the
formula correctly, 1 mark for rearranging to find the
correct value of h]*

20 a)

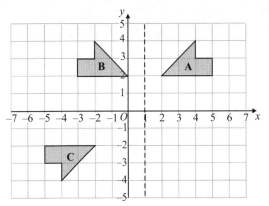

*[2 marks available — 2 marks for correct reflection,
otherwise 1 mark for shape reflected but in wrong position]*

b) Rotation of 180° about (0, 0) / about the origin.
*[3 marks available — 1 mark for using the word rotation,
1 mark for correct angle, 1 mark for correct centre of
rotation]*

21 Area of outside walls = 2(15 × 3) + 2(12 × 3) = 162 m²
Area of windows = 5(2 × 1) = 10 m²
Area of door = 3 × 2.5 = 7.5 m²
Area to paint = 162 – 10 – 7.5 = 144.5 m²
Number of litres needed = 144.5 ÷ 13 = 11.115... litres
Minimum number of tins needed = (2 × 5 litres) + (1 × 2.5 litres)
Cost = (2 × 20.99) + 12.99 = £54.97
*[6 marks available — 1 mark for finding the area of the outside
walls, 1 mark for finding the area of the windows and door,
1 mark for finding the area to paint, 1 mark for working out
how many litres of paint are needed, 1 mark for working out
how many tins are needed, 1 mark for cost]*

22 E.g.

x	$x^3 - x - 4$	Comment
1	–4	too small
2	2	too big
1.5	–2.125	too small
1.8	0.032	too big
1.7	–0.787	too small
1.75	–0.390625	too small

So x = 1.8 (to 1 d.p.)
*[4 marks available — 1 mark for any trial between 1 and 2,
1 mark for any trial between 1.5 and 2, 1 mark for a trial to
2 d.p., 1 mark for the correct answer]*

23 Shop A: 25% of £6.40 = 0.25 × 6.4 = £1.60, so with 25%
discount, 32 nappies cost £6.40 – £1.60 = £4.80 *[1 mark]*.
Price per nappy = £4.80 ÷ 32 = £0.15 (= 15p) *[1 mark]*.
Shop B: 56 nappies cost £7.84, so price per nappy = £7.84 ÷ 56
= 0.14 (= 14p) *[1 mark]*.
Nappies in Shop A cost 15p each after discount, but nappies in
Shop B cost 14p each, so Tom should buy nappies from Shop B
[1 mark].
[4 marks available in total — as above]

24 a) $a^4 \times a^5 = a^{4+5} = a^9$ *[1 mark]*

b) $b^9 \div b^3 = b^{9-3} = b^6$ *[1 mark]*

25 a) Pythagoras' theorem: $a^2 + b^2 = h^2$,
so $m^2 + 12.8^2 = 15.9^2$
$m^2 = 15.9^2 - 12.8^2 = 88.97$ *[1 mark]*
$m = \sqrt{88.97} = 9.432... = 9.4$ (1 d.p.) *[1 mark]*
[2 marks available in total — as above]

b) If it is a right-angled triangle, according to Pythagoras'
theorem, $a^2 + b^2 = h^2$, so $8.1^2 + 14.5^2$ should be equal to 15.6^2
[1 mark].
$8.1^2 + 14.5^2 = 275.86$ and $15.6^2 = 243.36$ *[1 mark]*.
Since $8.1^2 + 14.5^2$ does not equal 15.6^2, the triangle cannot
be a right-angled triangle *[1 mark]*.
[3 marks available in total — as above]

Index

Index